37.50

Statistical analysis of spherical data

Statistical analysis of spherical data

N.I.FISHER
Division of Mathematics and Statistics, CSIRO Australia

T.LEWIS
Department of Statistics, The Open University, UK

B.J.J.EMBLETON
Division of Mineral Physics and Mineralogy, CSIRO Australia

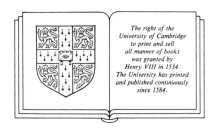

CAMBRIDGE UNIVERSITY PRESS
Cambridge
London   New York   New Rochelle
Melbourne   Sydney

Published by the Press Syndicate of the University of Cambridge
The Pitt Building, Trumpington Street, Cambridge CB2 1RP
32 East 57th Street, New York, NY 10022, USA
10 Stamford Road, Oakleigh, Melbourne 3166, Australia

© Cambridge University Press 1987

First published 1987

Printed in Great Britain by J. W. Arrowsmith Ltd

British Library cataloguing-in-publication data
Fisher, N. I.
Statistical analysis of spherical data.
1. Mathematical statistics   2. Sphere
I. Title   II. Lewis, Toby   III. Embleton, B. J. J.
519.5   QA276

Library of Congress cataloging-in-publication data
Fisher, N. I.
Statistical analysis of spherical data.

Bibliography
Includes index.
1. Mathematical statistics.   2. Spherical data.
I. Lewis, Toby.   II. Embleton, B. J. J. (Brian J. J.)   III. Title.
QA276.F489   1987   519.5   85-21363

ISBN 0 521 24273 8

Contents

	Preface	xi
1	**Introduction**	1
2	**Terminology and spherical coordinate systems**	17
2.1	Introduction	17
2.2	Spherical coordinate systems	17
2.3	Terminology	22
3	**Descriptive and ancillary methods, and sampling problems**	29
3.1	Introduction	29
3.2	Mathematical methods for unit vectors and axes in three dimensions	29
	3.2.1 Mean direction, resultant length and centre of mass	29
	3.2.2 Rotation of unit vectors and axes	32
	3.2.3 Some simple matrix calculations	33
	3.2.4 Eigenvectors, eigenvalues and the moment of inertia	33
	3.2.5 Calculation of points on an ellipse	34
3.3	Methods of data display	35
	3.3.1 Projections	35
	3.3.2 Sunflower plots	40
	3.3.3 Density contour and shade plots	41
	3.3.4 Directional/spatial plots	46
3.4	Exploratory analysis	46
3.5	Probability plots and goodness-of-fit tests	50
	3.5.1 Probability plots	50
	3.5.2 Goodness-of-fit tests	55
3.6	Simulation techniques	56
	3.6.1 Uses of simulation	56
	3.6.2 How to simulate data from various distributions	58
3.7	The jackknife method; permutation tests	60
3.8	Problems of data collection; sampling	63

4 Models — 67
- 4.1 Introduction — 67
- 4.2 Useful distributions on the line — 72
  - 4.2.1 The uniform distribution (on the line) — 72
  - 4.2.2 The normal distribution — 73
  - 4.2.3 The exponential distribution — 75
  - 4.2.4 The χ^2 (chi-squared) distribution — 76
  - 4.2.5 The F (or variance-ratio) distribution — 79
- 4.3 Useful distributions on the circle — 81
  - 4.3.1 The uniform distribution (on the circle) — 81
  - 4.3.2 The von Mises distribution — 81
  - 4.3.3 The wrapped normal distribution — 83
- 4.4 Distributions for modelling spherical data — 84
  - 4.4.1 The point "distribution" — 84
  - 4.4.2 The uniform distribution on the sphere — 84
  - 4.4.3 The Fisher distribution — 86
  - 4.4.4 The Watson distribution — 89
  - 4.4.5 The Kent distribution — 92
  - 4.4.6 The Wood distribution — 94
- 4.5 Other distributions on the sphere — 96

5 Analysis of a single sample of unit vectors — 101
- 5.1 Introduction — 101
- 5.2 Exploratory analysis — 106
  - 5.2.1 Graphical methods — 106
  - 5.2.2 Quantitative methods — 108
- 5.3 Analysis of a sample of unit vectors from a unimodal distribution — 110
  - 5.3.1 Simple procedures — 110
    - (i) Test for uniformity against a unimodal alternative — 110
    - (ii) Estimation of the median direction of a unimodal distribution — 111
    - (iii) Test for a specified value of the population median direction — 113
    - (iv) Test for rotational symmetry about the mean direction — 114
    - (v) Estimation of the mean direction of a symmetric unimodal distribution — 115
    - (vi) Test for a specified mean direction of a symmetric distribution — 116
  - 5.3.2 Parametric models: unimodal distributions with rotational symmetry — 117
    - (i) Test for uniformity against a Fisher $\mathbf{F}\{(\alpha, \beta), \kappa\}$ alternative — 117
    - (ii) Goodness-of-fit of the Fisher $\mathbf{F}\{(\alpha, \beta), \kappa\}$ model — 117
    - (iii) Outlier test for discordancy — 125
    - (iv) Parameter estimation for the Fisher distribution — 129
    - (v) Test for a specified mean direction — 133
    - (vi) Test for a specified concentration parameter — 134

Contents

	(vii) Other problems relating to the analysis of a single sample of data from the Fisher distribution	135
	5.3.3 Parametric models: unimodal distributions without rotational symmetry	136
	(i) Estimation of the parameters of the Kent distribution	136
	(ii) Elliptical confidence cone for the mean direction	138
	(iii) Test of whether a sample comes from a Fisher distribution, against the alternative that it comes from a Kent distribution	139
5.4	Analysis of a sample of unit vectors from a great-circle or small-circle distribution	140
	5.4.1 Introduction	140
	5.4.2 Simple estimation of the pole to a great- or small-circle distribution	141
	5.4.3 Parametric models for great-circle or small-circle distributions which are rotationally symmetric about the polar axis to the circle	143
5.5	Analysis of a sample of unit vectors from a multimodal distribution	144
	5.5.1 Introductory remarks	144
	5.5.2 Simple modal analysis	144
	5.5.3 A parametric model for a bimodal distribution of unit vectors	145
	(i) Estimation of the parameters of the Wood distribution	145
	(ii) Test for specified modal directions	148
5.6	General tests of uniformity and rotational symmetry	149
	5.6.1 General tests of uniformity	149
	5.6.2 General tests of rotational symmetry	150
6	**Analysis of a single sample of undirected lines**	**152**
6.1	Introduction	152
6.2	Exploratory analysis	156
	6.2.1 Graphical methods	156
	6.2.2 Quantitative methods	159
6.3	Analysis of a sample of undirected lines from a bipolar distribution	160
	6.3.1 Simple procedures	160
	(i) Test for uniformity against a bipolar alternative	160
	(ii) Estimation of the principal axis of a bipolar distribution	162
	(iii) Test for a specified value of the principal axis	163
	(iv) Test for rotational symmetry about the principal axis	164
	(v) Estimation of the principal axis of a symmetric bipolar distribution	165
	(vi) Test for a specified value of the principal axis of a symmetric bipolar distribution	167
	6.3.2 Parametric models: bipolar distributions with rotational symmetry	167
	(i) Test for uniformity against a Watson $\mathbf{W}\{(\lambda, \mu, \nu), \kappa\}$ bipolar alternative	168

viii   *Contents*

	(ii) Goodness-of-fit of the Watson $\mathbf{W}\{(\lambda, \mu, \nu), \kappa\}$ bipolar model	168
	(iii) Outlier test for discordancy	170
	(iv) Parameter estimation for the Watson bipolar distribution	175
	(v) Test for a specified value of the principal axis of the Watson bipolar distribution	177
	(vi) Test for a specified value of the concentration parameter of a Watson bipolar distribution	178
6.4	Analysis of a sample of undirected lines from a girdle (equatorial) distribution	178
	6.4.1 Introduction	178
	6.4.2 Simple procedures	179
	(i) Test for uniformity against a girdle alternative	179
	(ii) Estimation of the polar axis of a girdle distribution	180
	(iii) Test for a specified value of the polar axis	181
	(iv) Test for rotational symmetry about the polar axis	182
	(v) Estimation of the polar axis of a rotationally symmetric girdle distribution	182
	(vi) Test for a specified value of the polar axis of a rotationally symmetric girdle distribution	183
	6.4.3 Parametric models: girdle distributions with rotational symmetry	184
	(i) Test for uniformity against a Watson $\mathbf{W}\{(\lambda, \mu, \nu), \kappa\}$ girdle alternative	184
	(ii) Goodness-of-fit of the Watson $\mathbf{W}\{(\lambda, \mu, \nu), \kappa\}$ girdle model	184
	(iii) Outlier test for discordancy	186
	(iv) Parameter estimation for the Watson girdle distribution	189
	(v) Test for a specified value of the polar axis of a Watson girdle distribution	190
	(vi) Test for a specified value of the concentration parameter of a Watson girdle distribution	191
6.5	Analysis of a sample of measurements from a small-circle distribution	191
6.6	Analysis of a sample of axes from a multimodal distribution	192
6.7	A general test of uniformity	192
7	**Analysis of two or more samples of vectorial or axial data**	**194**
7.1	Introduction	194
7.2	Analysis of two or more samples of unimodal vectorial data	198
	7.2.1 Introduction	198
	7.2.2 Simple procedures	199
	(i) Test for a common median direction of two or more distributions	200
	(ii) Estimation of the common median direction of two or more unimodal distributions	201
	(iii) Test of whether two or more rotationally symmetric distributions have a common mean direction	204

Contents ix

	(iv) Estimation of a common mean direction of two or more rotationally symmetric distributions	206
	7.2.3 Parametric models – rotationally symmetric distributions	208
	(i) Test of whether two or more Fisher distributions have a common mean direction	208
	(ii) Estimation of the common mean direction of two or more Fisher distributions	211
	(iii) A test of whether the mean directions of several Fisher distributions are coplanar	213
	(iv) Estimation of the common plane of the mean directions of several Fisher distributions	216
	(v) A test of whether two or more Fisher distributions have a common concentration parameter	218
	(vi) Estimation of the common concentration parameter of two or more Fisher distributions	224
	(vii) Other problems of comparing two or more Fisher samples	224
7.3	Analysis of two or more samples of bipolar axial data	225
	7.3.1 Simple procedures	225
	(i) A test of whether two or more symmetric bipolar distributions with comparable dispersions have the same principal axis	225
	(ii) Estimation of the common principal axis of two or more symmetric bipolar distributions with comparable dispersions	226
7.4	Analysis of two or more samples of girdle axial data	227
	7.4.1 Simple procedures	228
	(i) A test of whether two or more symmetric girdle distributions with comparable dispersions have a common polar axis	228
	(ii) Estimation of the common polar axis of two or more symmetric girdle distributions with comparable dispersions	229
8	**Correlation, regression and temporal/spatial analysis**	230
8.1	Introduction	230
8.2	Correlation	231
	8.2.1 Introduction	231
	8.2.2 Correlation of two random unit vectors	232
	8.2.3 Correlation of two random unit axes	235
	8.2.4 Correlation between a random unit vector or axis and another random variable	237
8.3	Regression	238
	8.3.1 Introduction	238
	8.3.2 Simple regression of a random unit vector on a circular variable	239
	8.3.3 Regression of a random unit vector on a set of linear variables	240
8.4	Temporal and spatial analysis	242
	8.4.1 Introduction	242

8.4.2 Time series 242
 (i) A test for serial association in a time series of unit vectors 242
 (ii) A test for serial association in a time series of unit axes 243
8.4.3 Time-ordered sequences of unit vectors 244
8.4.4 Spatial analysis of unit axes 248

Appendix A Tables and charts 249

Appendix B Data sets 278

References 312

Index 313

Preface

1 The purpose of the book

The analysis of data in the form of directions in space, or equivalently of positions of points on a spherical surface, is required in many contexts in the Earth Sciences, Astrophysics and other fields. While the contexts vary, the statistical methodology required is common to most of these data situations. Some of the methods date back to the beginning of the century, but the main developments have been from about 1950 onwards. A large body of results and techniques is now disseminated throughout the literature. This book aims to present a unified and up-to-date account of these methods for practical use.

It is directed to several categories of reader:

(1) to the working scientist dealing with spherical data;
(2) to undergraduate or graduate students whose taught courses or research require an understanding of aspects of spherical data analysis, for whom it would be a useful supporting text or working manual;
(3) to statistical research workers, for reference with regard to current solved and unsolved problems in the field.

Because of the range of readership, priority has been given to providing a manual for the working scientist. Statistical notions are spelt out in some detail, whereas the statistical theory underlying the methods is, by and large, not included; for the statistician, references are given to this theory, and to related work. In particular, only one procedure is given for any specific problem. In some cases, the choice of procedure is dictated by certain optimality considerations, whereas in other cases a somewhat arbitrary choice has been made from several essentially equivalent procedures. As these matters are somewhat peripheral to the purpose of this book, we do not always give reasons for our choice.

The emphasis is on applications, with the statistical methods being illustrated throughout the book by data examples from many of the various relevant fields. A substantial number of these examples are drawn from Palaeomagnetism and Structural Geology, reflecting our experience and the state of the literature; however, the methods are readily applied in other areas. A knowledge of mathematics at, say, first year university level, and of basic probability and statistical methods, is assumed. However, mathematical presentation is only used where it is necessary for an understanding of the statistical technique being described.

Tables of all the main statistical functions required for practical use are given in an Appendix; some of these have been specially calculated for this book. It should be recognised that the analysis of spherical data has only got off the ground in a major way in the last 20 years, with the ready availability of computers. Particular care has been paid in the book to the computational and computer aspects of the procedures, so that there is no gap between the presentation of a technique and its implementation.

2 Survey of contents

Because we aim to present a comprehensive range of practical techniques within the limits of what is available at the time of writing, our account is accordingly extensive in some areas, incomplete in others, reflecting the current state of the art. Following a general survey in Chapter 1, questions of terminology, nomenclature and spherical coordinate systems are dealt with in Chapter 2. Chapter 3 gives a tool-kit for use in Chapters 5-8 on data analysis, and includes standard mathematical, statistical and computational methods, techniques for data display, and procedures for the exploratory analysis of data. The probability distributions employed in the book to represent various kinds of data are described in Chapter 4. Chapters 5 and 6 give extensive accounts of methods for analysing single samples of vectors and single samples of axes respectively. Chapter 7, also extensive, deals with the analysis of two or more samples of vectorial or axial data. Chapter 8 brings the reader to the current state of development of the subject. Dealing with problems of association between variables, correlation, regression, time series analysis and questions involving concomitant spatial information, it is necessarily incomplete and to some extent speculative.

3 How to use the book

By and large, there is a natural dichotomy of methods into those for *vectorial*, or directed, data, and those for *axial*, or undirected, data. Given this classification, there is a methodical development of statistical methods in each chapter. For example, Chapter 5 (Analysis of a single sample of unit vectors) progresses from exploratory analysis, to methods for unimodal data (simple methods, then parametric methods), to methods for great-circle or small-circle data, to methods for multimodal data (simple methods, then parametric methods), concluding with general tests of uniformity and symmetry. Chapters 5, 6 and 7 each give a figure early in the chapter, showing the development of that chapter. Having decided on the type of data (*vectorial* or *axial*) and the nature of the problem (one-sample, two-sample, correlation, etc.), the appropriate chapter and (except for Chapter 8) figure can be consulted to decide on the course of action. An example of how a complete analysis of a set of data might be carried out is given in Chapter 1.

4   Notation and conventions

The field of spherical data is distinguished by an extraordinary diversity of coordinate systems and methods of representing the data in two dimensions (i.e. projections). These matters are dealt with in some detail in §2.2 and §3.3.1. For data sets used in examples, the data are listed in their original coordinate system (i.e. that used when the data were published); all statistical analysis is done in terms of data (transformed if necessary) in polar coordinates or the related Euclidean coordinates; the results are then stated in terms of the original coordinate system. A particular method is used for displaying vectorial data which allows both hemispheres to be displayed simultaneously: this method is described in §3.3.1.

Many statistical terms will be unfamiliar to scientists from other disciplines, so a suitable list of definitions is given in §2.2.

5   Acknowledgements

We are indebted to many people for assistance in obtaining data sets and for permission to use tables and graphs. Acknowledgement of such assistance is given with each individual data set, table or graph.

Several typists at CSIRO were introduced to the delights of mathematical typing through preparation of the manuscript: we are most grateful to Diana Bridgewater, Colleen Sullivan and Lindy Apps for their skill and patience in this aspect of the work. The collective expertise of Vera Barhatov, Sue Clancy, Harold King and Jacki Ryan was invaluable for preparation of a number of figures and tables; our thanks also go to Chris Taylor and Dick Rattle for photographic help.

On the scientific side, we wish to record our particular gratitude to John Best, John Creasey, Peter Diggle, Chris Powell, Phil Schmidt and Geof Watson for a variety of forms of assistance.

Financial assistance, through the Overseas Visitors Programmes of the CSIRO Divisions of Mathematics and Statistics and of Mineral Physics, and through the Open University Statistics Research Visitors' Programme, was vital in enabling us to have several extended periods working together.

Finally, we thank Geoff Eagleson (Division of Mathematics and Statistics) and Ken McCracken (Division of Mineral Physics) for their continuing encouragement for the duration of this project.

Despite these various debts of gratitude, we must (selfishly) reserve for ourselves responsibility for any imperfections in the book, and would be most grateful to be advised of them.

The Greengate	Nicholas Fisher
Sydney	Toby Lewis
September 27, 1984	Brian Embleton

1
Introduction

What are *spherical data*? By a spherical measurement we simply mean the orientation of a straight line in space. In some scientific contexts we would wish to regard the line as directed, in others as undirected; in the first case we call it a *vector* and in the second case an *axis*.

Spherical data arise in many areas of scientific experimentation and observation. As examples of vectorial data from various fields, we instance from Astrophysics the arrival directions of showers of cosmic rays; from Structural Geology the facing directions of conically folded planes; from Palaeomagnetism the measurements of magnetic remanence in rocks; from Meteorology the observed wind directions at a given place; and from Physical Oceanography the measurements of ocean current directions. Examples of axial data from various fields include, from Crystallography the directions of the optic axes of quartz crystals in a sample of quartzite pebbles; from Astronomy the normals to the orbital planes of a number of comets; from Structural Geology the measurements of poles to joint planes or to axial plane cleavage surfaces; and from Animal Physiology the orientations of the dendritic fields at different sites in the retina of a cat's eye, in response to stimulus by polarised light.

Again, observations which are not in any way orientations can sometimes be usefully re-expressed in the form of orientations and analysed as spherical data; in Social Science, for example, it has been the practice to analyse data on occupational judgments by individuals as unit vectors.

Any vector can be represented by a point P on the surface of a unit sphere of centre O, $\overline{OP}$ being the directed line; any axis can be represented by a pair of points Q, P at opposite ends of a diameter QOP of the sphere, or equivalently by a point P on a unit hemisphere. Two quantities are required to define a point P on the sphere or hemisphere, and hence a spherical measurement, whether vectorial or axial. We discuss in the next chapter the various ways in which these quantities can be chosen, for example as the angles of colatitude and longitude of P, or the Cartesian coordinates x, y, z of P (amounting to two independent quantities since $x^2+y^2+z^2$ must equal 1).

Writing with specific reference to the Earth Sciences, Watson (1970) says:

> The most difficult problems ... are concerned with collecting the data – getting random samples from well-defined populations.

"This is not surprising", he goes on to say, "since the data of the earth sciences comes largely from field observation rather than from planned laboratory experiments", but his dictum holds good whatever the data context. We give some discussion of sampling problems and procedures in Chapter 3. However, our main purpose in this book is to present a unified and up-to-date account of methods for *analysing* data already collected.

A scientist who has measured or collected spherical data will have some questions in mind to which the data are intended to provide answers. For a single set of data – a single *sample* in statistical parlance (see Chapter 2) – can the observations reasonably be taken to be uniformly distributed? If not, what general qualitative features can be ascribed to the population represented by the sample? There must be clustering of some kind, but what is its nature? If the data are vectorial, is there a single preferred direction? Could there be more than one preferred direction – does the population appear multimodal, and if so how many modes are there? On the other hand, is there a preferred great circle – is the population of *girdle* form? or, again, could it be concentrated around a small circle, or other closed curve on the surface of the unit sphere? Does it have rotational symmetry – about the preferred direction in the unimodal case or the polar axis in the girdle case?

These questions about the main features of the population are essentially issues of modelling. Often they can be decided without the need to refer to the specific data set, from background knowledge of the nature, geological, biological or whatever, of the directional quantity under study. Often, on the other hand, the modelling has to be done on the basis of the data to hand. In either case, once they are decided, a further set of questions follows. What is the best estimate of the preferred direction in a unimodal population of vectors? Within what region around the best estimate can the true preferred direction reasonably be assumed to lie? How large a sample of observations is required if the preferred direction is to be estimated with a given precision? What parameter or parameters are needed to specify the *shape* of the population (in particular its degree of concentration or dispersion about the modal direction; and, if not rotationally symmetric, its asymmetry features); how can they be estimated, and between what limits can their true values reasonably be assumed to lie? Along with these there may be questions of data evaluation. Does it appear safe to use all the observations in the data set without question, or are there any suspect values which seem out of line with the rest of the observations? If so, what should be done about these *outliers*?

Analogous questions to all of these arise for vectorial data modelled by other types of population (multimodal, girdle etc.), and again for axial data. With axial data the main types of non-uniform model are bipolar and girdle distributions, whether rotationally symmetric or not; the role of the preferred direction in the above discussion of vectorial data will be played by the principal axis for an axial bipolar distribution and by the normal to the girdle plane, or polar axis, for an axial girdle distribution.

So far we have only been discussing the interpretation of a single sample of spherical data. In many cases, however, the data include a number of different samples. Suppose, to fix ideas, that there are two samples. A question that might well arise, where this makes sense in the data context, is whether the two samples are *consistent*, in other words whether it is reasonable to believe that they come from the same population. If so, how should any parameter of the population (e.g. modal direction, dispersion or concentration, principal axis etc.) be estimated from the combined information provided by the two samples? In other cases we may have samples from two populations with the same modal direction but different shapes; how then should the common modal direction be estimated, and within what region around the estimate can the true direction reasonably be taken to lie?

These are typical of the more straightforward questions of data interpretation which arise. Beyond them lie a whole range of more complex questions which can arise in various data situations. At this stage we will only mention one, by way of illustration; it arises in the context of comparing Apparent Polar Wander Paths in the course of palaeomagnetic studies. Given two sequences of time-ordered points on the surface of the unit sphere, what rotation of the second sequence brings it into "best" alignment with the first, with some suitable definition of "best"; and can the first sequence and the rotated second sequence then be regarded as consistent?

What we have been describing at some length are all *statistical* questions – *questions with uncertain answers*. We might find, for example, that *there is good reason to believe* that a data set comes from a non-uniform population. We might then assume that the data come from the well-known model for unimodal data, the Fisher distribution, whose shape is described by the *concentration parameter* κ. From further analysis it might emerge that *they are reasonably consistent* with the further assumption that $\kappa = 16$ (but this does not mean that κ has been proved equal to 16, since the data will also be consistent with other values for κ above and below 16). We might conclude from the analysis of two samples of data that *it is hardly credible* that they come from the same population; to believe this one would have to believe in an outside chance or long shot. In this case an alternative explanation for the data is required – an *alternative hypothesis* or rival model. This might be, for example, that the samples come from populations identical in form but differently oriented, so that they can be brought into coincidence by rotation; or again that the samples come from populations of different forms. The choice of alternative hypothesis is of particular importance when testing data for uniformity; if not uniform, are we to believe that there is a preferred direction, or that they are multimodal, or is there some other possibility? The efficiency of the analysis may well depend on this choice.

To take another type of statistical question: given a set of representative data from a population with unknown mean direction (colatitude α, longitude β), we may have to settle on a single *most plausible* value for (α, β). We calculate an *estimate*, say $(2.3°, 281.2°)$; possibly a *best estimate* in some sense, as there is usually more than one procedure for estimating, in a given data situation. The estimate $(2.3°, 281.2°)$ is worthless, however, without some indication of the error attaching to it. If the sample was so small and its information content so poor that all directions above the equator were reasonably believable as values of (α, β), the estimate $(2.3°, 281.2°)$ would be useless. With a larger data set we might find that, while $(2.3°, 281.2°)$ was our *best* estimate, any direction making an angle of up to, say, $8.4°$ with this direction would be *acceptable* as a value of (α, β) consistent with the data: we would have set up a *confidence cone*. The figure of $8.4°$ quantifies the error attaching to our estimate, and is an essential part of the answer.

All these statistical answers are statements not about certainties but essentially about probabilities. They require some assumption about where and in what way the data are *subject to uncertainty* instead of being exactly reproducible; in other words, some assumption about the probability mechanism or *model* underlying the data. It may be quite strong or specific, e.g. that the observations come from a Fisher distribution with known mean direction; or it may be fairly weak, e.g. that the observations all come from a unimodal distribution (giving rise to the range of "non-parametric" methods which are useful when there is no basis for any strong assumption about distributional form etc.), or even

Introduction

simply that the observations all come from the same homogeneous population, whichever its form may be. Assessment of what is reasonably believable or barely credible, of what is the range of acceptable values of a parameter and which one of these is most plausible, can then be derived in terms of probabilities calculated on the basis of the assumptions.

These considerations are exemplified in Chapters 5, 6, 7, and 8, in the discussion of the various facets of statistical analysis of spherical data. It is, perhaps, profitable to view the way they blend together for the complete analysis of a data set, so we consider below an example of such an analysis (although, since the techniques required have not yet been introduced, the discussion will be somewhat qualitative). It is convenient here to use artificial data to allow a number of aspects of data analysis to be highlighted.

> ***Example* 1.1.** The data in Figure 1.1(a) constitute a sample of 75 unit vectors supposedly drawn from some common population. The coordinate system used is (Colatitude, Longitude) (**§2.2(a)**), and the method of display is an *equal-area projection* (**§3.3.1(b)**). The problem is to estimate a preferred direction for the population, and to provide some assessment of the error in this estimate.
>
> In the initial, *exploratory* phase of analysis, we make a qualitative assessment of the data. Figure 1.1(a) shows that the data are largely concentrated in a single group; however, an equal-area projection does not provide us with an appreciation of the shape of the pattern of points unless the data are "centred" in the projection. Accordingly, we make a rotation of the data (**§3.2.2**) so that the *sample mean direction* (**§3.2.1**) plots at the centre, as shown in Figure 1.1(b). (The small cross in the lower centre indicates the position of the South Pole, relative to the data set.) We now see that the sample exhibits some degree of *rotational symmetry* about its mean direction, and that there is one *outlying point* (that is, a point noticeably distant from the main data mass). At this stage, we may seek to enhance the shape of the sample distribution by computing contours of an estimated probability density (**§3.3.3**). The superimposed contours for the rotated data set are displayed in Figure 1.1(c). The inner contours are rather more elliptical than circular; nevertheless, it is worth exploring further the hypothesis that the underlying distribution is rotationally symmetric about its mean direction. So we conclude, after exploratory analysis, that the data appear to come from a single population, possibly rotationally symmetric about its mean direction, although there is some question about how to explain or handle the outlying point.
>
> For the next phase of the analysis, we have a choice of two paths to follow:
>
> (a) We can use a method requiring very few assumptions about the underlying distribution (a "simple" method as termed in this book) to estimate the preferred direction. This would involve a preliminary

6   1. *Introduction*

test of whether the hypothesis of rotational symmetry was tenable (§**5.3.1(iv)**), and depending on whether the hypothesis was or was not adjudged reasonable, calculation of the mean direction (§**5.3.1(v)**) or *median direction* (§**5.3.1(ii)**) and an associated estimate of error.
(b) We can attempt to fit a probability distribution to the data, the Fisher distribution (§**4.4.3**) if the hypothesis of rotational symmetry is reasonable, otherwise the Kent distribution (§**4.4.5**) which has elliptical contours of equal density.

Figure 1.1. (a) Equal area projection of 75 unit vectors (artificial data). The coordinate system is (colatitude, longitude). See Example 1.1.
(b) Plot of data in Figure 1.1(a) rotated to centralise the sample mean direction.
(c) Contour plot of the data in Figure 1.1(b).

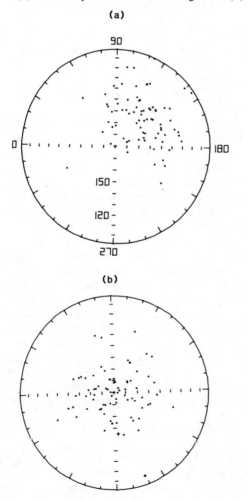

Introduction

It is worthwhile pausing for a few moments in our analysis to elaborate the reasons for preferring the second option:

(A) If we find that a particular parametric model (i.e. probability distribution) is a reasonable fit to the data, then the particular distribution – say, the Fisher distribution – that we fit, together with the estimated parameters of the distribution, gives us a complete summary of the data in condensed form. We will also have available a range of inferential procedures (such as a confidence cone for the mean direction, tests concerning the model parameters, and the methods in (B) below) which are valid even for small sample sizes; by contrast, the procedures in (a) are valid only for moderate to large sample sizes (at least 25, say).

(B) A satisfactory general approach to the treatment of outlying observations is only possible in the context of a parametric model. With such a model, it may be feasible to test formally, for example, whether one or more outliers can reasonably be taken to be drawn from the same distribution as the rest of the data, or whether there are data from, say, two distinct populations present.

Finally, if we cannot fit a parametric model satisfactorily, we still have recourse to the methods in (a).

Returning to the data set in hand, suppose we decide to examine the Fisher distribution as a suitable model. We begin with some graphical procedures (§**5.3.2(ii)**) designed (i) to compare the observed angular deviations of the observations from their sample mean direction with what might be expected for a Fisher distribution, and (ii) to test for rotational symmetry. (A third graphical procedure is given in §**5.3.2(ii)** but we omit it here for simplicity.) The so-called *colatitude* and *longitude* plots for these data are given in Figures 1.2(a) and 1.2(b) respectively. If the data are "Fisherian", the colatitude plot of points should be reasonably approximated by a straight line through the origin, whose slope then provides a graphical estimate of the *concentration parameter* κ of the distribution ($\hat{\kappa}_{gr} = 1/\text{slope}$). The points of the longitude plot

Figure 1.1 (*cont.*)

(c)

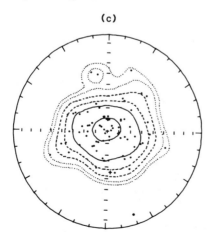

should be scattered about the 45° line shown. We see that the longitude plot is well-behaved in this respect, indicating that the assumption of rotational symmetry is reasonable. The colatitude plot also looks reasonable, except for the most extreme point which lies some distance away from an otherwise best-fitting line. It corresponds to the outlier evident in Figure 1.1(b). The question is: can we reasonably assume that this point is consistent with the main data mass, so that by chance we have obtained a rather wayward observation; or is it *statistically unlikely* that this observation comes from the same population as the rest? To answer this, we test the outlier for *discordancy* (§**5.3.2(iii)**). It turns out that the *significance probability* from the test is about 0.065, perhaps not strong enough evidence for us to classify the outlier as discordant, but perhaps sufficient for us to accord it some special treatment.

Had the agreement between the plotted points and appropriate straight lines been rather more dubious, we could have performed formal tests of goodness-of-fit (§**5.3.2(ii)**) to quantify the evidence for fit. However, this is unnecessary in these cases. From Figure 1.2(a) we obtain a graphical estimate of κ from the slope of the best-fitting line (ignoring the outlier), namely $\hat{\kappa}_{gr} = 1/(0.5/4.25) = 8.5$.

Figure 1.2. Probability plots for checking goodness-of-fit of the Fisher distribution to the data in Figure 1.1. (a) Colatitude plot, (b) longitude plot. See Example 1.1.

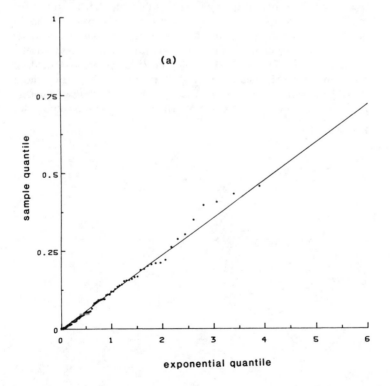

Introduction

Thus far, we have found that the Fisher distribution is a suitable model for the data. In the final phase, we estimate the parameters of the model and calculate appropriate estimates of error. The Fisher distribution is characterised by its mean direction and its concentration parameter. We have already had to calculate the sample mean direction to perform the rotation and the goodness-of-fit tests; its value is (Colatitude 145.2°, Longitude 135.2°). In estimating κ, we may wish to downweight the effect of the outlying point, in which case we use a *robust* method of estimation (**§5.3.2(iv)**), to obtain $\hat{\kappa} = 8.6$. (In this example, the robust estimate is only about 3% different from the non-robust estimate, because the outlier was not substantially distant from the main data mass.) Finally, the confidence cone (i.e. error) associated with our estimate of the mean direction has semi-angle 6.1°. Were it of interest, we could now proceed to test hypotheses about whether the mean direction of the underlying distribution could reasonably be taken to have a specified value using the methods in **§5.3.2(v)**, and similarly for the concentration parameter (**§5.3.2(vi)**).

The treatment of spherical data in the early literature on the subject does not in general meet these requirements of a satisfactory statistical analysis, since the methodology effectively dates from R.A. Fisher's 1953

Figure 1.2 (*cont.*)

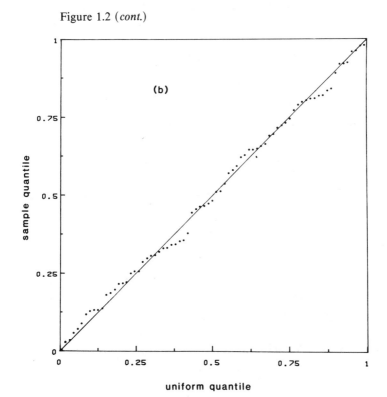

paper. Attempts were made, certainly, to use standard statistical methods not specific to spherical data. Winchell (1937), for example, proposed testing for non-uniformity of the orientations of a set of crystal units by plotting them on a Lambert equal-area projection, dividing the circular area into 148 equal grid squares (with allowance for incomplete squares at the periphery), and testing the goodness-of-fit of a Poisson distribution to the 148 counts of points in the squares. The alternative hypothesis in this test is simply that the expected density of points is not the same for all the squares, so a significantly high value of the goodness-of-fit statistic would give no clue as to what type of non-uniform model was appropriate for the data. Moreover, in most applications one would not expect the data points to be distributed over the whole area of the Lambert circle, so that routine application of Winchell's procedure would give a high number of empty squares conveying no information but being interpreted as evidence of non-uniformity. (Incidentally, while questions of attribution are mentioned occasionally in this book, conscientiousness in making acknowledgment can be carried too far! Winchell in his paper expresses his "special obligation" to the chairman of the department, who "in spite of considerable time and energy spent upon [a certain statistical] problem ... was unable to find the solution."). Flinn (1958) commented on the defects of a number of statistical tests then current, including Winchell's tests and some tests given by Chayes (1949) in a chapter on the statistical analysis of three-dimensional fabric diagrams which he contributed to a book on Structural Petrology; for a modern critique see Dudley, Perkins & Giné (1975). We may mention Chayes's enterprising though fallacious "empty space test" of uniformity, in which the plotted points on the Lambert projection, n in all, say, are contoured and the area inside the zero contour estimated. A χ^2 goodness-of-fit test is then carried out on the two frequencies n and 0 in the areas outside and inside the zero contour, using expected values proportional to these areas.

In this and other such tests, there is no concept of a distribution to represent the pattern of the spherical data in the event that they are not uniform. But, as will be shown in later chapters, the appropriate test depends on this alternative *model*.

When it comes to estimation, the specification of a model is essential for the calculation of accurate measures of the error of the estimate (though with large data sets there is the possibility, as a fall-back position, of getting approximate error values by the "bootstrap" methods developed in the late 1970's and described in Chapter 3).

A model can be chosen in the light of exploratory analysis or reference to previous experience with similar data. For example, the Watson girdle distribution is often found to be a good fit for a sample of normals to bedding planes measured around a cylindrical fold; but is this model *valid* for the data set in hand? An important part of the analysis is to check that the model is reasonably consistent with the data. Statistical procedures such as goodness-of-fit tests are available for this purpose, and will be described where appropriate in later chapters.

What if the model is not quite right – if, for example, a sample of vectors assumed to come from a Fisher population with its in-built rotational symmetry, in fact comes from a population with some noticeable degree of asymmetry? This situation is known in the trade as one of "departures from the model" (though really models depart from data rather than the other way around). Depending on the nature and extent of the departures, a test or estimation procedure may be affected severely or only to a small extent. In the former case the procedure is sensitive to the model assumptions, in the latter it is insensitive and, if satisfactory when the model holds good, will continue to be satisfactory under the indicated departures from the model. Such a *robust* procedure is obviously advantageous. Of course, some payment has to be made for the benefit of robustness in using such an all-purpose procedure, since it will tend to be rather less efficient when the data are in full conformity with the model and will give results of somewhat lower precision.

The reign of the *ad hoc* methods of analysis of spherical data could not continue indefinitely; there was an increasing demand for soundly based statistical techniques. In the context of vectorial data, the main motivation for statistical advance came from Palaeomagnetism, a fast-expanding science. The breakthrough came in 1953 with Sir Ronald Fisher's seminal paper "Dispersion on a sphere". In this paper he put forward as a model for angular errors on the sphere the fundamental vectorial distribution with density $C \exp(\kappa \cos \theta)$ which bears his name, developed methods of inference for the mean direction and the concentration parameter, calculated the distribution of the sum of a set of Fisher-distributed vector observations, and applied his results to a set of palaeomagnetic data due to J. Hospers, viz., directions of remanent magnetism measured in specimens from Icelandic lava flows. Actually Hospers, in a paper on remanent magnetism of rocks published in 1951, had thanked Fisher "for the calculation of the estimates of precision" of the true direction of the remanent magnetism, so Fisher had already been applying his theory to Hospers' data several years before he published his 1953 paper.

We quote for later reference the terse opening of the main part of Fisher's paper (following a brief Introduction), in which he introduces the distribution, with no mathematical explanation or preamble:

> 1.1 *The fundamental distribution*
> We may take as our fundamental distribution of elementary errors over the surface of the unit sphere, which is the field of possible observations, that in which the frequency density is proportional to
> $$e^{\kappa \cos\theta},$$
> where θ is the angular displacement from the true position...

The modern methodology of spherical data analysis undoubtedly dates from Fisher's 1953 paper, with its innovative results on inference methods for spherical data. The spherical distribution with density $C \exp(\kappa \cos\theta)$ had, it is true, been introduced twelve years previously together with a number of other spherical and circular distributions by K. J. Arnold in his unpublished Ph.D. thesis "On Spherical Probability Distributions" (1941), and in a generalised form by Langevin in 1905 in the context of the theory of magnetism. Arnold, whose thesis contains a number of important results, studied some of the properties of what is now called the Fisher distribution, and could claim to have his name associated with the distribution; Batschelet in his book *Circular Statistics in Biology* (1981) calls it the von Mises-Arnold-Fisher distribution. One wishes that Arnold had published his work in a journal. Actually, there is a fascinating historical twist to this question of priorities, which we have learned about during the preparation of this book from the British statistician G.A. Barnard. It turns out that Fisher had in fact devised the distribution in the early 1920's and had put his notes away in a drawer, there to remain for some thirty years until resuscitated for his 1953 paper! Professor Barnard has very kindly given us an account of this episode and his permission to reproduce it:

Mill House
Hurst Green
Brightlingsea
Colchester
Essex CO7 0EH

30 June 1981

Dear Nick,

Fisher sent me an offprint of his paper 'Dispersion on a sphere' when it came out, and I recall remarking to him that the really hard part of the work consisted in the first sentence of his section 1.1 (introducing the $e^{\kappa \cos\theta}$ density). I had had some directional data brought to me (by two different sets of people in our physics department at Imperial one suspecting anisotropic decay of particles in HE physics, the other on rock magnetism) and while I saw you could use the length of the vector sum to test isotropy, I had not seen how to do estimation. He replied that it was easier for him, since he had, in the twenties, asked himself what would be the analogue, on the sphere, of the normal density in the plane, and had made some notes on it. When approached by (I presume) Hospers (- though there were others working in that line at the time, including Runcorn, at Cambridge) he was able to go to his filing cabinet and pull out the notes, and answer the question on the spot. We agreed that being able to do something like that gave one particular pleasure - I had myself done something of the sort when asked about the dead time problem for a Geiger counter.

When trying to think of a precise year, the figure 1922 comes into my head, for Fisher's first work. This seems rather early; but he was in touch with Eddington, the astronomer, in 1920, and may well have gone on thinking about the problems he had raised.

Thanks for reminding me about the spherical BF problem. I must get down to it when I get back from Canada in August.

Best wishes
George A. Barnard

P.S. In case you are wanting to put in more history, I have a date which I happen to be able to fix precisely. I gave my inaugural lecture at Imperial on the 7th June, 1955, and both Fisher and Blackett came to the dinner which followed. They got into conversation about rock magnetism, in the course of which Fisher said to Blackett that not only was he going to have to admit polar reversal, but, Fisher suspected, Blackett would have to accept continental drift also.

Blackett had been very slow to accept polar reversal, not only because of his mistrust of statistics ('If I need statistics to analyse my data it means I haven't got enough data'), but also because he had some attachment to a conjecture he had made a bit earlier, to the effect that any massive rotating body would, by that fact alone, generate a magnetic field.

Following Fisher's paper the stage was set for the development of a full-going methodology of orientation statistics, both on the sphere and the circle. The leading name in this development is G.S. Watson, whose published work in this field started in 1956 and whose steady flow of major contributions continues unabated nearly three decades later. Watson's interest in the statistics of circular data was stimulated by problems in animal behaviour, in particular Schmidt-Koenig's experiments on the homing ability of pigeons, while the motivation for his interest in the statistics of spherical data came from problems in Palaeomagnetism. Here his collaboration with E. Irving, a pioneering figure in the development of Palaeomagnetism, was most fruitful. In 1966 Watson wrote:

> Many people contributed to the construction of distributions to represent directional data. No real progress was made with the inference questions until the problem was drawn to the attention of the late Sir Ronald Fisher (1953) by geophysicists interested in palaeomagnetism. His paper showed how progress towards methods of statistical analysis

could be made. The writer had the great good fortune to be introduced, at that happy moment, to this subject by E. Irving, whose book (1964) could serve as a model for the application of statistics to the earth sciences. As a result, a series of simple practical methods were devised... (Watson, 1966).

Within a few years of Fisher's paper, many of the problems of inference for Fisher-distributed data had been solved, in particular by Watson and E. J. Williams in a classic paper (1956). The ensuing period saw progress by a number of statisticians, notably M. A. Stephens, R. J. Beran and Watson himself, on further theoretical problems of inference and their exploitation in a variety of practical applications, particularly in the Earth Sciences.

Statistical techniques for axial data did not start being developed until the 1960's, several distributional models being published at this period. Much as the Fisher distribution with density $C \exp(\kappa \cos \theta)$ is the basic distribution representing vectorial data, the most important distribution representing axial data is the one with density $C \exp(\kappa \cos^2 \theta)$. First proposed as a model by Arnold (1941) and later by Dimroth (1962) and Breitenberger (1963), its properties and applications were studied in some detail by C. Bingham (1964) in his Ph.D. thesis "Distributions on the sphere and on the projective plane", and it was also used by Scheidegger (1964) for modelling normals to fault planes, and by Scheidegger (1965) in the analysis of sedimentological data on the orientations of bedding planes and grain axes. A key contribution was by Watson (1965), who presented significance tests and confidence regions for the polar axis and the shape parameter κ for the girdle form of the distribution ($\kappa < 0$), and applied these to the analysis of a set of normals to foliation planes in Structural Geology.

There is an interesting question of nomenclature connected with the $C \exp(\kappa \cos^2 \theta)$ distribution. Following Mardia's usage in his influential book (1972), it is widely called the *Dimroth-Watson distribution*; Watson himself calls it the *Scheidegger-Watson distribution* (Watson, 1983a). If there is to be joint attribution with Watson, a strong case can also be made for using Bingham's name (leaving aside Arnold's priority). In this book we have taken the view that the requirements of proper recognition and simplicity are both well met by entitling it the *Watson distribution*.

A new phase of expansion in the development of the subject arrived with the publication in 1972 of K.V. Mardia's book *Statistics of Directional Data*, a systematic account of theory and methodology for both circular and spherical data, with emphasis on the applicability of the methods in many fields. As H.E. Daniels said at a meeting of the Royal Statistical

Society in 1975,

> Professor Mardia did us all a service by publishing the right book at the right time, so bringing an important but hitherto specialized field to the attention of all statisticians. It is a young and growing field... (Daniels, 1975).

With the increased awareness among statisticians of the importance of spherical data analysis and the challenging problems that it offers, there has been progress along new lines in recent years. New probability distributions have been devised as models for vectorial data having non-Fisherian characteristics such as asymmetry and multimodality; a particularly useful development here is the *Fisher-Bingham* family of distributions introduced by J.T. Kent (1982). The topic of outlying or suspect observations in spherical data has attracted attention in the last few years; methods for identifying and testing them and for estimating parameters by robust procedures not unduly affected by their presence have been constructed for various data situations. Then there is the wide and relatively unexplored field of problems concerned with relationships between variables, including the analogues for spherical data of topics such as correlation, regression and time-series analysis for ordinary linear variables. Research activity in these areas has been developing rapidly. The statistical analysis of Apparent Polar Wander Paths, with its problems specific to the spherical context, is a noteworthy area of current progress in this general field.

Spherical statistics, "a young and growing field" in Professor Daniels's words, is as we have seen a phenomenon almost entirely of the last 30 years, and largely of the last 20. It has taken root at the right time; without the general advent of computers in the 1960's much of the methodology would not have been feasible, and the available techniques would have been quite limited in scope. The use of the computer is vital in several ways. It makes feasible a range of essential procedures for the display, manipulation and exploratory analysis of data - for example rotation, smoothing, density estimation. Large sets of data can be processed and studied interactively. Again, many of the inference procedures for spherical data - estimation of parameters, setting up of confidence cones, etc. - involve heavy calculations which necessitate the use of a computer; without it they would be impracticable. Other examples of essential techniques which the computer makes feasible are simulation, permutation testing, and bootstrap assessment of variability.

What is the present state of the art? Methods for exploratory analysis and procedures for the display, plotting, mapping and contouring of data are well established, and the requisite computer programs available. Progress has also been made in recent years with the development of techniques for testing and estimation which do not require the assumption of a specific probability model for the data, such as a Fisher distribution or a Watson distribution. As regards probability distributions for modelling data, quite a number have been proposed in the literature; not all of these, however, are suitable for practical data analysis, since the procedures required for the various forms of inference are either unknown or intractable. With vectorial data, useful distributions are available to represent symmetric and asymmetric unimodal populations, bimodal populations with modes of equal weight, and girdle-type populations. With axial data, the range of really useful distributions is at present limited to the case of populations with rotational symmetry, whether girdle or bipolar. There is a need for a convenient model for asymmetric axial data. The analysis of single samples of vectorial or axial data, whether on the basis of an assumed model or by nonparametric methods, has been extensively investigated. Procedures for comparing several samples or for combining the information they contain are also well understood for vectorial data and currently under development for axial data. A new and important book by Watson (1983a) sets out much of the underlying theory and places it in a more general framework. The treatment of outliers in spherical data has been explored in some cases, but much remains to be done. The study of problems of relationship between a spherical random variable and another variable or variables, or of the dependence of a spherical random variable on time, is in a relatively early stage.

2

Terminology and spherical coordinate systems

2.1 Introduction

Many different ways of representing a three-dimensional unit vector or axis have been developed over the centuries, due not only to the requirements of different disciplines (Astronomy, Geodesy, Geology, Geophysics, Mathematics, ...) but also to diverse needs within a discipline: in Geology, for example, there appear to be five or six systems in current use. In this book, we shall use either polar coordinates or the corresponding direction cosines for all purposes of statistical analysis. The following sub-section (§**2.2**) defines several of the coordinate systems and gives the mathematical relationship of each to polar coordinates.

Later chapters of this book, concerned with statistical analysis, abound with words and phrases which have particular meanings in Statistics, and, possibly, rather different meanings in other areas. A good example of this is the word "sample", which for our purposes is loosely taken to mean a collection of measurements of a particular characteristic, but which has a general scientific meaning of an observational or sampling unit (e.g. a drill-core specimen on which a single measurement may be made). §2.3 gives definitions of a number of such words and phrases.

2.2 Spherical coordinate systems

The type of data we shall be dealing with will be either directed lines or undirected lines. For the former, the measurements will be unit *vectors*, such as the direction of magnetisation of a rock specimen, or the direction of palaeocurrent flow. For the latter, which we shall term *axes* (cf. §**2.3**), the line measured might be the normal to a fracture plane, and so have no sense (direction) unless this is ascribed on some other basis. Sometimes the coordinate system refers to angles defining the plane itself, rather than to the normal to the plane. These points are clarified in the diagrams accompanying the definitions.

2. Terminology and spherical coordinate systems

(a) **Polar coordinates** Figure 2.1 shows a sphere of unit radius, centred at O. P is a point on its surface, and can be identified with the unit vector $\overrightarrow{OP}$. Also drawn are three orthogonal axes (x, y, z). The *colatitude* θ is the angle between $\overrightarrow{OP}$ and $\overrightarrow{Oz}$; the *longitude* ϕ is the angle, measured anticlockwise, between $\overrightarrow{Ox}$ and $\overrightarrow{OP^*}$, the projection of $\overrightarrow{OP}$ onto the x-y plane. Thus $0° \leq \theta \leq 180°$, $0° \leq \phi < 360°$. The coordinates of P in terms of x, y and z are the *direction cosines*

$$x = \sin\theta \cos\phi, \qquad y = \sin\theta \sin\phi, \qquad z = \cos\theta$$

for θ, ϕ in radian measure ($\theta_{\text{rad}} = \theta° \times \pi/180$).

For an *axis*, $\overrightarrow{OP}$ is replaced by the diameter passing through P and O to P' (Figure 2.1). Since axial data are usually recorded in the lower hemisphere, the coordinates of P' would be used; in terms of θ and ϕ they are $(180° - \theta, \phi + 180°)$.

(b) **Geographical coordinates** The point of intersection of $\overrightarrow{Oz}$ with the surface of the Earth is the North Pole. The variation of P along the Oz, or North-South, axis is measured as the angle θ' (the *latitude*) between OP and the x-y plane (the equatorial plane), with angles above the plane being positive and those below negative. The *longitude* ϕ' is

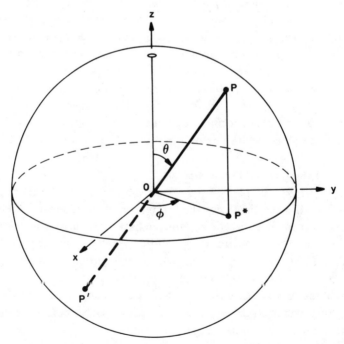

Figure 2.1 Definition of polar coordinates: θ = colatitude, ϕ = longitude.

2.2. Spherical coordinate systems

measured as the angle between the meridian containing P and the zero reference meridian or, alternatively, the angle between the projection OP^* of OP onto the x-y plane and the x-axis, in an Easterly (anti-clockwise) direction (Figure 2.2). In terms of polar coordinates,

$$\theta' = 90° - \theta, \qquad \phi' = \phi$$

(c) Geological coordinates Refer to Figure 2.3. The orientation of a planar feature, for example an inclined stratum such as a bedding plane or a sheet, is defined by its strike and dip, or its dip and direction of dip (dip azimuth). The *strike S* is the bearing or direction of a horizontal line in the plane, and is often measured in an Easterly sense from North. The *dip D* is the angle of inclination of a planar feature from the horizontal; it is measured at right angles to the strike. The *direction of dip A* is thus $S + 90°$. In terms of geological coordinates,

$$\theta = D + 90°, \qquad \phi = 360° - A$$

An upward directed normal (or *pole*) to the plane will have direction cosines

$$\cos \theta \cos \phi, \qquad \cos \theta \sin \phi, \qquad -\sin \theta$$

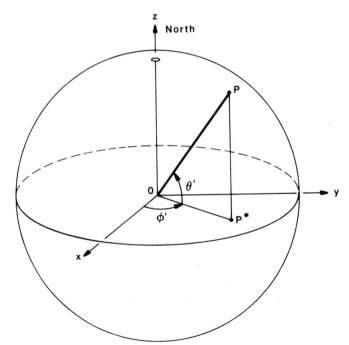

Figure 2.2 Definition of geographical coordinates: θ' = latitude, ϕ = longitude.

20   2. Terminology and spherical coordinate systems

relative to orthogonal axes x, y (counterclockwise) and z positive upwards. Thus, the endpoint or pole in the lower hemisphere corresponding to these direction cosines has polar coordinates $(270° - \theta, \phi + 180°)$.

A unique orientation of a plane may be obtained by recording the dip and dip azimuth, whereas a strike and dip record should be complemented with a description of the direction of dip (e.g. to the South-South-East) to avoid ambiguity. With the increasing use of in-the-field electronic recording, the shorter notation is often preferred.

Linear features with directional significance (directed lines) are commonly treated as unit vectors; magnitude may be unimportant, e.g. glacial striae, or cannot be measured, e.g. palaeocurrent velocities. A directed line may be specified by its *inclination* (Inc.), the angle between the vector and the horizontal plane, and its *declination* (Dec.) - or azimuth

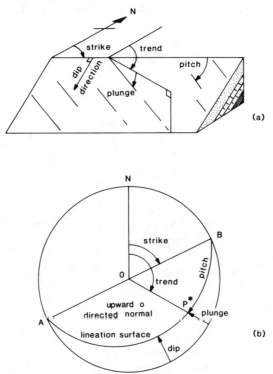

Figure 2.3 Definition of geological coordinates. Figure 2.3(a) shows variables used to describe the orientation of linear features; Figure 2.3(b) shows a stereographic representation of a directed line in a plane. The curve *APB* is the intersection of the plane with the lower hemisphere. The upward directed normal is plotted with an open circle to denote intersection with the upper hemisphere. *OP** is the orientation of a line in the plane.

2.2. Spherical coordinate systems

– taken to be the angle between North and the projection of the vector onto the horizontal plane. Inclination is positive [negative] for a downward [upward] pointing vector, and declination is measured in an Easterly sense. Thus

$$\theta = \text{Inc.} + 90°, \qquad \phi = 360° - \text{Dec.}$$

The orientation of a unit vector in terms of θ and ϕ is covered in (*a*) above.

Directed lines, and undirected lines such as fault lines, joints and foliations, often occur in a plane. The variables commonly used in Geology to describe the orientation of linear features are shown diagrammatically in Figure 2.3(a). The angle between North and the vertical projection of the feature onto a horizontal plane is the *trend*. The *plunge* is the angle between the feature and the horizontal surface. *Pitch* is measured in the plane and is the angle between the strike direction and the linear feature. Watson (1983a, pages 17–19) discusses six combinations of strike, dip, trend, plunge and pitch which may be used to describe the orientation of a linear feature in the plane. To specify the orientation uniquely, it may again be necessary to qualify the measurements with a description of the feature's approximate bearing (e.g. South-East). Figure 2.3(b) shows a stereographic representation of these orientation parameters. The particular combination to be used in practice will often depend on a geologist's intuitive judgment as to which of the measurements can be made most accurately (cf. §3.8).

(*d*) *Astronomical coordinates* There are at least five coordinate systems used in Astronomy, detailed discussion and illustration of which are beyond the scope of this Section. We confine ourselves to the definition of five systems, and refer the reader to Kraus (1966, Chapter 2) and Smart (1977, Chapters 1 and 2) for amplification. Following the description in Kraus (*loc. cit.*, page 33) the systems are:

1. *The horizon system* (azimuth, altitude), which is based on a plane through the observing point parallel to the horizon (Figure 2.4): in terms of polar coordinates,

   azimuth $= 360° - \phi$

   altitude (elevation angle) $= 90° - \theta$

2. *The equatorial system* (Declination, Right Ascension), in which the Earth's equator is the plane of reference (Figure 2.5); in terms of polar coordinates,

   Declination $= 90° - \theta$

   Right Ascension $= \phi$

3. *The ecliptic system* (celestial longitude, celestial latitude) in which the plane through the Earth's orbit is taken as reference; in terms of polar coordinates,

   celestial longitude = ϕ

   celestial latitude = $90° - \theta$

4. *The galactic system* (galactic longitude, galactic latitude), which is based on a plane parallel to the plane of our galaxy; in polar coordinates,

   galactic longitude = ϕ

   galactic latitude = $90° - \theta$

5. *The super-galactic system*: as for the galactic system, but based on a plane parallel to a plane in which there is an apparent local concentration of galaxies.

2.3 Terminology

ALTERNATIVE HYPOTHESIS – see NULL HYPOTHESIS.

AXIAL DISTRIBUTION – A probability distribution of an axial random variable (see AXIS).

AXIS – A direction in three dimensions, without a sense (for example, a dipole axis, or the Earth's axis of rotation). Hence RANDOM AXIS, AXIAL OBSERVATION, AXIAL RANDOM VARIABLE.

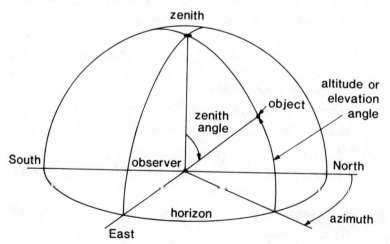

Figure 2.4 Definition of the horizon system of coordinates.

2.3. Terminology

BOOTSTRAP METHOD - A statistical procedure which provides an assessment of the variability of an estimate using only the data from which the estimate was computed (see §**3.6**).

CONFIDENCE CONE, CONFIDENCE REGION - As for CONFIDENCE INTERVAL, except that the region computed to cover the unknown parameter is not an interval but a conical or other specified shape.

CONFIDENCE INTERVAL - Suppose θ is a parameter to be estimated. A confidence interval for θ is an interval of values computed from a sample, which includes the unknown value of θ with some specified probability. Thus, if we construct a 95% confidence interval for θ for each of a series of independent samples, in the long run θ will be contained in 95% of these intervals.

CONFIDENCE LEVEL - The probability (e.g. 95% or 99%) that a confidence interval will cover the unknown parameter value. [More generally in this book, we shall refer to a $100(1-\alpha)$% confidence level, where $\alpha = 0.05$ or 0.01 for 95% or 99% confidence].

Figure 2.5 Definition of equatorial system of coordinates. (Based on Kraus (1966, Fig. 2-11)).

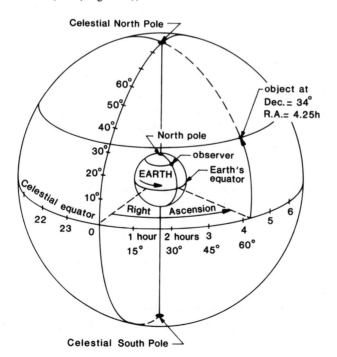

2. Terminology and spherical coordinate systems

CONTOUR – A line joining up contiguous points representing variate values for which the probability density function has a given value. (This is the only sense in which "contour" will be used in this book; of course, the notion of a contour has far wider applicability.)

CRITICAL REGION – The set of possible values of a test statistic which will result in rejection of a hypothesis being tested, at a specified significance level.

CUMULATIVE DISTRIBUTION FUNCTION – see DISTRIBUTION FUNCTION.

DISCORDANT OBSERVATION – In relation to a set of observations (main data set) generated by some prescribed probability model (e.g. the observations belong to a normal distribution with mean 50 and standard deviation 3), a *discordant* observation is one which is generated by a *different* probability model (e.g. it belongs to a normal distribution with mean 60 and standard deviation 3). An observation is *judged* discordant if it is *statistically unreasonable* on the basis of the prescribed probability model. (For example, an outlying unit vector in a sample supposedly drawn from a Fisher distribution on the sphere is judged discordant if it is rejected by the test described in §**5.3.2(iii)**).

DISTRIBUTION FUNCTION (d.f.) – The distribution function $F(x)$ of a random variable is the proportion of members in the population with values less than or equal to x.

EQUATORIAL DISTRIBUTION – See GIRDLE DISTRIBUTION.

GIRDLE DISTRIBUTION – An axial distribution which is concentrated around a great or small circle (rather than at the end points of some fixed axis). If there is rotational symmetry about the axis normal to the plane of the great circle, the distribution is called *symmetric girdle*. A girdle distribution concentrated on a great circle is also termed an *equatorial* distribution.

GOODNESS-OF-FIT TEST – A statistical test which assesses the agreement between an observed set of values and a hypothesised probability distribution (see §**3.5**).

GREAT CIRCLE – A closed curve on the surface of the sphere created by the intersection of the sphere and a plane passing through the centre of the sphere.

INDEPENDENT – Two random variables X and Y are statistically independent if knowledge that X has taken a particular value, say x_0, does not change one's probability statement about possible values of Y.

2.3. Terminology

ISOTROPIC – A circular or spherical random variable has an *isotropic distribution* if all possible unit directions (or axes) are equally likely to occur. (In this sense, *isotropic* is the same as *uniform*.) A spherical random variable has a distribution *isotropic with respect to a reference direction* (or *axis*) if its azimuth relative to the reference direction has an isotropic distribution (being a circular random variable).

JACKKNIFE METHOD – A method which reduces bias in estimation and which can be used to provide approximate confidence intervals in cases where ordinary distribution theory is intractable (see §3.7).

LEVEL OF SIGNIFICANCE – In hypothesis testing, it is usual to obtain from a given sample of data a test statistic calculated for the purposes of the test. If the test statistic falls in a range of values (the CRITICAL REGION) which, in total, have a small probability of occurrence under the hypothesis, the hypothesis is rejected. This small probability is the *level of significance* or *significance level*.

LINEAR (or REAL) RANDOM VARIABLE – A random variable specified by a single number (cf. VECTOR RANDOM VARIABLE).

MEAN – The mean value, or first MOMENT, of a random variable X is the average value of X over all its possible values x, each x being weighted by the value of the probability density function $f(x)$ of X at x. The mean value of a function $g(X)$ of a random variable X is the average value of $g(X)$ computed similarly. For a sample of measurements $x_1, \ldots, x_n$, the *sample* mean is $\sum_{i=1}^{n} x_i/n$, and the mean value of $g(x)$ is $\sum_{i=1}^{n} g(x_i)/n$.

MOMENT – A moment of a random variable X is the MEAN value of some power of X. (See MEAN, VARIANCE).

NULL HYPOTHESIS – The particular hypothesis under test. Usually, it will be tested against a competitor, the ALTERNATIVE HYPOTHESIS.

OBSERVATION – An observation is an observed value of a RANDOM VARIABLE; a SAMPLE comprises a set of observations of (or on) a random variable.

ORDER STATISTIC – When a sample of variate values are arranged in ascending order of magnitude, these ordered values are known as the order statistics of the sample.

ORTHOGONAL MATRIX – A square matrix **H** is orthogonal if $\mathbf{H'H} = \mathbf{I}$, the identity matrix (which has 1's on its main diagonal, and 0's elsewhere).

OUTLIER – An observation in a sample of variate values which is suspiciously far from the main data mass. (An outlier is not necessarily a DISCORDANT OBSERVATION).

PARAMETER – A numerical characteristic of a population, which may be known or may require estimation.

PERCENTILE – The percentiles are the set of partition values of a random variable which divide the total frequency into 100 equal parts. See also QUANTILE.

PERMUTATION DISTRIBUTION – Suppose T is a statistic calculated from a sample S, for testing a hypothesis H. Under the assumptions of H, we may be able to construct a family $S_1, \ldots, S_m$ of samples (one of which is S), each of which was just as likely before sampling commenced, by re-arranging the observations in S. The permutation distribution of T is the set of values $T_1, \ldots, T_m$ calculated from these samples (see §**3.7**).

PERMUTATION TEST – A method of testing a hypothesis, in which the test statistic T calculated from a sample is compared with the possible values in the PERMUTATION DISTRIBUTION of T. The hypothesis is rejected if T falls in an appropriate CRITICAL REGION of the permutation distribution (see §**3.7**).

POPULATION – The collection (or set) of all individual units possessing some characteristic of interest.

PROBABILITY DENSITY FUNCTION – For a univariate random variable, an expression which, when multiplied by an elemental range dx, gives the proportion of variate values between x and $x+dx$. Corresponding notions apply for bivariate or multivariate probability density functions.

PROBABILITY PLOT – A graphical technique for comparing two probability distributions to investigate their similarity or otherwise (see §**3.5.1**).

PSEUDO-RANDOM NUMBER – Numbers which are produced by a definite process, but which nonetheless are designed to have properties of randomness.

QUANTILE – The class of (n) partition values $A_1, \ldots, A_n$ say which divide the total frequency of a population into a given number, $n+1$, of equal parts. (See §**3.5.1** and Figure 3.16(b)). A PERCENTILE is a particular kind of quantile.

RANDOM SAMPLE – A set of statistically INDEPENDENT OBSERVATIONS from a common distribution (see INDEPENDENT, OBSERVATION, SAMPLE).

2.3. Terminology

RANDOM VARIABLE – A quantity (linear or vector) which may take any of the values of a specified set with a specified relative frequency or probability. Sometimes called VARIATE.

RANDOMISATION DISTRIBUTION – see PERMUTATION DISTRIBUTION.

RANDOMISATION TEST – See PERMUTATION TEST.

REFLECTION MATRIX – An ORTHOGONAL MATRIX whose determinant is −1.

ROBUST (STATISTICAL PROCEDURE) – A procedure which is not very sensitive to specified departures from the assumptions on which it depends.

ROTATION MATRIX – An ORTHOGONAL MATRIX whose determinant is +1.

SAMPLE – A part of a POPULATION, or a subset from that set of units, which is provided by some process or other, usually by deliberate selection with the object of investigating a characteristic or characteristics of the parent population or set.

SIGNIFICANCE LEVEL – see LEVEL OF SIGNIFICANCE.

SMALL CIRCLE – A closed curve on the surface of a sphere resulting from the intersection of the sphere and a plane not passing through the centre of the sphere (see GREAT CIRCLE).

SPLINE – A continuous function whose graph is composed of segments each of which is a polynomial.

STANDARD DEVIATION – The (positive) square root of the variance of a random variable.

STATISTIC – A function of the sample values.

TRACE – The trace of a square matrix is the sum of the terms on its main diagonal.

VARIANCE – The second moment of a random variable X when measured relative to its MEAN value. For a random sample of observations $x_1, \ldots, x_n$, with sample mean $\bar{x} = \sum_{i=1}^{n} x_i / n$, the *sample* variance is usually defined as

$$s^2 = \sum_{i=1}^{n} (x_i - \bar{x})^2 / (n-1)$$

(although in some books the divisor n is used instead of $n-1$).

2. Terminology and spherical coordinate systems

VARIATE – See RANDOM VARIABLE.

VECTOR RANDOM VARIABLE (or RANDOM VECTOR) – A random variable specified by two or more numbers (e.g. a direction).

Throughout this book, the natural logarithm of a variable x will be denoted $\log x$.

3

Descriptive and ancillary methods, and sampling problems

3.1 Introduction

The contents of this chapter constitute a tool-kit for use in the subsequent chapters on data analysis. §**3.2** deals with some basic mathematical methods for vectors and matrices; §**3.3** and §**3.4** are concerned with methods of data display, and qualitative (or descriptive) features of spherical data sets. *In particular, the basic method we have adopted for displaying vectorial data, which may cover both hemispheres, is explained in* §**3.3.1**. §**3.5** describes some standard statistical methods for deciding whether a given random sample of observations is adequately fitted by some specified probability distribution, and whether two independent samples have been drawn from the same (unspecified) distribution; §**3.6** describes the use of simulation as an aid in complicated analyses; §**3.7** describes jackknife procedures and permutation tests; and §**3.8** is a brief discourse on problems of data collection.

The mathematical results presented in §**3.2** are purely for reference purposes, and no derivations are given; most, if not all, of the results are available in standard texts.

3.2 Mathematical methods for unit vectors and axes in three dimensions

3.2.1 *Mean direction, resultant length and centre of mass*

Consider a collection of points $P_1, \ldots, P_n$ on the surface of the unit sphere centred at O, with P_i corresponding to a unit vector with polar coordinates (θ_i, ϕ_i) and direction cosines $x_i = \sin \theta_i \cos \phi_i$, $y_i = \sin \theta_i \sin \phi_i$, $z_i = \cos \theta_i$, $i = 1, \ldots, n$. The direction cosines of P_i can be written as a vector

$$\begin{pmatrix} x_i \\ y_i \\ z_i \end{pmatrix} \quad \text{or its transpose} \quad \begin{pmatrix} x_i \\ y_i \\ z_i \end{pmatrix}' = (x_i \, y_i \, z_i).$$

The angle ψ between the unit vectors $\overrightarrow{OP_1}$ and $\overrightarrow{OP_2}$ is then given (in radians) by

$$\cos\psi = x_1 x_2 + y_1 y_2 + z_1 z_2 = (x_1\ y_1\ z_1)\begin{pmatrix} x_2 \\ y_2 \\ z_2 \end{pmatrix} \quad (3.1)$$

with ψ in the range $0° \leq \psi \leq 180°$ (see Figure 3.1).

The *resultant vector* of the n unit vectors is a vector with direction $(\hat{\theta}, \hat{\phi})$ and length R say, obtained as follows (see Figure 3.2). Let

$$S_x = \sum_{i=1}^{n} x_i, \quad S_y = \sum_{i=1}^{n} y_i, \quad S_z = \sum_{i=1}^{n} z_i. \quad (3.2)$$

Then

$$R^2 = S_x^2 + S_y^2 + S_z^2 \quad (3.3)$$

so that $(\hat{\theta}, \hat{\phi})$ has direction cosines $(\hat{x}, \hat{y}, \hat{z}) = (S_x/R, S_y/R, S_z/R)$.

Thus

$$\begin{aligned} \sin\hat{\theta}\cos\hat{\phi} &= \hat{x} \\ \sin\hat{\theta}\sin\hat{\phi} &= \hat{y} \\ \cos\hat{\theta} &= \hat{z} \end{aligned} \quad (3.4)$$

whence

$$\hat{\theta} = \arccos(\hat{z}), \quad \hat{\phi} = \arctan(\hat{y}/\hat{x}), \quad (3.5)$$
$$0° \leq \hat{\theta} \leq 180°, \quad 0° \leq \hat{\phi} < 360°$$

Figure 3.1 P_1, P_2, P_3, P_4 and P_5 are points representing unit vectors in 3-dimensional space. c is the angle between the vectors $\overrightarrow{OP_1}$ and $\overrightarrow{OP_2}$.

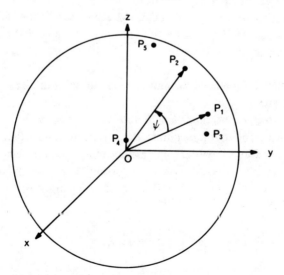

3.2.1. Mean direction and resultant length

where

$$0° < \hat{\phi} < 90° \quad \text{if } \hat{x} > 0, \hat{y} > 0$$
$$90° < \hat{\phi} < 180° \quad \text{if } \hat{x} < 0, \hat{y} > 0 \quad (3.6)$$
$$180° < \hat{\phi} < 270° \quad \text{if } \hat{x} < 0, \hat{y} < 0$$
$$270° < \hat{\phi} < 360° \quad \text{if } \hat{x} > 0, \hat{y} < 0.$$

(Care is required in programming the exceptional cases $\hat{x} = 0$ and $\hat{y} = 0$).

$(\hat{\theta}, \hat{\phi})$ is called the *mean direction* of the n unit vectors, R the *resultant length*, and $\bar{R} = R/n$ the *mean resultant length*. Note that the centre of mass of $P_1, \ldots, P_n$ (considered as unit masses) has coordinates $(\sum x_i/n, \sum y_i/n, \sum z_i/n))$ in the direction $(\hat{\theta}, \hat{\phi})$ at a distance $\bar{R}$ from the origin. Obviously the centre of mass of distinct points on the surface of the unit sphere must be interior to the sphere, so $0 \le \bar{R} \le 1$, $\bar{R} = 1$ corresponding to all points being coincident. If the points are concentrated close together, $\bar{R}$ will be close to 1, whereas increasing scatter results in

Figure 3.2 The vector resultant of $\overrightarrow{OP}_1$, $\overrightarrow{OP}_2$, $\overrightarrow{OP}_3$, $\overrightarrow{OP}_4$ and $\overrightarrow{OP}_5$ is $\overrightarrow{OP}$, with length R. The mean direction is a unit vector in the direction $\overrightarrow{OP}$. The centre of mass of P_1, P_2, P_3, P_4 and P_5 is $\hat{P}$, (a point interior to the unit sphere unless P_1, P_2, P_3, P_4 and P_5 all coincide); the length of $\overrightarrow{O\hat{P}}$ is the mean resultant length $\bar{R}$.

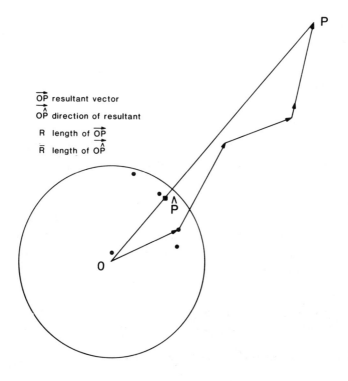

smaller values of $\bar{R}$. However, $\bar{R}$ is not always a reliable indicator of scatter: symmetric placement of the points, for example in two identical groups at opposite ends of an axis, can result in $\bar{R}=0$.

3.2.2 Rotation of unit vectors and axes

Let (θ, ϕ) be the polar coordinates of a unit vector measured relative to a pole in the direction $(0, 0)$. It will frequently be necessary, in subsequent chapters, to find the coordinates (θ', ϕ') say relative to some other direction (θ_0, ϕ_0) as pole. This can be achieved by the following rotation. Let ψ_0 be an arbitrary angle. The general form of a matrix giving a rotation of the x, y and z axes with (θ_0, ϕ_0) as new z direction is

$$\begin{pmatrix} \cos\theta_0 \cos\phi_0 \cos\psi_0 - \sin\psi_0 \sin\phi_0, & \cos\theta_0 \sin\phi_0 \cos\psi_0 + \cos\phi_0 \sin\psi_0, & -\sin\theta_0 \cos\psi_0 \\ -\cos\theta_0 \cos\phi_0 \sin\psi_0 - \sin\phi_0 \cos\psi_0, & -\cos\theta_0 \sin\phi_0 \sin\psi_0 + \cos\phi_0 \cos\psi_0, & \sin\theta_0 \sin\psi_0 \\ \sin\theta_0 \cos\phi_0, & \sin\theta_0 \sin\phi_0, & \cos\theta_0 \end{pmatrix}$$

$$= \mathbf{A}(\theta_0, \phi_0, \psi_0) \qquad (3.7)$$

Then the direction cosines of (θ', ϕ') are given by

$$\begin{pmatrix} \sin\theta' \cos\phi' \\ \sin\theta' \sin\phi' \\ \cos\theta' \end{pmatrix} = \mathbf{A}(\theta_0, \phi_0, \psi_0) \begin{pmatrix} \sin\theta \cos\phi \\ \sin\theta \sin\phi \\ \cos\theta \end{pmatrix} \qquad (3.8)$$

from which (θ', ϕ') can be calculated using (3.5) and (3.6). The angle ψ_0 is a rotation about the polar axis through (θ_0, ϕ_0), and is unimportant in many cases. If it is arbitrarily set to zero, we can use a simplified rotation matrix

$$\mathbf{A}(\theta_0, \phi_0, 0) = \begin{pmatrix} \cos\theta_0 \cos\phi_0 & \cos\theta_0 \sin\phi_0 & -\sin\theta_0 \\ -\sin\phi_0 & \cos\phi_0 & 0 \\ \sin\theta_0 \cos\phi_0 & \sin\theta_0 \sin\phi_0 & \cos\theta_0 \end{pmatrix}. \qquad (3.9)$$

Note that θ' is the angle between (θ, ϕ) and (θ_0, ϕ_0).

For axial data, suppose that the axis (x, y, z) is measured relative to the polar axis $(0, 0, 1)$ and we seek to find its coordinates (x', y', z') relative to (x_0, y_0, z_0). For convenience, we first determine one end of the (x_0, y_0, z_0) axis in polar coordinates using (3.5) and (3.6), to obtain (θ_0, ϕ_0) say. Then proceed as for vectors, to obtain $\mathbf{A}(\theta_0, \phi_0, \psi_0)$ or $\mathbf{A}(\theta_0, \phi_0, 0)$, and hence

$$\begin{pmatrix} x' \\ y' \\ z' \end{pmatrix} = \mathbf{A}(\theta_0, \phi_0, \psi_0) \begin{pmatrix} x \\ y \\ z \end{pmatrix}. \qquad (3.10)$$

z' is the cosine of the smaller of the angles between (x, y, z) and (x_0, y_0, z_0).

3.2.3 Some simple matrix calculations

(i) Matrix required to move a unit 3×1 vector or axis **x** to another such vector or axis **y**.

Calculate the 3×3 symmetric matrix

$$\mathbf{H} = (\mathbf{x}+\mathbf{y})(\mathbf{x}+\mathbf{y})'/(1+\mathbf{x}'\mathbf{y}) - \mathbf{I}_3 \tag{3.11}$$

where $\mathbf{I}_3$ is a 3×3 matrix with 1's on the main diagonal and zeroes elsewhere. Then $\mathbf{Hx} = \mathbf{y}$ and $\mathbf{Hy} = \mathbf{x}$ (Watson 1983a, page 28).

(ii) Inverse of a symmetric 2×2 matrix.

Let

$$\mathbf{Z} = \begin{pmatrix} a & b \\ b & c \end{pmatrix}, \qquad \Delta = ac - b^2 \neq 0.$$

Then

$$\mathbf{Z}^{-1} = \frac{1}{\Delta}\begin{pmatrix} c & -b \\ -b & a \end{pmatrix} \tag{3.12}$$

(iii) Calculation of $\mathbf{Z}^{1/2}$ for a symmetric positive definite 2×2 matrix **Z**.

Compute the eigenvalues and eigenvectors of **Z** as in Step 1 of §**3.2.5**, obtaining t_1 and t_2 as eigenvalues from (3.23), and the matrix **Y** in (3.22) whose columns are the eigenvectors. Define

$$\mathbf{X}^{1/2} = \begin{pmatrix} t_1^{1/2} & 0 \\ 0 & t_2^{1/2} \end{pmatrix} \tag{3.13}$$

Then

$$\mathbf{Z}^{1/2} = \mathbf{Y}\mathbf{X}^{1/2}\mathbf{Y}' \tag{3.14}$$

3.2.4 Eigenvectors, eigenvalues and the moment of inertia

Let $(x_1, y_1, z_1), \ldots, (x_n, y_n, z_n)$ be a collection of unit vectors or axes, and define the *orientation matrix*

$$\mathbf{T} = \begin{pmatrix} \sum x_i^2 & \sum x_i y_i & \sum x_i z_i \\ \sum x_i y_i & \sum y_i^2 & \sum y_i z_i \\ \sum x_i z_i & \sum y_i z_i & \sum z_i^2 \end{pmatrix}. \tag{3.15}$$

If we think of a unit mass being placed at each point (x_i, y_i, z_i), $i = 1, \ldots, n$ then the moment of inertia I of the n points about some axis (x_0, y_0, z_0) is given by

$$n - (x_0 \; y_0 \; z_0)\mathbf{T}\begin{pmatrix} x_0 \\ y_0 \\ z_0 \end{pmatrix}.$$

Accordingly, the variation of this moment of inertia, as the choice of axis varies, gives information about the scatter of the points.

The axis about which the moment is least is called the *principal axis*; to complete the set of three orthogonal coordinates there are two *minor axes*. These axes correspond to the eigenvectors (or *latent vectors* or *characteristic vectors*) of **T**, which we denote $\hat{\mathbf{u}}_1$, $\hat{\mathbf{u}}_2$ and $\hat{\mathbf{u}}_3$. Associated with the eigenvectors are the eigenvalues $\hat{\tau}_1$, $\hat{\tau}_2$, $\hat{\tau}_3$ respectively, which satisfy

$$\hat{\tau}_1 \geq 0,\ \hat{\tau}_2 \geq 0,\ \hat{\tau}_3 \geq 0,\ \hat{\tau}_1 + \hat{\tau}_2 + \hat{\tau}_3 = n. \tag{3.16}$$

We shall assume that $0 \leq \hat{\tau}_1 \leq \hat{\tau}_2 \leq \hat{\tau}_3$. If $\hat{\tau}_1 = \hat{\tau}_2 = \hat{\tau}_3 = n/3$, then there is no axis with greater moment of inertia than any other; otherwise, $\hat{\mathbf{u}}_3$ is the principal axis. The eigenvalues and eigenvectors provide the spectral decomposition of the orientation matrix:

$$\mathbf{T} = \hat{\tau}_1 \hat{\mathbf{u}}_1 \hat{\mathbf{u}}_1' + \hat{\tau}_2 \hat{\mathbf{u}}_2 \hat{\mathbf{u}}_2' + \hat{\tau}_3 \hat{\mathbf{u}}_3 \hat{\mathbf{u}}_3'. \tag{3.17}$$

Computer programs to calculate the eigenvalues and eigenvectors of a symmetric positive definite matrix are available in most standard mathematical software packages. Alternatively, a suitable algorithm is listed in the program described by Diggle & Fisher (1985).

We shall often use the normalized eigenvalues

$$\bar{\tau}_1 = \hat{\tau}_1/n,\ \bar{\tau}_2 = \hat{\tau}_2/n,\ \bar{\tau}_3 = \hat{\tau}_3/n,\ \bar{\tau}_1 + \bar{\tau}_2 + \bar{\tau}_3 = 1. \tag{3.18}$$

3.2.5 Calculation of points on an ellipse

In subsequent chapters, we shall occasionally obtain the confidence region for an unknown reference direction or axis in the form of an elliptical cone. Let the estimate of this direction or axis be denoted by $\mathbf{h}_3$, and let $\mathbf{h}_1$ and $\mathbf{h}_2$ be two other mutually orthogonal unit vectors, such that

$$\mathbf{H} = (\mathbf{h}_1, \mathbf{h}_2, \mathbf{h}_3) \tag{3.19}$$

is an orthogonal matrix. Typically, the elliptical confidence cone will be centred on $\mathbf{h}_3$ rotated to the axis $(0, 0, 1)$, and points (x, y, z) on the rotated ellipse will satisfy an equation of the form

$$Ax^2 + 2Bxy + Cy^2 = D, \qquad x^2 + y^2 + z^2 = 1 \tag{3.20}$$

The problem is to find a set of points on this ellipse relative to the original, unrotated axes.

Step 1 Calculate as follows the eigenvectors and eigenvalues of the matrix

$$\mathbf{Z} = \begin{pmatrix} A & B \\ B & C \end{pmatrix}. \tag{3.21}$$

The eigenvectors are the two columns of

$$\mathbf{Y} = \begin{pmatrix} a & -b \\ b & a \end{pmatrix} \tag{3.22}$$

where
$$a = \alpha/(1+\alpha^2)^{1/2}$$
$$b = 1/(1+\alpha^2)^{1/2}$$
and
$$\alpha = (A-C)/(2B) + \sqrt{[(A-C)^2/(4B^2)+1]} \quad \text{(provided } B \neq 0\text{)}.$$

The eigenvalues are the diagonal elements of **X**, where
$$\mathbf{X} = \mathbf{Y'ZY}$$
$$= \begin{pmatrix} t_1 & 0 \\ 0 & t_2 \end{pmatrix}. \tag{3.23}$$

Step 2 Calculate $g_1 = \sqrt{(D/t_1)}$, $g_2 = \sqrt{(D/t_2)}$. Then $\beta_1 = \arcsin g_1$ and $\beta_2 = \arcsin g_2$ are the semi-vertical angles (in radians) of the major and minor semi-axes of the ellipse.

Step 3 For any given angle ψ between 0 and 2π, let $v_1 = g_1 \cos\psi$, $v_2 = g_2 \sin\psi$. Calculate
$$x = av_1 - bv_2, \quad y = bv_1 + av_2, \quad z = \sqrt{(1-x^2-y^2)}.$$
Then
$$\mathbf{H} \begin{pmatrix} x \\ y \\ z \end{pmatrix}$$
is a point on the perimeter of the desired elliptical confidence region.

Step 4 Repeat *Step* 3 for a range of ψ values between 0 and 2π, to obtain the desired set of points.

3.3 Methods of data display
3.3.1 *Projections*

A projection of spherical data is a representation of the points in the plane. In many cases, the data turn out to be confined to a single hemisphere. If this is not the case, separate projections can be made for the data in each hemisphere (see below) or data from one hemisphere can be projected as small discs (●) and from the other hemisphere as open circles (○).

There is a remarkable profusion of projections, spawned by the diverse requirements of astronomers, cartographers, geologists and others. We shall give brief descriptions of those few most relevant to our purpose, and give references where possible to other descriptions of projections. Let P be a point on the upper hemisphere of a unit sphere, with colatitude θ and longitude ϕ. We denote by P^* its projection onto the x-y plane, with coordinates (x^*, y^*). It is sometimes convenient to work with polar

coordinates (ρ, ψ) in the plane. These are shown in Figure 3.3. We have
$$x^* = \rho \cos \psi, \quad y^* = \rho \sin \phi$$
and conversely,
$$\rho = \sqrt{(x^{*2} + y^{*2})}, \quad \psi = \tan^{-1}(y^*/x^*).$$

The choice of projection depends, of course, on the properties which one wants preserved after projecting the data. For density estimation, an equal-area (Schmidt) projection is appropriate: this projection preserves the density of points, although the shapes of projected groups of points will vary according to their original position on the sphere. On the other hand, if one wants small circles on the sphere plotting as circles in the plane, an equal-angle (Wulff) projection should be used; generally, as noted by Hoek and Brown (1980, page 61), "Engineers tend to prefer the equal angle projection because geometrical constructions required for the solution of engineering problems are simpler and more accurate on this projection than for the equal area projection".

Graph paper is available for some of these projections, which allows the data to be plotted directly. Such background graphs are known as *nets*.

A detailed reference book on the subject is by Richardus and Adler (1972). Phillips (1971) is a specialist reference for Structural Geology; see also Turner and Weiss (1963, Chapter 3) and Hoek and Brown (1980, Chapter 4). Berry and Mason (1959, Chapter 2) discuss projections in

Figure 3.3 Representation of a point P by its projection P^* in a plane, in polar coordinates (ρ, ψ) and rectangular coordinates (x^*, y^*).

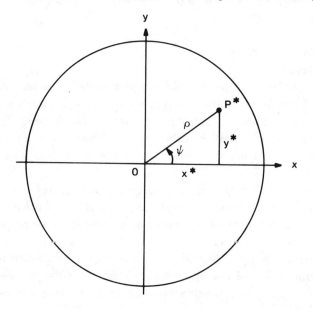

3.3.1. Projections

the context of Crystallography, Bomford (1980, Chapter 4) in the context of Geodetic Astronomy, and Lafeber (1965) in the context of Soil Science.

If the data are spread over both hemispheres, rather than being confined to either the upper or the lower hemisphere, it is highly desirable to be able to display both hemispheres simultaneously, for contouring and other purposes. The method adopted in this book is to consider the sphere cut at the equator and hinged at colatitude $\theta = 90°$, longitude $\phi = 0°$. The upper hemisphere is displayed normally; the lower hemisphere appears reversed from left to right, as shown in Figure 3.8 (see §**3.3.2**). Even if all the data are confined to the lower hemisphere, this convention will be used for consistency. As will be seen, this form of display allows contour plots to be utilised rather better than they might otherwise have been. As axial data are conventionally displayed in the lower hemisphere, they will always appear in the usual (i.e. correct, non-reversed) form. If the data or contours have been rotated prior to display, the rotated position of the original North or South pole is marked with a + sign.

(a) Wulff, or "stereographic", projection

Definition $\rho = \tan(\frac{1}{2}\theta)$, $\psi = \phi$, $0 \leq \rho \leq 1$, $0 \leq \psi < 2\pi$. See Figure 3.4.

Properties This is an equal-angle projection: great and small circles project as circular areas, hence, for example, a contour plot of a unimodal data set which exhibits circular contours following projection onto a

Figure 3.4 Definition of the equal-angle projection P^* of P.

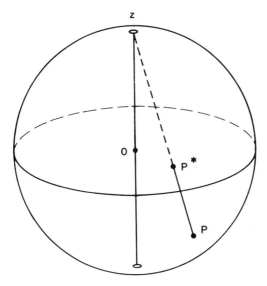

Wulff net indicates that the data are isotropic about their preferred direction (see §**3.3.3**). The Wulff projection is commonly used in engineering work, Crystal Morphology, Palaeomagnetism and Structural Geology.

(b) Lambert, or Schmidt, projection

Definition $\rho = 2\sin(\frac{1}{2}\theta)$, $\psi = \phi$, $0 \le \rho \le \sqrt{2}$, $0 \le \psi < 2\pi$. See Figure 3.5.

Properties This is an equal-area projection: equal areas on the surface of the sphere map to equal areas on the plane of projection. Hence, densities of points are preserved, an important consideration when using contoured densities (see §**3.3.3**).

(c) Orthographic projection

Definition $x^* = \sin\phi$, $y^* = \cos\theta$. See Figure 3.6.

Properties This projection has primarily aesthetic interest when displaying data. However, it is the view seen by astronomers of heavenly bodies.

(d) Gnomonic or central projection

Definition For $\frac{1}{2}\pi < \theta \le \pi$, $\rho = -\tan\theta$, $\psi = \phi$. See Figure 3.7.

Properties This projection can only be used for data confined to one hemisphere, and then it is unsatisfactory for points elevated more than

Figure 3.5 Definition of the equal-area projection P^* of P.

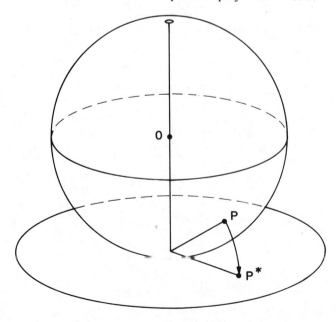

3.3.1. Projections

45° from vertically down (in terms of Figure 3.7). However, it has the property that all great circles are projected as straight lines.

These projections are so-called polar projections. For data clustered around an equator, a pole can be located on that equator and a suitable projection made. Further details of such *equatorial* projections may be found in the general references listed above, and in Mardia (1972, page 216).

Figure 3.6 Definition of the orthographic projection P^* of P.

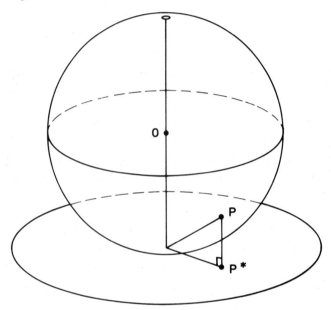

Figure 3.7 Definition of the gnomonic projection P^* of P.

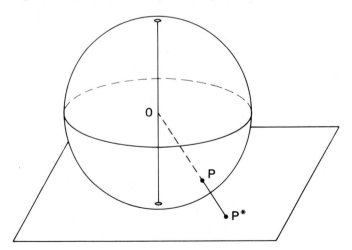

40   3. *Descriptive and ancillary methods; sampling problems*

3.3.2   *Sunflower plots*

In some situations, a sample of observations may either fall into natural sub-groups, or be divided into sub-groups (see e.g. §5.4), as in Figure 3.8. It may then be desirable to summarise each sub-group by some representative quantity, such as the mean direction of its members, and to display these summary quantities. It is possible to do this easily, while still retaining information about the numbers of observations in each sub-group, using the notion of sunflowers (see e.g. Chambers *et al.* 1983). We use the symbols ·, |, ⋏, +, ... to denote 1, 2, 3, 4, ... observations, or, in large samples, · might be 1-5, | 6-10, ⋏ 11-15, and so on. The symbols are then plotted out at the points corresponding to the projections (Schmidt, Wulff, ...) of the mean directions of the sub-groups, each representing the sample size of its sub-group. For data distributed over the whole sphere, either both hemispheres can be plotted, or sunflowers on the hidden side printed in a circle: ⊙, ⓘ, ⊗, etc. As an example, the information in Figure 3.8 is shown in condensed form as a sunflower plot in Figure 3.9.

Information about the degree of concentration of each subgroup about its mean direction can also be included in an individual sunflower by altering its size, or overprinting with circles, but the diagrams start to become more difficult to interpret. An alternative to sunflowers is the use of circles of varying radii - · ∘ ○ - which may be more effective for conveying information about density (cf. §3.3.4).

This form of data presentation is designed for a flat-bed plotter or other high-quality graphics device; however it is also feasible to produce

Figure 3.8 Multimodal sample of data plotted using an equal-area projection. See Example 5.6 and §5.5 for information about the data.

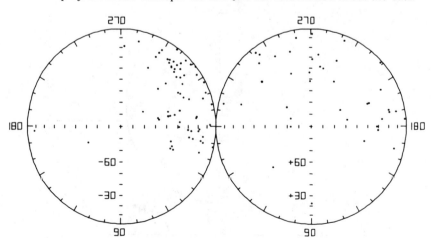

sunflower plots on a line-printer by overstriking various combinations of ·, /, \, -, +, ×.

3.3.3 Density contour and shade plots

Whilst basic projections (and to some extent sunflower plots) have the virtue that they can be performed by hand, allowing rapid visual assessment of a data set, they may not be effective in highlighting qualitative features of the data, such as symmetry or multimodality. An appropriate way to enhance these features is to compute some (nonparametric) estimate of the true density giving rise to the data, and to display either the contours of this estimated density, or a shade plot of it. This necessitates the use of a computer, and a high-quality plotter (at least, for contour plots).

Suitable estimates can be computed in various ways, as discussed by Diggle & Fisher (1985). The key point about any form of nonparametric density estimation is that it involves *some* degree of smoothing of the raw data, *and the amount of such smoothing must be related to the size of the data set*: essentially, the larger the data set, the less smoothing is required (i.e. the spherical cap over which we average gets smaller) so that in the limit, with an enormous data set, we would do no smoothing at all.

Suppose then, that $P_1, \ldots, P_n$ are points on the surface of the unit sphere, with direction cosines $(x_1, y_1, z_1), \ldots, (x_n, y_n, z_n)$ respectively, and that we wish to estimate the density $f(x, y, z)$ at $P = (x, y, z)$. A suitable way to do this is to compute a weighted average of the n data points, with largest weights being given to those values closest to (x, y, z). Thus we estimate $f(x, y, z)$ by

$$\hat{f}(x, y, z) = \sum_{i=1}^{n} W_n(P, P_i) \qquad (3.24)$$

Figure 3.9 Sunflower plot summarising the data in Figure 3.8. Each petal represents 5 data points (see text).

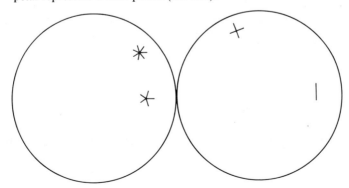

where $W_n(P, P_i)$ is the weight given to the ith point P_i when estimating the density at P, and W_n depends on n. An appropriate choice for W_n is a function of the form

$$W_n(P, P_i) = [C_n/(4\pi n \sinh(C_n))] \exp[C_n(xx_i + yy_i + zz_i)] \tag{3.25}$$

(Diggle & Fisher, 1985), which is the form of the Fisher density (see §4.4.3). The weight depends only on the angular distance or deviation of P_i from P and not on its direction, and the amount of smoothing is controlled by the value of C_n: the larger the value of C_n, the smaller the amount of smoothing. Since we would like to smooth less and less the more data we have, C_n should increase with n.

Diggle and Fisher give two recommendations for choosing C_n. Firstly, if there are reasons to believe that the sample is drawn from a Fisher distribution or at least, is unimodal and approximately isotropic about its polar axis (see §3.4), then use $C_n = \hat{\kappa} n^{1/3}$, where $\hat{\kappa}$ is an estimate of κ in (4.19) computed as in §5.3.2(iv), say. More generally, C_n can be chosen by a cross-validation method, as described by Watson (1983a), a précis of which we give here.

For a given value of C_n, let $\hat{f}_j(P_j)$ be an estimate of the density at P_j for a sample of size $n-1$ obtained by omitting P_j from the original sample:

$$\hat{f}_j(P_j) = \sum_{\substack{i=1 \\ i \neq j}}^{n} W_{n-1}(P_j, P_i). \tag{3.26}$$

Then calculate a *pseudo-log-likelihood* $L(C_n)$ of the data as

$$L(C_n) = \sum_{j=1}^{n} \log\{\hat{f}_j(P_j)\}. \tag{3.27}$$

The function L should increase from its value at $C_n = 0$ to some maximum, and then decrease. This maximum can be found by some simple maximisation procedure.

The computer program listed in Diggle & Fisher (1985) computes a suitable value of C_n in this fashion.

To calculate a density estimate for axial data, the only alteration to (3.24) is that the function W_n is replaced by the Watson density (§4.4.4):

$$W_n(P, P_i) = h(C_n) \exp[C_n(xx_i + yy_i + zz_i)^2] \tag{3.28}$$

where

$$h(C_n) = 1 \Big/ \left[4\pi \int_0^1 \exp(C_n u^2)\, du\right]. \tag{3.29}$$

If there are good reasons to believe that the underlying distribution is Watson bipolar or at least a bipolar distribution approximately isotropic about its principal axis (cf. §3.4) then a suitable choice of smoothing

3.3.3. Density contour and shade plots

constant is $C_n = \hat{\kappa} n^{1/3}$ where $\hat{\kappa}$ is an estimate of κ in (4.31) computed as in §6.3.2(iv). Otherwise, cross-validation can be used, as described for unit vectors.

Given an estimate of the density at any desired point, we may now produce either a *contour plot* or a *shade plot*. For a contour plot, a convenient form of representation is to project the density estimate at suitable points using a Schmidt (equal-area projection), and to contour the projected density; this is done in the program of Diggle and Fisher (1985), and requires a plotting device for output. An alternative, as yet unavailable, would be to produce a three-dimensional picture of the density as mountains and valleys on a spherical surface.

An important consideration when using contour plots is the choice of contour levels. Suppose we have computed values of the function (f say) to be contoured at a reasonably large number N of points on the unit disc, obtaining (in increasing order)

$$f_1 \le f_2 \le \ldots \le f_N. \tag{3.30}$$

For example, with vectorial data we might evaluate f at those gridpoints of a regular grid lying inside the two discs corresponding to projections of the upper and lower hemispheres: see Figure 3.10. A 10×10 grid for each disc yields about 80 points on each disc. Three basic methods for choosing say, k contour levels $c_1, \ldots, c_k$ are:

(a) k values equally spaced between the smallest and largest f-values, f_1 and f_N respectively; thus

$$c_i = f_1 + i \times (f_N - f_1)/(k+1), i = 1, \ldots, k. \tag{3.31}$$

(b) k values dividing the N f-values into $k+1$ approximately equal groups (equal probability contours) each of approximately $N/(k+1)$ values; thus

$$c_i = f_{j(i)}, j(i) = \text{nearest integer to } (i - \tfrac{1}{2})N/k, i = 1, \ldots, k. \tag{3.32}$$

Figure 3.10 Regular grid containing points at which the density $\hat{f}$ is evaluated for choice of contour levels.

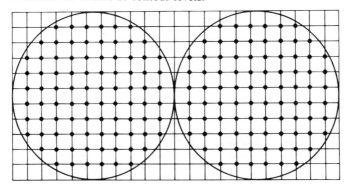

(c) k values dividing the N f-values into $k+1$ groups decreasing in size as the f-values increase. For example, if the smallest interval (which contains the largest f-values) contains M values, we might choose the next interval to contain rM values, $r > 1$, and each successive interval to have r times more values than the preceding one. This requires the choice

$$M = N(r-1)/(r^k - 1).$$

To appreciate the differences between these three alternatives, consider the data in Figure 3.8. Using a 10×10 grid for each disc, we obtain 160 f-values, plotted in increasing order in Figure 3.11. Two key features immediately appear:

- there are a lot of zero or near-zero f-values corresponding to patches of the sphere where there are few if any data

- there are a few f-values which are large relative to the other non-zero f-values.

Clearly it is not desirable to have contours in the bald patches; method (a) will not have this problem, whereas methods (b) and (c) will. On the

Figure 3.11 A plot of function values of the density estimate for the data of Figure 3.8, evaluated at the grid points falling inside the discs in Figure 3.10.

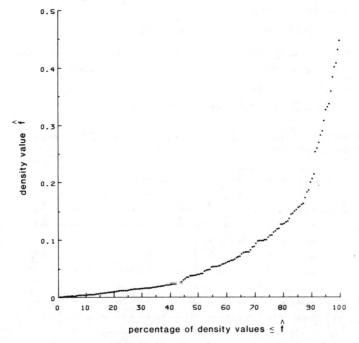

3.3.3. Density contour and shade plots

other hand, since there are only a few high f-values, we would hope to have only one contour at a high level, and the rest delineating details of the great proportion of the data corresponding to non-zero f-values. For the reason that method (a) avoids the earlier problem, it may fail here; method (b) might succeed if the number of contours is large; and some form of method (c) can virtually guarantee one high contour level.

A reasonable approach which is a compromise between (b) and (c) is as follows:

Step 1 Produce a density plot of the function values, like Figure 3.11, and decide approximately what percentage of them are essentially zero. For these data, 30-40% would be appropriate.

Step 2 Discard these small f-values, leaving N_0 values, say.

Step 3 Use method (b) for the N_0 f-values, with the modification that the partition of the N_0 values into $k+1$ equal groups each of $N_0/(k+1)$ values is displaced to the right, with the right-most interval (containing the highest f-values) being only 1/4 of the size (i.e. $N_0/[4(k+1)]$) and the left-most interval being increased correspondingly.

> A contour plot for the data in Figure 3.8 is shown in Figure 3.12, using the smoothing value $C_n = 13.6$ from the cross-validation method. The dotted lines correspond to the lowest contours and the continuous lines to the highest. The contours were obtained by using method (a), which in this case appeared to give the best results.

A *shade plot*, which can be produced either on a plotter or on a lineprinter, is obtained by placing, at the Schmidt projection of (x, y, z) onto a disc, a mark whose density is proportional to the estimated density

Figure 3.12 A contour plot of the data in Figure 3.8. See text.

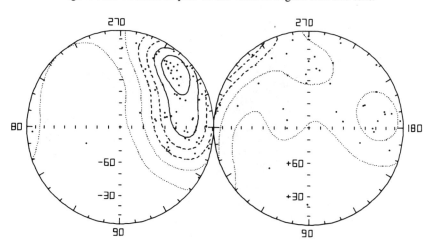

$\hat{f}(x, y, z)$, for a suitably large number of points (x, y, z). On the lineprinter, the variation in intensity of the mark can be obtained by overprinting various keyboard characters (see e.g. Macleod (1970)). This can also be done on a plotter. (The program of Diggle and Fisher has the facility to produce a file containing a set of projected points together with density estimates, for this purpose).

A shade plot for the data in Figure 3.8, using a 30×30 grid for each disc, is shown in Figure 3.13.

3.3.4 Directional/spatial plots

In some situations, data have a temporal or spatial component in addition to being directional. Methods for smoothing such data are considered in §**8.4** including some techniques for graphical display. However, for display of raw data comprising samples taken from a number of sampling sites (where the sites may be geographical, or may be different times, or different stages of treatment), it may suffice to display sunflower plots of the samples at each site. An example of this sort of display is given in Figure 3.14, for hypothetical examples of axes at a number of localities. It shows a strong modal group present in the South-West quadrant of each sample, another shallower group progressing from the North-West to the North-East quadrant as we move across the diagram, a fairly strong mode present only in the two left-most samples, and a few isolated modes.

3.4 Exploratory analysis

In this subsection, we draw together some of the techniques in §**3.2** and §**3.3** to assist us in making inferences about qualitative features of a data set. Typical questions of interest in this phase of analysis are:

Figure 3.13 A shade plot of the data in Figure 3.8. See text.

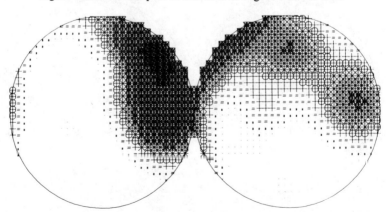

3.4. Exploratory analysis

(i) Are the data unimodal or multimodal (where, for axial data, unimodal means bipolar axial); if multimodal, how many modes are there?

(ii) If the data are unimodal, do they appear to be isotropic about the polar axis, or approximately isotropic except for one or two points, or is the sample basically non-isotropic?

(iii) (Particularly for axial data). Do the data lie approximately about a great or small circle, and if so, are they isotropic with respect to the polar axis?

These questions can often be answered by studying (a) a plot of the raw data, (b) a contour plot of the data, and (c) the mean resultant length $\bar{R}$ (for vectors) and normalised eigenvalues $\bar{\tau}_1$, $\bar{\tau}_2$, $\bar{\tau}_3$ of the orientation matrix **T** of the data (see §**3.2.4**). The information about the distribution of the data, as conveyed by $\bar{R}$ and the relative magnitudes of $\bar{\tau}_1$, $\bar{\tau}_2$ and $\bar{\tau}_3$, is set out in Table 3.1, based on Table 8.3 in Mardia (1972).

Woodcock (1977) has discussed in detail several methods of graphing the eigenvalues to identify the nature of the clustering, in the context of

Figure 3.14 A schematic diagram of several sunflower plots for samples of data with spatial variation. See text.

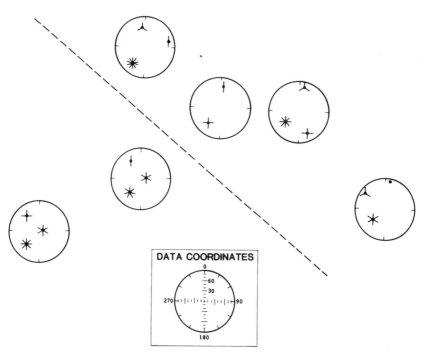

specification of fabric shapes. The following analysis is based on the first graphical method in Woodcock's paper. The reader interested in further detail on this and other eigenvalue methods and critical commentary on their interpretation may wish to refer to Woodcock's paper. (Woodcock & Naylor (1983) also give a brief discussion of these graphical methods, and references to other papers on this general topic).

Because $\bar{\tau}_1 + \bar{\tau}_2 + \bar{\tau}_3 = 1$, there are only two independently varying $\bar{\tau}$'s so it is convenient to work with two ratios, $\bar{\tau}_2/\bar{\tau}_1$ and $\bar{\tau}_3/\bar{\tau}_2$; in fact, for some distributions, these ratios may be as large as 1000 or more, so we use $\log(\bar{\tau}_2/\bar{\tau}_1)$ and $\log(\bar{\tau}_3/\bar{\tau}_2)$ instead. With these variables as axes, various configurations of points may be plotted as shown in the displays in Figure 3.15, which are based on Woodcock (1977, Figure 1) and Woodcock & Naylor (1983, Figure 3). The empirical *shape* criterion $\hat{\gamma} = \log(\bar{\tau}_3/\bar{\tau}_2)/\log(\bar{\tau}_2/\bar{\tau}_1)$ is useful in discriminating girdle-type distributions from clustered distributions; lines of constant $\hat{\gamma}$-value radiate from the origin in Figure 3.15 with increasing *strength* $\hat{\zeta} = \log(\bar{\tau}_3/\bar{\tau}_2)$. Distributions of the girdle type plot below the line $\hat{\gamma} = 1$, and of the cluster type plot above it. Transitional distributions (partly girdle, partly cluster) plot around the line $\hat{\gamma} = 1$. Uniform, or isotropic, distributions plot at the origin ($\hat{\zeta} = 0$). Note that this method does not necessarily help with multimodal data: the mean resultant length can be used as a further discriminator for vectorial data as shown in Table 3.1, or multimodality can be recognised from a contour plot. Of course, care should be exercised in making inferences based on $\hat{\gamma}$ and $\hat{\zeta}$ for small samples (say n less than 25-30).

Table 3.1. *Descriptive interpretation of the shape of a spherical distribution in terms of the normalised eigenvalues of the orientation matrix and, for vectorial data, the mean resultant length.*

Relative magnitudes of eigenvalues	Type of distribution	Other features
$\bar{\tau}_1 = \bar{\tau}_2 = \bar{\tau}_3 = \frac{1}{3}$	uniform	
$\bar{\tau}_3$ large; $\bar{\tau}_1, \bar{\tau}_2$ small		
(i) $\bar{\tau}_1 \neq \bar{\tau}_2$	unimodal if $\bar{R}$ near 1 bimodal otherwise	concentration at one end of $\hat{\mathbf{u}}_3$ concentration at both ends of $\hat{\mathbf{u}}_3$
(ii) $\bar{\tau}_1 \approx \bar{\tau}_2$	unimodal if $\bar{R}$ near 1 bipolar otherwise	rotational symmetry about $\hat{\mathbf{u}}_3$
$\bar{\tau}_1$ large; τ_2, τ_3 large		
(i) $\bar{\tau}_2 \neq \bar{\tau}_3$	girdle	girdle plane spanned by $\hat{\mathbf{u}}_2, \hat{\mathbf{u}}_3$
(ii) $\bar{\tau}_2 \approx \bar{\tau}_3$	symmetric girdle	rotational symmetry about $\hat{\mathbf{u}}_3$

3.4. Exploratory analysis

Figure 3.15 Classification of spherical distributions according to the shape parameter γ and strength parameter ζ, being functions of the eigenvalues of the orientation matrix. (a) theoretical variation of γ and ζ in terms of the two axes $\log(\tau_2/\tau_1)$ and $\log(\tau_3/\tau_2)$ (Based on Woodcock & Naylor (1983, Figure 3)). (b) Typical patterns of samples for given observed values of $\bar{\tau}_1, \bar{\tau}_2$ and $\bar{\tau}_3$ and hence estimated shape and strength parameters $\hat{\gamma}$ and $\hat{\zeta}$. (Based on Woodcock (1977, Figure 1)).

(a)

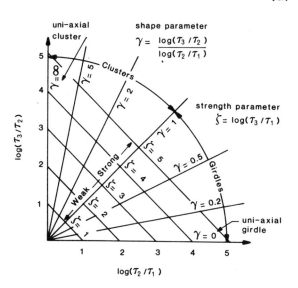

(b)

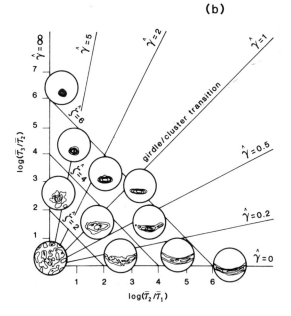

3.5 Probability plots and goodness-of-fit tests

3.5.1 Probability plots

Let $X_1, \ldots, X_n$ be a random sample of *linear* (as distinct from directional) observations, and denote the observations re-arranged in ascending order by $X_{(1)} \leq \ldots \leq X_{(n)}$: $X_{(i)}$ is the *i*th *order statistic*, $i = 1, \ldots, n$.

An important property of the order statistics is that they divide the area under their underlying probability density function into $n+1$ areas $A_1, \ldots, A_{n+1}$ (Figure 3.16(a)), each of which has *average* value $1/(n+1)$. So an informal method of deciding whether a particular distribution (with distribution function $F(x)$ and density function $f(x)$) fits the data, subject to change of origin and scale, is to compare $X_{(1)}, \ldots, X_{(n)}$ with the numbers $a_1, \ldots, a_n$ which divide the area under $f(x)$ into $n+1$ equal areas (Figure 3.16(b)). To do this, we plot the points $(a_i, X_{(i)}), i = 1, \ldots, n$. If the plot is approximately linear, the distribution may be regarded as a satisfactory fit; the slope and intercept of the best-fitting line then provide information about the mean and variance of the distribution. Such a plot is known as a *quantile-quantile* (*Q-Q*) plot.

Figure 3.16 The basis of a probability plot for checking goodness-of-fit of a distribution with pdf $f(x)$ to a sample of data. (a) The order statistics $X_{(1)}, \ldots, X_{(n)}$ of a random sample divide the area under the probability density function $f(x)$ into $n+1$ areas $A_1, \ldots, A_{n+1}$ each with mean value $1/(n+1)$. (b) The values $a_1, \ldots, a_n$ divide the area under $f(x)$ into $n+1$ areas each exactly $1/(n+1)$.

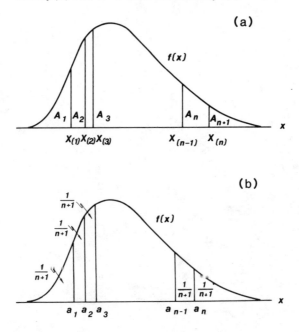

3.5.1. Probability plots

The numbers $a_1, \ldots, a_n$ are calculated from the formula
$$F(a_i) = (i - \tfrac{1}{2})/n, \; i = 1, \ldots, n, \tag{3.33}$$
whence
$$a_i = F^{-1}((i - \tfrac{1}{2})/n), \; i = 1, \ldots, n, \tag{3.34}$$
where F^{-1} is the inverse function to F. We list in Table 3.2 some common distribution functions $F(x)$, together with their inverse functions $F^{-1}(x)$, which will be required later in Chapters 5, 6 and 7 for probability plotting purposes.

In general, if the plot is reasonably linear, of the form $y = a + bx$, the distribution from which the sample was drawn will have as an estimate of its average value μ, $\hat{\mu}_{gr} = a + b \times$ (average value of the F-distribution), and as an estimate of its standard deviation σ, $\hat{\sigma}_{gr} = b \times$ (standard deviation of the F-distribution).

To fix ideas, we give examples of these plots, taken from Lewis & Fisher (1982), using the following artificial data:

47.5, 32.4, 38.1, 43.9, 31.2, 49.5, 43.5, 38.5, 55.9, 40.0, 23.6, 33.7

Thus $X_1 = 47.5, \ldots, X_{12} = 33.7$ ($n = 12$), and the ordered values are $X_{(1)} = 23.6$, $X_{(2)} = 31.2, \ldots, X_{(12)} = 55.9$.

Plotting procedure for the normal model Suppose we wish to examine the normal distribution as a suitable model. Plot $X_{(i)}$ against $Q_i = F^{-1}((i - \tfrac{1}{2})/n)$, $i = 1, \ldots, n$, where F^{-1} is the inverse of the unit normal distribution function in Table 3.2. An approximately linear plot will indicate the plausibility of a normal $N(\mu, \sigma^2)$ model for the data; the intercept and slope of the line will then give rough estimates of μ and σ respectively.

Table 3.2. *Some useful distribution functions $F(x)$ and their inverse functions $F^{-1}(x)$.*

Distribution	$F(x)$	$F^{-1}(x)$
Uniform	$x, \; 0 \leq x \leq 1$	x
(Unit) Exponential	$1 - \exp(-x)$	$-\log(1-x)$
(Unit) Normal	$(2\pi)^{-1/2} \int_{-\infty}^{x} \exp(-\tfrac{1}{2}y^2) \, dy$	See Abramowitz & Stegun (1970, §**2.6.2.22-23**) for approximate formula, or Beasley & Springer (1977) for computer code
χ_1^2	$(2\pi)^{-1/2} \int_0^x y^{-1/2} \exp(-\tfrac{1}{2}y) \, dy$	Use e.g. IMSL subroutine MDCHI

3. Descriptive and ancillary methods; sampling problems

For the artificial data, the observed and theoretical quantities are shown in Table 3.3.

The Q-Q plot of the points $(Q_i, X_{(i)})$ is shown in Figure 3.17. It seems reasonably linear, suggesting that the normal distribution is a plausible model. A straight line fitted by eye yields graphical estimates $\hat{\mu}_{gr} = 40$ for the average value, $\hat{\sigma}_{gr} = 8.5$ for the standard deviation.

Table 3.3. *Order statistics $X_{(1)}, \ldots, X_{(12)}$ of artificial data, and theoretical quantiles of the uniform distribution $((i-\frac{1}{2})/n)$, the normal distribution (Q_i) and the exponential distribution (E_i)*

i	$X_{(i)}$	$(i-\frac{1}{2})/n$	Q_i	E_i
1	23.6	1/24	−1.73	0.04
2	31.2	3/24	−1.15	0.13
3	32.4	5/24	−0.81	0.23
4	33.7	7/24	−0.55	0.34
5	38.1	9/24	−0.32	0.47
6	38.5	11/24	−0.10	0.61
7	40.0	13/24	0.10	0.78
8	43.5	15/24	0.32	0.98
9	43.9	17/24	0.55	1.23
10	47.5	19/24	0.81	1.57
11	49.5	21/24	1.15	2.08
12	55.9	23/24	1.73	3.18

Figure 3.17 A normal probability plot for the artificial data. a_i is the *i*th unit normal quantile, $X_{(i)}$ the *i*th sample quantile. See text and Table 3.3.

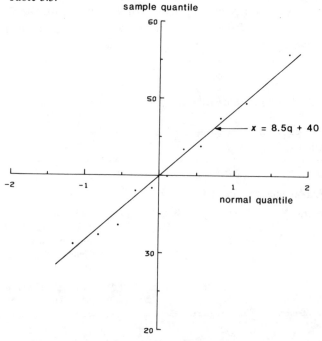

3.5.1. Probability plots

Plotting procedure for the exponential model Suppose that we were interested in the exponential distribution as a model for the data.

Plot $X_{(i)}$ against $E_i = \log(n/(n-i+\frac{1}{2}))$, $i = 1, \ldots, n$. An approximately linear plot through the origin will indicate the plausibility of an exponential model for the data; the slope of the line will then give a rough estimate of $1/\lambda$, the mean of the distribution.

> If we perform a Q-Q plot of $(E_i, X_{(i)})$, $i = 1, \ldots, 12$ for the artificial data, we obtain Figure 3.18. The plot is clearly curved, and not passing through the origin, indicating the unsuitability of the exponential model.

Plotting procedure for the uniform model If the assumed model is uniform with range $(a, a+h)$, we can define its $(i-\frac{1}{2})/n$ quantile in terms of the quantile U_i of the standardized uniform distribution on $[0, 1]$, namely $U_i = (i-\frac{1}{2})/n$. Then the inverse function is $F^{-1}((i-\frac{1}{2})/n) = a + hU_i$.

Plot $X_{(i)}$ against $(i-\frac{1}{2})/n$. An approximately linear plot will indicate the plausibility of a uniform model for the data. The intercept and slope should agree reasonably with the values a and h, if they are known, or will estimate them roughly if they are not known.

Q-Q plots can also be used to check whether two independent random samples seem to come from the same distribution (see Chapter 7). Suppose the samples are $X_1, \ldots, X_m$ and $Y_1, \ldots, Y_n$, with respective sets

Figure 3.18 An exponential probability plot for the artificial data. See text and Table 3.3.

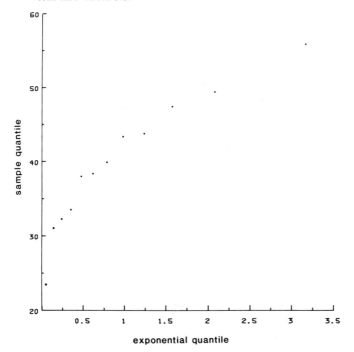

of order statistics $X_{(1)} \leq \ldots \leq X_{(m)}$ and $Y_{(1)} \leq \ldots \leq Y_{(n)}$. *We take first the case* $m = n$. Then plot $(X_{(i)}, Y_{(i)})$, $i = 1, \ldots, n$. If the samples are drawn from the same distribution, the plot should be reasonably well approximated by a straight line passing through the origin with slope 45°. More generally, if the plot is linear and the "best-fitting" line has the form $y = a + bx$, the implication is that the distributions of X and Y have the same shape; also that the average value of the Y-distribution is approximately $a + b \times$ (average value of the X-distribution), and the standard deviation of the Y-distribution is approximately b times that of the X-distribution. *Now suppose that* $m < n$. Then we have to choose a subset of m points from $Y_{(1)}, \ldots, Y_{(n)}$ for plotting purposes. The appropriate points to plot are

$$X_{(i)} \text{ against } Y_{(j_i)}, \text{ where } j_i = \text{integral part of } n(i - \tfrac{1}{2})/m + 1.$$
(3.35)

Again, to fix ideas, we give a simple example using artificial data.

Table 3.4 gives two ordered samples, of sizes $m = 20$ and $n = 25$, and a suitable subset of 20 of the second sample. A Q-Q plot of $(X_{(i)}, Y_{(j_i)})$,

Figure 3.19 A quantile-quantile (Q-Q) plot comparing the two samples of artificial data in Table 3.4.

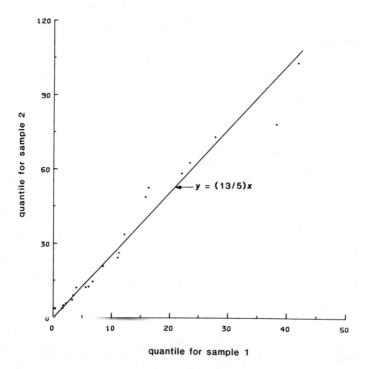

$i = 1, \ldots, 20$, is shown in Figure 3.19. The plot appears reasonably linear, approximately of the form $y = (13/5)x$, suggesting that X and Y have distributions of the same shape, with

average value of $Y = (13/5) \times$ (average value of X)

and

standard deviation of $Y = (13/5) \times$ (standard deviation of X)

3.5.2 Goodness-of-fit tests

The Q-Q plot introduced in the previous subsection allows us to make an *informal* assessment of some distribution function $F(x)$ as model for the data. Often it is desirable to perform a formal test. We describe two such tests, which are of similar type, but which have different applications. As in §3.5.1, let $X_1, \ldots, X_n$ be a random sample hypothesised to be drawn from a distribution $F(x)$, and let $X_{(1)} \leq \ldots \leq$

Table 3.4. *Order statistics* $X_{(1)}, \ldots, X_{(20)}$ *and* $Y_{(1)}, \ldots, Y_{(25)}$ *of two samples of artificial data, to be compared using a Q-Q plot;* $X_{(i)}$ *is plotted against* $Y_{(j_i)}$

i	$X_{(i)}$	$Y_{(i)}$	$j_{(i)}$	$Y_{(j_i)}$
1	0.2	4.0	1	4.0
2	1.5	4.1	2	4.1
3	1.6	5.1	4	5.3
4	2.1	5.3	5	6.1
5	3.2	6.1	6	7.5
6	3.4	7.5	7	9.4
7	3.9	9.4	9	12.4
8	5.5	10.9	10	12.6
9	6.0	12.4	11	12.8
10	6.7	12.6	12	15.0
11	8.4	12.8	14	21.1
12	10.9	15.0	15	24.5
13	11.1	17.7	16	26.4
14	12.1	21.1	17	34.0
15	15.7	24.5	19	48.9
16	16.2	26.4	20	52.7
17	21.9	34.0	21	58.5
18	23.3	41.9	22	63.0
19	38.1	48.9	24	68.3
20	41.8	52.7	25	103.0
21		58.5		
22		63.0		
23		63.3		
24		78.3		
25		103.0		

$X_{(n)}$ be the order statistics of the sample. Calculate

$$D_n^+ = \text{maximum of } [i/n - F(X_{(i)})], \quad i = 1, \ldots, n \qquad (3.36)$$

and

$$D_n^- = \text{maximum of } [F(X_{(i)}) - (i-1)/n], \quad i = 1, \ldots, n. \qquad (3.37)$$

The Kolmogorov-Smirnov statistic is defined by

$$D_n = \text{maximum } (D_n^+, D_n^-) \qquad (3.38)$$

The Kuiper statistic is defined by

$$V_n = D_n^+ + D_n^-. \qquad (3.39)$$

In subsequent applications, the distribution functions $F(x)$ may have to be estimated (in part) from the data. Because of this, the distributions of D_n and V_n may depend on the particular situation. However, if $F(x)$ is known completely, *the distributions of D_n and V_n do not depend on $F(x)$*. Provided $n \geq 8$ we may then calculate the modified statistics

$$D_n^* = D_n(n^{1/2} + 0.12 + 0.11/n^{1/2}) \qquad (3.40)$$

and

$$V_n^* = V_n(n^{1/2} + 0.155 + 0.24/n^{1/2}). \qquad (3.41)$$

The distributions of D_n^* and V_n^* are essentially independent of sample size n; selected critical values for D_n^* and V_n^* are given in Appendix A6.

3.6 Simulation techniques

3.6.1. Uses of simulation

In a book concerned with Data Analysis, it may strike the (non-statistical) reader as paradoxical to find a section on how to simulate samples of data. However, simulation methods have an important role to play in at least two areas of Statistics:

(i) *Evaluating a new statistical procedure* Suppose we develop a new statistical test for the hypothesis that a sample of unit vectors has been drawn from, say, the uniform distribution on the sphere. We may wish to compare it with existing tests in respect of ability to detect departure from the null hypothesis - for example, that the distribution is unimodal. It may be possible to investigate this possible superiority theoretically (i.e. mathematically). Alternatively, one can generate a large number of samples from some unimodal distribution by *simulation*, and look at the performances of the various competing tests over the many samples to see which is most successful in rejecting the null hypothesis when it is not true. In this way, we *know* what the true situation is, and by taking a sufficiently large number of samples, we can determine the behaviour of the various tests for this precisely-controlled "environment".

3.6.1. Uses of simulation

Use of the simulation technique for precisely this example is described by Diggle, Fisher & Lee (1985).

(ii) *Assessing the variability of an estimate from data* Suppose we have estimated some parameters, and have no obvious way of assessing the error or variability of the estimating process. This can often be done using simulation techniques. To illustrate this first on a simple problem (for which a satisfactory solution already exists), suppose we have a random sample of n unit vectors from a Fisher distribution (a unimodal distribution on the sphere determined completely by its mean direction (α, β) and concentration parameter κ - see §4.4.3) and seek to estimate the mean direction, and to assign some measure of error to this estimate. Let $(\hat{\alpha}, \hat{\beta})$ and $\hat{\kappa}$ be the usual parameter estimates (**§5.3.2(iv)**). Now *simulate* a large number N of samples of size n from the Fisher distribution with mean direction $(\hat{\alpha}, \hat{\beta})$ and concentration parameter $\hat{\kappa}$, and for each sample i compute an estimate $(\hat{\alpha}_i, \hat{\beta}_i)$ of the mean direction. Then plot the N estimates $(\hat{\alpha}_1, \hat{\beta}_1), \ldots, (\hat{\alpha}_N, \hat{\beta}_N)$: the pattern of points so obtained gives an idea of the error in $(\hat{\alpha}, \hat{\beta})$. Further, a cone centred on $(\hat{\alpha}, \hat{\beta})$ and containing 95% of the means (counting the actual mean $(\hat{\alpha}, \hat{\beta})$) would be an approximate 95% confidence cone for the unknown (α, β). (For example, if $N = 99$, a cone centred on $(\hat{\alpha}, \hat{\beta})$ and containing 94 of the simulated mean directions would be used).

Provided that a reasonable estimate of κ is used, the simulation method will give satisfactory results.

An even simpler version of this problem is obtained by dropping the specific assumption that the underlying distribution is Fisherian, and assuming simply that it is unimodal. In this case, the *only* knowledge we have about the distribution is that contained in the data. To assess the variability of our estimate, we proceed as follows. Generate n pseudo-random numbers $r_1, \ldots, r_n$ between 0 and 1, and define m_i = integer part of $(n \times r_i + 1)$, $i = 1, \ldots, n$. This gives n pseudo-random integers $m_1, \ldots, m_n$ between 1 and n, with some m_i's probably being the same. Form the sample $(\theta_{m_1}, \phi_{m_1}), \ldots, (\theta_{m_n}, \phi_{m_n})$ (some of whose members will probably be the same) and compute the sample mean direction. Repeat this whole procedure say 200 times, giving 200 pseudo-estimates of the mean direction, and plot them. This will give an indication of the variability of the original estimate.

This method of estimating errors is one of a collection known in Statistics as "bootstrap techniques" (see Efron (1982), Efron & Gong (1983)).

Of the two sorts of applications listed above, the latter is of particular value to us. While the example given was simple, and a perfectly good alternative procedure for it exists already, a bootstrap method may be

58   3. Descriptive and ancillary methods; sampling problems

the only feasible approach in more complex situations. This is exemplified by the following problem (which is discussed further in §**8.4.3**):

> Consider samples of unit vectors collected at a number of different locations corresponding to different periods of time. It is required to fit a smooth curve through the data, as a function of time, and to provide some assessment of the variability of this curve.

In this problem, errors occur in estimating sample mean directions and in specifying ages. Further, one may have only a point estimate of each age and a hazy idea of the error of this estimate (e.g. 350 my as the estimate, with maximum error ±50 my). In such a case, in simulating the error of the fitted path we would be particularly interested in the greatest effect of the error in age.

3.6.2   How to simulate data from various distributions

The first requirement is a generator of so-called *pseudo-random* numbers, that is, a process which will supply a sequence of numbers $R_1, R_2, R_3, \ldots$ of any desired length, such that each R_i is uniformly distributed in the interval [0, 1] and such that the sequence is free from pattern (for example, serial association). There are many methods available for producing pseudo-random numbers, and their virtues and faults reasonably well-known. Almost all computing systems have a machine function which produces these numbers. However, many of these generators create sequences with considerable pattern, so the user should seek advice before employing the generator offered by a particular computer. On the other hand, a package such as IMSL produces "reasonable" sequences. Alternatively, a piece of FORTRAN code published by Wichmann & Hill (1982) can be used.

In the sequel, $R_0, R_1, R_2, R_3, \ldots$ is taken to be a sequence of pseudo-random variates.

Uniform distribution $U(a, a+h)$

$$U = a + hR_1$$

Exponential distribution $E(\lambda)$

$$E = -\lambda \log(R_1)$$

Normal distribution $N(\mu, \sigma^2)$

1. $U = R_1$, $V = R_2$
2. Set $X = 1.71553(U - 0.5)/V$
3. If $X^2 \leq 4(1 - U)$ or $X^2 \leq -4 \log(U)$ go to 6
4. $U = R_3$, $V = R_4$ (or $U = R_5$, $V = R_6$ etc)
5. Go to 2
6. $Z = \sigma X + \mu$ is the required deviate

This is based on the method of Kinderman & Monahan (1977).

3.6.2. How to simulate data

Uniform distribution on the sphere U_S

1. Simulate independent variates N_1, N_2, N_3 from the standard normal $N(0, 1)$ distribution
2. $X = N_1/(N_1^2 + N_2^2 + N_3^2)^{1/2}$
   $Y = N_2/(N_1^2 + N_2^2 + N_3^2)^{1/2}$
   $Z = N_3/(N_1^2 + N_2^2 + N_3^2)^{1/2}$
3. The colatitude Θ and longitude Φ are then obtained from the direction cosines (X, Y, Z) by relations of the type (3.4)–(3.6).

Fisher distribution $F\{(\alpha, \beta), \kappa\}$

0. Set $\lambda = \exp(-2\kappa)$
1. Colatitude $\Theta = 2 \arcsin \sqrt{\{-\log[R_1(1-\lambda)+\lambda]/2\kappa\}}$
2. Longitude $\Phi = 2\pi R_2$
3. Rotate (Θ, Φ) to (Θ', Φ') using (3.8) with $A(\alpha, \beta, 0)$ as in (3.9)
4. (Θ', Φ') is the required pseudo-random variate.

This method was given by Fisher, Lewis and Willcox (1981).

Watson distribution $W\{(\lambda, \mu, \nu), \kappa\}$

(a) *Bipolar distribution* $(\kappa > 0)$

0. Set $C = 1/(\exp(\kappa) - 1)$
1. $U = R_1$, $V = R_2$
2. Set $S = (1/\kappa)\log(U/C + 1)$
3. If $V \leq \exp(\kappa S^2 - \kappa S)$ go to 6
4. $U = R_3$, $V = R_4$ (or $U = R_5$, $V = R_6$ etc.)
5. Go to 2
6. Colatitude $\Theta = S$, longitude $\Phi = 2\pi R_0$
7. Rotate $(X = \sin\Theta\cos\Phi$, $Y = \sin\Theta\sin\Phi$, $Z = \cos\Theta)$ to (X', Y', Z') using (3.8) with $A(\alpha, \beta, 0)$ as in (3.9), where $\sin\alpha\cos\beta = \lambda$, $\sin\alpha\sin\beta = \mu$, $\cos\alpha = \nu$, and (α, β) are found using equations such as (3.4)–(3.6).
8. (X', Y', Z') are the direction cosines of the required pseudo-random variate.

(b) *Girdle distribution* $(\kappa < 0)$

0. Set $C_1 = \sqrt{|\kappa|}$, $C_2 = \arctan(C_1)$
1. $U = R_1$, $V = R_2$
2. Set $S = (1/C_1)\tan(C_2 U)$
3. If $V < (1 - \kappa S^2)\exp(\kappa S^2)$ go to 6
4. $U = R_3$, $V = R_4$ (or $U = R_5$, $V = R_6$, etc.)
5. Go to 2
6. Proceed as in 6–8 for bipolar case.

60   3. *Descriptive and ancillary methods; sampling problems*

(In large simulations, it may be desirable to increase the speed of the Watson generators by using so-called squeeze techniques, that is, by using simple bounds on the exponential function occurring in Step 3, and testing against those first).

The methods for generating pseudo-random Watson variates are due to Best & Fisher (1986).

3.7 The jackknife method; permutation tests

(*i*) *The jackknife method.* The jackknife method can sometimes be used, with a large sample of data, to calculate an approximate confidence region for a parameter of interest. Suppose that we wish to estimate the parameter ψ from the sample $X_1, \ldots, X_n$, and that $\hat{\psi}$ is a biased estimate of ψ, which is asymptotically normally distributed. The first step is to find an estimate which is less biased (in large samples), the so-called *jackknife estimate* $\hat{\psi}_J$. To do this, calculate the n pseudo-values

$$\hat{\psi}_i = n\hat{\psi} - (n-1)\hat{\psi}_i^*, \quad i = 1, \ldots, n, \tag{3.42}$$

where $\hat{\psi}_i^*$ is computed from $X_1, \ldots, X_n$ with X_i omitted. Then

$$\hat{\psi}_J = (1/n) \sum_{i=1}^{n} \hat{\psi}_i. \tag{3.43}$$

Next, calculate the *standard deviation* $n^{-1/2} s_J$ of $\hat{\psi}_J$, where

$$s_J^2 = [n(n-1)]^{-1} \sum_{i=1}^{n} (\hat{\psi}_J - \hat{\psi}_i)^2. \tag{3.44}$$

An approximate $100(1-\alpha)\%$ confidence interval for ψ is given by

$$(\hat{\psi}_J - s_J z_{\alpha/2}, \hat{\psi}_J + s_J z_{\alpha/2}) \tag{3.45}$$

where $z_{\alpha/2}$ is the upper $100(\alpha/2)\%$ point of the standard normal distribution (Appendix A1).

If $\hat{\psi}$ is an unbiased estimate, $\hat{\psi}$ can be used instead of $\hat{\psi}_J$ in the confidence interval in (3.45).

(*ii*) *Permutation or randomisation procedures.* We give a brief discussion of these procedures in the context of testing a hypothesis, and more particularly, of testing a hypothesis of correlation between two variates. Application to estimation problems will not arise explicitly in this book.

Suppose we are interested in testing for correlation between two real (i.e. linear) random variables X and Y, using a sample of independent measurements $(X_1, Y_1), \ldots, (X_n, Y_n)$. The kernel of the reasoning behind a permutation test is that, *under the null hypothesis that X and Y are independent, the pairing of X_1 with Y_1, X_2 with $Y_2, \ldots, X_n$ with Y_n is just one of many equally likely pairings that could have occurred, given the specific X-values $X_1, \ldots, X_n$ and Y-values $Y_1, \ldots, Y_n$.* For example, another equally likely sample is $(X_1, Y_2), (X_2, Y_3), \ldots, (X_{n-1}, Y_n)$,

3.7. The jackknife method; permutation tests

(X_n, Y_1). The total number of different possible samples corresponds to the number of different permutations of the Y-sequence $Y_1, \ldots, Y_n$ relative to the fixed X-sequence $X_1, \ldots, X_n$, that is, $n!$ possible samples.

To perform a permutation test, we need to find some statistic which is a sensible measure of correlation: the usual statistic is

$$r = \frac{\sum (X_i - \bar{X})(Y_i - \bar{Y})}{\{\sum (X_i - \bar{X})^2 \sum (Y_i - \bar{Y})^2\}^{1/2}}, \quad (3.46)$$

where $\bar{X} = \sum X_i/n$, $\bar{Y} = \sum Y_i/n$. The first step is to calculate r for the sample in hand; call it r^*. Next, obtain a different pairing of the X's and Y's by permuting the Y's, and calculate the sample correlation for this new sample. Repeat for all possible samples. We obtain a total of $n!$ correlations $r_1, \ldots, r_N$ ($N = n!$).

The next step is to see whether our actual value r^* is an extreme value compared with the other $N - 1$ correlation values. Re-arrange the correlations so that they are in increasing order:

$$r_{(1)} \leq r_{(2)} \leq \ldots \leq r_{(N)}. \quad (3.47)$$

We suppose, for definiteness, that the alternative hypothesis of interest is that X and Y are positively correlated. In such a case, we would tend to reject the hypothesis of independence if r^* were "significantly large" compared with the other r-values; for example, to test at the $100\alpha\%$ level, reject the hypothesis if r^* is among the $100\alpha\%$ largest values (the *critical region*). The interpretation of this is that our basic assumption (that all $n!$ possible samples were equally likely) was wrong, that there is in fact something special about the specific pairing of X_1 with Y_1, X_2 with $Y_2, \ldots, X_n$ with Y_n, and that it represents (in this case) positive association.

The computational aspect of a permutation test is a matter of considerable importance. For all except the smallest sample sizes, it is not feasible to produce all $n!$ possible samples. This problem can be surmounted satisfactorily by generating a large number M (say 1000 or 2000) *random* permutations of $Y_1, \ldots, Y_n$, and calculating the correlations for the corresponding M samples. Algorithms for enumerating all $n!$ permutations of n objects, or for generating a random permutation of n objects, are given by Nijenhuis & Wilf (1978; page 54; page 62).

Another computational point is that it may be possible to calculate a simplified version of the statistic. For the correlation r, certain quantities do not change from sample to sample, specifically $\bar{X}$, $\bar{Y}$, $\sum (X_i - \bar{X})^2$ and $\sum (Y_i - \bar{Y})^2$. On this basis, it can be shown that instead of (3.46) we can use

$$s = \sum X_i Y_i \quad (3.48)$$

as our measure of correlation.

3. Descriptive and ancillary methods; sampling problems

To fix ideas, we use a trivial set of data: suppose our sample is just four points (3,1), (8,3), (11,7), (12,6). Under the hypothesis of independence, given these observed X's and Y's there are 23 other samples (i.e. $4!-1$) that we could have obtained, as listed in Table 3.5. From (3.48), the observed value of s is $3+24+77+72=176$; s-values for the other samples are in Table 3.5. When arranged in increasing order, the resulting sequence is shown at the foot of Table 3.5. Taking as our alternative the hypothesis that X and Y are positively associated, for a test at the 5% level (approximately) our observed value would have to be in the top 5% of values – in this case, just the largest value, which it is not. (Note that it is not always possible to get a test at exactly the 100α% level: the possible levels are multiples of $(100/n!)$% or, if we have used M random permutations, multiples of $(100/M)$%. In this case, the exact level of the test is $(100/24)$% ≈ 4%.

Table 3.5. *Calculations for permutation test for correlation between X and Y; see text*

	Permutation No.	Permuted Y-values				$S=\sum X_i Y_i$
(original Y-data)	1	1	3	7	6	176
	2	3	1	7	6	166
	3	7	1	3	6	134
	4	1	7	3	6	164
	5	3	7	1	6	148
	6	7	3	1	6	128
	7	6	3	1	7	137
	8	3	6	1	7	152
	9	1	6	3	7	168
	10	6	1	3	7	143
	11	3	1	6	7	167
	12	1	3	6	7	177
	13	1	7	6	3	161
	14	7	1	6	3	131
	15	6	1	7	3	139
	16	1	6	7	3	164
	17	7	6	1	3	116
	18	6	7	1	3	121
	19	6	7	3	1	119
	20	7	6	3	1	114
	21	3	6	7	1	146
	22	6	3	7	1	131
	23	7	3	6	1	123
	24	3	7	6	1	143
X-data		3	8	11	12	

Ordered value #	1	2	3	4	5	6	7	8	9	10	11	12
Ordered S-value	114	116	119	121	123	128	131	131	134	137	139	143
Ordered value #	13	14	15	16	17	18	19	20	21	22	23	24
Ordered S-value	143	146	148	152	161	164	164	166	167	168	176	177

Permutation tests can be applied in any situation in which it is possible to identify a set of possible samples which, under the null hypothesis, and conditional on the observed values, were just as likely as the observed sample. To mention briefly one other situation without pursuing it in depth, suppose we wish to compare the average values of two real (i.e. linear) random variables X and Y, assumed to have identical distributions except, possibly, for a difference in their average values. We have available a sample of m independent measurements $X_1, \ldots, X_m$ of X, and a similar sample of n measurements $Y_1, \ldots, Y_n$ of Y. Under the null hypothesis that the average values are the same, X and Y have identical distributions. Thus, the $m + n$ observations $X_1, \ldots, X_m, Y_1, \ldots, Y_n$ constitute a random sample from this common distribution, and given their specific values, there are many allocations of m observations to one sample and n to the other sample which could have occurred and which would have been equally likely. Specifically, there are $\binom{m+n}{m}$ possible, equally likely, samples. A suitable measure of difference between the population averages is the difference between the sample averages, $\sum_{i=1}^{m} X_i/m - \sum_{j=1}^{n} Y_j/n$, and we can proceed analogously to the correlation example.

A statistical treatment of such procedures is given, for example, in Maritz (1981).

3.8 Problems of data collection; sampling

The most important aspect of statistical analysis of a set of data is knowledge of the sampling methods used to obtain the data, for without this there is no valid way of assessing the variability. For example, in estimating a mean palaeomagnetic direction from a sample of measurements of n specimens, it would be important to know that the time span represented by the specimens was sufficient for averaging out of secular variation, but not so great that substantial change in the direction of the Earth's magnetic field had taken place. It is useful to distinguish three types of sampling problems:

(i) instrumental problems
(ii) choice-of-coordinate-system problems
(iii) problems of spatial or temporal variation in the variate being studied

(i) The difficulties which arise here are caused by instrument "drift", that is, systematic variation of the measuring instrument with time. The effect of drift is to cause serial association (association between consecutive members of the sequence of measurements in the order they were made) in an otherwise independent set of data. A test for serial association is given in §**8.4.2**.

(ii) This sort of problem has been noted by Watson (1970; 1983a, page 19). To quote from the description given in these references on coordinate systems:

> There are six combinations {(Strike, dip, trend), (Strike, dip, pitch), (Strike, dip, plunge), (Strike, trend, plunge), (Dip, trend, plunge), (Dip, trend, pitch)} which could be used to specify the orientation of a directed line in a plane. The combination which gives the greatest accuracy depends on the practical situation. The measurement errors in some of these quantities are correlated and dependent on the values of other quantities, e.g. when a plane is almost horizontal (low dip) its strike is hard to determine. While these problems are well known and intuitive "solutions" are part of the practical training of geologists, there do not appear to be any formal studies in print.

(iii) Problems due to spatial or temporal variation are very common, particularly in the Earth Sciences. In some situations we are interested in a "representative" value for a given region. Instances of this are:

> *Example* 3.1 A mean palaeocurrent direction for measurements of palaeocurrents made in a number of rock units in a defined area.
>
> *Example* 3.2 (mentioned above). A mean palaeomagnetic direction for measurements of directions made in samples of sufficient range of ages to average out the effect of secular variation.

In each of these examples, the difficulties are overcome by using some form of representative sampling scheme. A basic reference is Cochran (1977).

On the other hand, it may be the spatial or temporal variation which is of interest, in which case it is equally important to sample in a representative manner. Thus, in Example 3.1, we may wish to obtain smoothed estimates of the various palaeocurrent directions in the area (the statistical methods for which are discussed in §**8.4.3**); and in Example 3.2, to estimate Apparent Polar Wander as expressed in data collected in a continent over an *extended* time period (see the statistical methodology in §**8.4.3**). Two points to note in such cases are

(a) the hierarchical structure of variability

and

(b) the need to make more measurements at sites (or ages) of greater variability

> For Example 3.1, the lowest level of variability will be that due to repeated sampling at a single site (within a bed or stratum) to estimate the local mean direction. The next level of variability is the site-to-site (between beds or between strata) variation, which will usually be of far greater magnitude (otherwise there would be no detectable spatial structure); and then perhaps area-to-area variation. Typically, some strata will exhibit far greater variability than others and should be

3.8. Problems of data collection; sampling

sampled more abundantly. Some guidance on how to do this can be obtained from a pilot study, as described briefly below. In Example 3.2, the lowest level of variability is provided by the sample of independent measurements made on a given rock unit, from which a single pole position is estimated with appropriate cone of confidence. The semi-angle of each confidence cone should be rather smaller than the angular separation of two consecutive pole positions, the next level of variability.

(Statistical methods for studying hierarchical error structures are considered in §7.2.3).

It is often possible to make a *pilot study* (or *reconnaissance survey*) of an area later to be studied extensively. A prototype for such a study for the palaeocurrent example might be the following:

1. Suppose the area displayed in Figure 3.20(a) is chosen for study. Resources are available to select 20 locations and to sample from each. A sampling grid is shown in Figure 3.20(b). Choose a site in each square (either at random, having pre-chosen randomly two coordinates in the square; or use the midpoint of each square) and sample there (25 measurements, for example). If unable to sample exactly at the location specified by the sampling plan, then use a pre-determined strategy for finding a suitable site nearby.

2. For each (statistical) sample of 25 measurements, compute the mean direction and spherical standard error $\hat{\sigma}$ (see §**5.3.1(v)**). In the follow-up study, sample in each area proportionally to the value of $\hat{\sigma}^2$ for that area. (Of course, a follow-up study may well involve rather more sampling sites; the several sampling sites falling into one of the original sampling squares would each be sampled according to this rule).

Figure 3.20 (a) A hypothetical region to be sampled. (b) A possible sampling frame for a pilot study; see text.

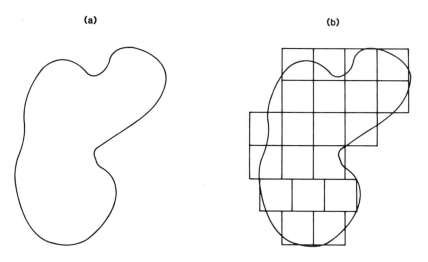

3. Descriptive and ancillary methods; sampling problems

Sampling problems have been discussed by several writers, in the context of their respective subjects. We review this material briefly.

Potter and Pettijohn (1977), in the context of palaeocurrent analysis, consider many of the issues raised above in far more detail, and also point out the merit of sampling an area rather than collecting *all available* data. As well, they consider the problem of tilt correction, and make specific recommendations about sample sizes when collecting palaeocurrent data.

Turner & Weiss (1963) quote from the geological literature on the subject of sampling problems, It is appropriate to give their own distillation of these problems as they relate to fabric analysis (*op. cit.*, page 153):

> The main restrictions to the application of rigorous statistical methods stem from the necessarily somewhat fortuitous nature of the sampling. Except in unusually completely exposed bodies, data can be gathered only at localities decided by the whims of subaerial erosion. Areas of excellent exposure may be surrounded by areas of poor or no exposure; and even where exposed, rocks may be inaccessible to the geologist, as on vertical cliffs. The statistical evaluation of data from macroscopic domains is made questionable also by the extent of necessary extrapolation into unsampled parts of the body. Data are measured in a relatively thin skin of exposed material of large areal extent. As the size of this area increases the depth to which structural observations must be extrapolated increases, if full geologic use is to be made of the data. Added to these difficulties is an almost complete ignorance of the sources and relative magnitude of errors in measurement and observation. For example, are the errors involved in measuring the attitude of a dipping surface with a Brunton compass of the same magnitude as those involved in measuring a plunging lineation? Are either of these errors as large, in a given situation, as the errors associated with superficial gravitational creep of rocks exposed on steep slopes?

The reader is referred to this book for other information pertinent to fabric analysis.

Irving (1964, pages 57-58) and McElhinny (1973, pages 68-70) discuss sampling problems specifically related to palaeomagnetic data, including sources of error, hierarchical structure of error, and temporal effects; for other information on the hierarchical structure of errors, see Watson & Irving (1957).

4
Models

4.1 Introduction

In this chapter we give an account of the main probability distributions at present available for modelling spherical data. A distribution can be specified equivalently by its probability density function (pdf) or by its probability density element (pde); we begin by clarifying this point as a necessary preliminary.

Probability density functions and probability density elements Consider first a random variable X on the line with probability density function (pdf) $f(x)$; for example, X distributed exponentially with pdf $f(x) = \lambda \exp(-\lambda x)$ $(x > 0)$. The proportion of X-values in the population which lie in the *elemental range* between x and $x + dx$, in other words the probability $\text{Prob}(x \leq X \leq x + dx)$, is $f(x)\,dx$; we can call this the probability density *element* (pde) of X. The distribution can be specified equivalently by the pdf $f(x)$ (e.g. $\lambda \exp(-\lambda x)$) which is *not* a probability, or by the pde $f(x)\,dx$ (e.g. $\lambda \exp(-\lambda x)\,dx$) which *is*. The same applies to a vectorial random variable (Θ) on the circle with pdf $f(\theta)$, and equivalently with pde $f(\theta)\,d\theta$, the proportion of Θ-values in the *elemental range* between θ and $\theta + d\theta$. An example is the von Mises distribution (see §**4.3** below) with pdf

$$f(\theta) = C \exp(\kappa \cos \theta),$$

where

$$C = 1/[2\pi I_0(\kappa)] \qquad (0 \leq \theta < 2\pi)$$

and equivalently pde $C \exp(\kappa \cos \theta)\,d\theta$.

So far, the distinction between pdf and pde only involves multiplying by the elemental range dx or $d\theta$. By contrast, with a spherical random variable (Θ, Φ) a different situation arises. Suppose (Θ, Φ) (which may be either vectorial or axial) has pdf $f(\theta, \phi)$. The elemental area of the

67

unit spherical surface having Θ between θ and $\theta + d\theta$ and Φ between ϕ and $\phi + d\phi$ is *not* $d\theta\, d\phi$ but

$$dS = \sin\theta\, d\theta\, d\phi \tag{4.1}$$

which we write dS as it is an element of surface. The proportion of (Θ, Φ)-values in the population which lie in the elemental *joint range* between

$$\theta \text{ and } \theta + d\theta; \quad \phi \text{ and } \phi + d\phi,$$

i.e. which lie in the elemental *surface region* dS, is certainly

$$f(\theta, \phi)\, d\theta\, d\phi;$$

but this is *not* $f(\theta, \phi)\, dS$, it is

$$h(\theta, \phi)\, dS$$

where $h(\theta, \phi)$ is a function related to the pdf $f(\theta, \phi)$ by

$$f(\theta, \phi) = h(\theta, \phi) \sin\theta \tag{4.2}$$

The distribution of (Θ, Φ) can be specified either by the pdf $f(\theta, \phi)$ or by the pde $h(\theta, \phi)\, dS$; we will make use of both forms.

Figure 4.1(a) Probability density function $f(\theta, \phi)$ of the Fisher distribution referred to its mean direction, as a function of θ.

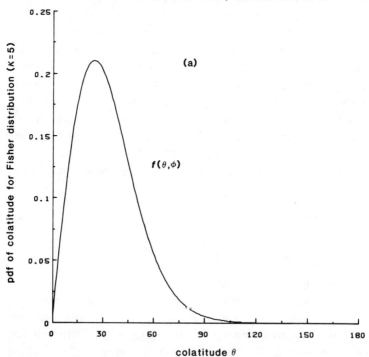

4.1. Introduction

For an example of the importance of this distinction, see Mardia (1972, page 229) relating to the Fisher distribution (discussed in §**4.4** below) referred to its axis of symmetry. Mardia gives graphs of the pdf for various values of the parameter κ in a form depending only on θ (his $g(\theta, \phi)$ is our $f(\theta, \phi)$); these show the pdf to be zero when $\theta = 0$, as is inevitable from the factor $\sin \theta$ in (4.2). The graph for $\kappa = 5$ is shown in Figure 4.1(a) as an illustration. Yet the distribution has its *greatest concentration* where $\theta = 0$, as illustrated by the "profile" graph of $h(\theta, \phi)$ against θ for the same case $\kappa = 5$ in Figure 4.1(b). From this point of view the h-function, reflecting the pde, describes the distribution more directly than the f-function (the pdf).

We can now move on to our account of distributional models. For each distribution, we give a description with the following information where available:

- **Main features** e.g. vectorial or axial; unimodal, multimodal, girdle or other; rotationally symmetric or asymmetric; etc. These features relate directly to the question: "Can this distribution be a useful model for our particular data set?"

Figure 4.1(b) Profile graph of $h(\theta, \phi)$ for the Fisher distribution referred to its mean direction, as a function of θ.

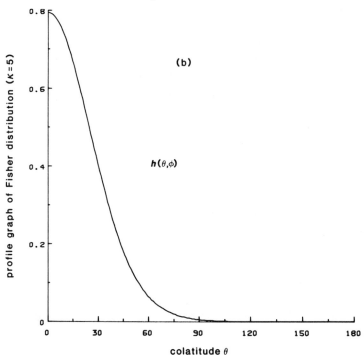

- **General form of the distribution** as specified by pdf or pde.

- **Parameters of the distribution** e.g. mean direction; concentration parameter; etc. These will appear in the expression (pdf or pde) for the form of the distribution. In the case of a *direction* parameter (vectorial or axial), e.g. the mean direction of a unimodal distribution, we shall make use of any or all of the following equivalent specifications:

    (i) angular coordinates of colatitude and longitude, typically (α, β).

    (ii) the equivalent direction cosines, typically (λ, μ, ν), where $\lambda = \sin \alpha \cos \beta$, $\mu = \sin \alpha \sin \beta$, $\nu = \cos \alpha$.

    (iii) a single vector symbol, typically $\boldsymbol{\lambda}$ where $\boldsymbol{\lambda} = (\lambda, \mu, \nu)'$.

In the same way when giving formulae for the distribution (pdf, pde etc), we shall specify a random spherical observation in whichever of the following forms is most convenient:

    (i) angular coordinates (θ, ϕ)
    (ii) direction cosines (x, y, z)
    where
    $x = \sin \theta \cos \phi$, $y = \sin \theta \sin \phi$, $z = \cos \theta$
    (iii) the single vector symbol $\boldsymbol{x} = (x, y, z)'$.

- **Basic form of the pde $C e^T \, dS$** All the spherical distributions described have pde's of the form $C e^T \, dS$, where C is a constant (of the appropriate value to make the total probability equal to 1), dS is the elemental surface area $\sin \theta \, d\theta \, d\phi$, and T is a function of the random direction (θ, ϕ) or equivalently of (x, y, z) or $\boldsymbol{x}$ (see above). Generally T depends only on the angle ψ which the random direction (θ, ϕ) makes with some fixed direction (α, β), or again on the angles ψ_1, ψ_2, ψ_3 which it makes with three such fixed directions; typically T is a simple function such as $\kappa \cos \psi$ or $\kappa \cos^2 \psi$ or $\kappa_1 \cos^2 \psi_1 + \kappa_2 \cos^2 \psi_2 + \kappa_3 \cos^2 \psi_3$. The function T contains in basic form all the information about the shape and properties of the distribution; e.g. if T depends only on a single angle ψ the distribution has rotational symmetry, and if $T(-\psi) = T(\psi)$ an axial distribution is indicated.

- **Notation** for use in this book. Except for the usual Greek χ^2 symbol for the chi-squared distribution, each distribution is denoted by a capital letter or letters (e.g., **N** for the normal distribution, **VM** for the von Mises distribution, **F** for the Fisher distribution). Bold face letters (e.g.,

4.1. Introduction

F) are used for distributions on the sphere and circle; these are the distributions of primary interest in the book, their presence being essentially in the role of models for data. Italicised letters (e.g., N) are used for distributions on the line, whose role here is in nearly all cases as sampling distributions. The parameters of a distribution are shown either in brackets (e.g., $N(\mu, \sigma^2)$, $F\{(\alpha, \beta), \kappa\}$ or as subscripts (e.g. χ^2_ν). In accordance with common practice the same letter F is used both for the Fisher distribution on the sphere, $F\{(\alpha, \beta), \kappa\}$, and the variance-ratio distribution on the line, F_{ν_1, ν_2}; there is no risk of confusion. Uniform distributions occur on the line, the circle and the sphere: these are denoted respectively by the symbols $U(a, b)$, $\mathbf{U_C}$ and $\mathbf{U_S}$.

In employing the notation, $N(\mu, \sigma^2)$, for example, will be used to denote both the normal distribution with mean μ and variance σ^2 and a random variable having this distribution; so we can write "X is distributed $N(10, 9)$", or again "Prob $(N(10, 9) > 16) = 0.023$".

- *Moments* of interest; e.g., resultant length, spherical variance.

- *Standardised form of the distribution* for particular values of the parameters; e.g., choosing the mean direction to be $(0, 0, 1)$.

- *Limiting forms of the distribution* where appropriate; e.g., as some parameter tends to 0 or ∞.

- *Special properties*

- *Applications and data contexts*

- *Tables* References to tables in this book or in other books and journals. When giving tabulated values of a parameter the notation 4(2)12, for example, is an abbreviation for 4, 6, 8, 10, 12.

- *References* to other published material on the distribution.

Analysis of spherical data necessarily involves some standard distributions on the line (e.g., the normal distribution) and on the circle (e.g., the von Mises distribution); descriptions of the main linear and circular distributions required in this book are given, for completeness, in §**4.2** and §**4.3**. The survey of spherical distributions for practical use is given in §**4.4**, starting with a summary Table. Other distributions for modelling spherical data have been proposed in the literature, which are at present of less practical applicability and which have not been used in this book. A list of these, with references, is given in §**4.5**.

4.2 Useful distributions on the line

Distribution	Notation	Section
Uniform	$U(a, b)$	4.2.1
Normal	$N(\mu, \sigma^2)$	4.2.2
Exponential	$E(\lambda)$	4.2.3
χ^2 (Chi-squared)	χ^2_ν	4.2.4
F (Variance ratio)	F_{ν_1, ν_2}	4.2.5

4.2.1 The uniform distribution (on the line)

All values x of the random variable X between a lower limit a and an upper limit b are equally likely - the population of values is spread uniformly over the range $a \le x \le b$.

Form of the distribution The pdf (see Figure 4.2) is
$$f(x) = \text{constant} = 1/(b-a) \text{ for } a \le x \le b, \text{ 0 otherwise.}$$
There are two *parameters*, a and b.

Notation $U(a, b)$.

Moments The distribution has mean $(a+b)/2$ and variance $(b-a)^2/12$.

Standardised form This is $U(0, 1)$ with pdf $f(x) = 1$ for $0 \le x \le 1$, 0 otherwise. If Z is distributed $U(0, 1)$, and $X = a + (b-a)Z$, then X is distributed $U(a, b)$.

Typical data context If a point P is chosen at random on the line joining two fixed points A and B at distance L apart, the random length AP is distributed $U(0, L)$.

Observations from $U(0, 1)$ can be directly *simulated* from sequences of random digits, and then converted to observations from $U(a, b)$; e.g. random digits ... 2 5 5 0 9 3 2 4 ... give 0.2550, 0.9324 as two values (each

Figure 4.2 Probability density function of the uniform $U(a, b)$ distribution.

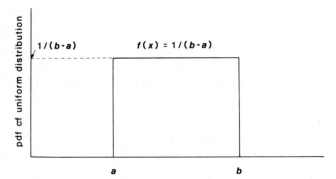

4.2.2 The normal distribution

to 4 decimal places) from $U(0, 1)$; if values from, say, $U(0, 360°)$ are required, these are at once available as 360×0.2550, 360×0.9324, i.e. 91.8°, 335.7°. See §3.6.2.

4.2.2 The normal distribution

This is a symmetric unimodal distribution, serving as a general probability model for errors of measurement or deviations from an underlying pattern, and also as a plausible model for many types of measurement. It is often called the *Gaussian distribution*.

Form of the distribution The pdf (see Figure 4.3) is

$$f(x) = [\sigma\sqrt{(2\pi)}]^{-1} \exp[-\tfrac{1}{2}(x-\mu)^2/\sigma^2], \quad -\infty < x < \infty \quad (4.3)$$

There are two *parameters*, the mean μ and the variance σ^2. All values of x are possible, but this in no way precludes the use of a normal distribution to model a random variable whose range of possible values is limited (e.g. weights of pieces of coal which cannot be negative, nor greater than some realistic upper limit), because the normal distribution used to model the data has effectively zero probability for values far enough from the mean.

Figure 4.3 Probability density function of the normal distribution (with mean μ).

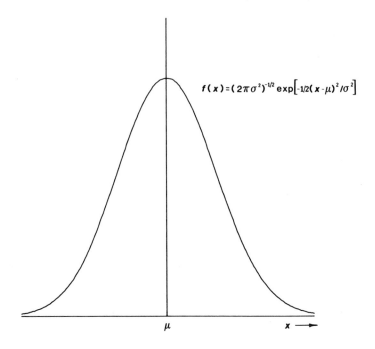

Notation $N(\mu, \sigma^2)$

Standardised form This is $N(0, 1)$, the normal distribution with zero mean and unit variance; its pdf is

$$f(x) = [\sqrt{(2\pi)}]^{-1} \exp[-\tfrac{1}{2}x^2], \quad -\infty < x < \infty \tag{4.4}$$

If Z is distributed $N(0, 1)$ and $X = \mu + \sigma Z$, then X is distributed $N(\mu, \sigma^2)$; conversely, if X is distributed $N(\mu, \sigma^2)$ and $Z = (X - \mu)/\sigma$, then Z is distributed $N(0, 1)$. This enables tables of $N(0, 1)$ [see below] to be used for any normal distribution $N(\mu, \sigma^2)$.

Special properties For a very wide variety of distributions it is a fact that sums of independent observations, and means of independent observations, from a common distribution are distributed approximately normally (the *central limit theorem*, a remarkable result). That is to say, if X is a random variable with mean μ and variance σ^2 and distribution (possibly unknown) of any form belonging to this very wide variety, and if $X_1, X_2, \ldots, X_n$ are n independent observations of X, then if we write

$S = X_1 + X_2 + \ldots + X_n =$ sum of the observations

$\bar{X} = S/n =$ mean of the observations,

S has mean $n\mu$ and variance $n\sigma^2$, and its standardised form $(S - n\mu)/(\sigma\sqrt{n})$ is distributed approximately $N(0, 1)$; equivalently $\bar{X}$

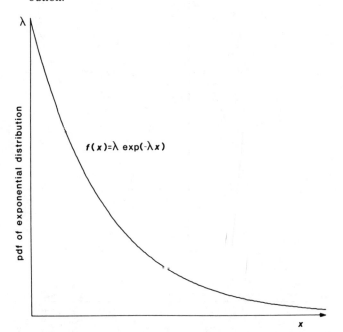

Figure 4.4 Probability density function of the exponential $E(\lambda)$ distribution.

has mean μ and variance σ^2/n, and its standardised form $(\bar{X} - \mu)/(\sigma/\sqrt{n})$ is distributed approximately $N(0, 1)$. The closeness of the approximation depends of course on the number of observations (i.e., sample size) n and on the shape of the X-distribution.

The above property enables the normal distribution to be used as a universal model for large-sample means and sums, and this is one of its most important applications.

Another special feature of the normal distribution is the 'additive property': the sum of any number of normally distributed random variables (whether independent or not) is itself normally distributed.

Tables As explained above, only tables of $N(0, 1)$ are needed. A table of percentage points for $N(0, 1)$ is given in this book as Appendix A1, and a table of the distribution function is given as Appendix A2. Tables of percentage points, df values and pdf values for $N(0, 1)$ are available in many texts and reference books; see, e.g., Pearson & Hartley (1970, Tables 1 and 3), Fisher & Yates (1974, Tables 1, 2, 3), Owen (1962, Tables 1.1 and 1.2).

4.2.3 The exponential distribution

An exponential random variable X takes positive values x only. The distribution of X is skew, the probability density $f(x)$ decreasing steadily as x increases; see Figure 4.4.

Form of the distribution The pdf is

$$f(x) = \lambda \exp(-\lambda x) \text{ for } x \geq 0, \quad 0 \text{ otherwise,} \tag{4.5}$$

and the distribution function (df) is

$$F(x) = \text{Prob}(X \leq x) = \begin{cases} 0 & (x < 0), \\ 1 - \exp(-\lambda x) & (x \geq 0). \end{cases} \tag{4.6}$$

There is one *parameter*, λ.

Notation $E(\lambda)$

Moments The distribution has mean $1/\lambda$ and variance $1/\lambda^2$.

Standardised form This is $E(1)$, the exponential distribution with unit mean. Its pdf is $f(x) = \exp(-x)$ ($x \geq 0$), 0 otherwise. Its df is $1 - \exp(-x)$ for $x \geq 0$; equivalently, the upper tail probability $\text{Prob}(X \geq x)$ is equal to $\exp(-x)$ for $x \geq 0$. If Z is distributed $E(1)$, and $X = Z/\lambda$, then X is distributed $E(\lambda)$; conversely, if X is distributed $E(\lambda)$, and $Z = \lambda X$, then Z is distributed $E(1)$. This enables random values from the $E(1)$ distribution to be used for simulating values from the $E(\lambda)$ distribution with any value of the parameter λ (see below).

The exponential distribution has a wide variety of *applications*, e.g. to lifetime and interval phenomena. In the analysis of spherical data it is important as the effective distribution of (1−cosine of colatitude) in a Fisher distribution referred to its axis of symmetry (see **§4.4.3**), hence giving easily the effective distribution of the colatitude itself.

No special *tables* are necessary for $E(\lambda)$, since from (4.6) Prob $(X \geq x)$ has the easily calculated form $\exp(-\lambda x)$. If we require the percentage point of $E(\lambda)$ corresponding to a given value α of the upper tail area Prob $(X \geq x)$, its value x_α is given by

$$\exp(-\lambda x_\alpha) = \alpha$$

i.e. $x_\alpha = -(1/\lambda) \log \alpha.$ (4.7)

Random values from $E(1)$ are available for simulation purposes, see e.g. **§3.6.2**. In the context of spherical data analysis, random values (θ, ϕ) from a Fisher distribution can be constructed from simulated random values from $E(1)$ together with simulated uniform values from $U(0, 360°)$ (**§4.2.1**); see **§4.4.3**.

4.2.4 The χ^2 (*chi-squared*) distribution

The χ^2-distribution is, along with the normal distribution, the most important statistical sampling distribution, and an essential tool in statistical data analysis. Basically, if $X_1, X_2, \ldots, X_\nu$ are ν independent random variables each distributed $N(0, \sigma^2)$ (see **§4.2.2**), then

$$(X_1^2 + X_2^2 + \ldots + X_\nu^2)/\sigma^2 \quad (4.8)$$

is a random variable whose distribution, which does not depend on σ^2, is called the χ^2-*distribution with parameter* ν, or, in common statistical parlance, '*with ν degrees of freedom*'. From this definition many applications follow; we mention two of the most important:

(i) If s^2 is the variance of a sample of n observations from a normal distribution $N(\mu, \sigma^2)$, then

$$(n-1)s^2/\sigma^2$$

is distributed as χ^2 with $n-1$ degrees of freedom. So the χ^2-distribution can be used to make inferences about the value of the unknown population variance σ^2 from the observed sample variance s^2.

(ii) Suppose that any individual in a population must belong to one out of k given categories, and that a sample of n individuals is classified according to these categories. If the probability that an individual belongs to category j $(j = 1, 2, \ldots, k)$ is believed to be p_j, and if the actual number of individuals in the sample of n which are observed to fall in category j is $n_j (n_1 + n_2 + \ldots + n_k = n)$, then the measure of goodness of fit of the

4.2.4. The χ^2 distribution

probability model $(p_1, p_2, \ldots, p_k)$ to the data $(n_1, n_2, \ldots, n_k)$

$$\sum_{j=1}^{k} (n_j - np_j)^2/(np_j) \tag{4.9}$$

is distributed approximately as χ^2 with $k-1$ degrees of freedom when the model is valid.

General form of the distribution We will call the χ^2 random variable Y to avoid confusion with the normal random variables X mentioned above. Its pdf is

$$f(y) = \begin{cases} Cy^{m-1} \exp(-\tfrac{1}{2}y) & \text{for } y \geq 0, \\ 0 & \text{otherwise} \end{cases} \tag{4.10}$$

where $m = \tfrac{1}{2}\nu$.

The constant C is equal to $1/(2^m \Gamma(m))$, where $\Gamma(m)$ (the *gamma function*) is equal to

$(m-1)!$ if m is an integer

$$\frac{(2m-1)! \sqrt{\pi}}{(m-\tfrac{1}{2})! 2^{2m-1}} \quad \text{if } m - \tfrac{1}{2} \text{ is an integer.} \tag{4.11}$$

There is *one parameter*, viz ν, the number of *degrees of freedom*. The random variable takes positive values y only. The pdf has different shapes according to the value of ν. For $\nu = 1$ the pdf is infinite at $y = 0$ and decreases steadily as y increases (Figure 4.5). For $\nu = 2$ the pdf (Figure 4.5) has a similar exponential shape to Figure 4.4; see below. For larger ν the pdf is unimodal, approaching normal form as ν increases (Figure 4.5).

Notation χ^2_ν where ν denotes the degrees of freedom.
We shall denote the upper $100\beta\%$ point of the χ^2_ν distribution by $\chi^2_\nu(\beta)$, i.e.

Prob $(\chi^2_\nu \geq \chi^2_\nu(\beta)) = \beta$.

By the same token, the lower $100\beta\%$ point of the distribution will be denoted by $\chi^2_\nu(1-\beta)$.

Moments χ^2_ν has mean ν and variance 2ν.

Special properties The χ^2_2-distribution is the same as $E(\tfrac{1}{2})$, as is seen by writing $m = 1$ in (4.10). As one of the consequences, it follows from (4.7) that the percentage point of χ^2_2 corresponding to an upper tail area α is $-2 \log \alpha$.

Since (4.8) is a sum, it follows (**§4.2.2**) that χ^2_ν is distributed approximately normally for large ν; compare Figure 4.3 with Figure 4.5 ($\nu = 10$).

The χ^2-distribution has an additive property of importance in statistical analysis: if S is the sum of independent random variables distributed as:

$$\chi^2_a, \chi^2_b, \chi^2_c, \ldots,$$

then S is distributed as $\chi^2_{a+b+c+\ldots}$.

An important application of this property is to the pooling of separate estimates of the same unknown variance σ^2.

Approximations to χ^2 If a random variable Y has the χ^2-distribution with ν degrees of freedom, there are a number of useful approximations to Y in terms of normal random variables. We shall use the following reasonably simple approximation, which gives satisfactory results for values of ν down to $\nu = 3$ (Lewis, 1986):

$1/\{1 - \frac{1}{6}\log(Y/\nu)\}$ is approximately distributed

$$N(1 - (9\nu)^{-1}, \sigma^2) \tag{4.12}$$

where $\sigma = (18\nu)^{-1/2} + (18\nu)^{-3/2}$.

Tables A table of percentage points for χ^2_ν for $\nu = 1(1)30(2)40(10)100$ is given in Appendix A3, and a table of the distribution functions of χ^2_1, χ^2_2 and χ^2_3 is given in Appendix A4. For other values of ν, percentage

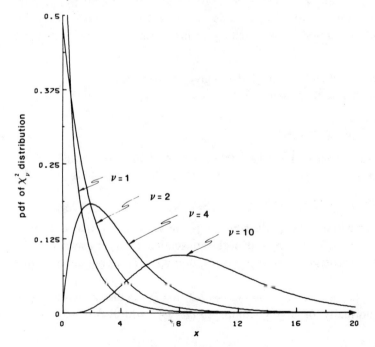

Figure 4.5 Probability density function of the χ^2_ν distribution for $\nu = 1$, 2, 4 and 10.

4.2.5. The F distribution

points of χ^2 can be found from the normal approximation (4.12) in conjunction with the normal distribution tables A1 or A2 respectively, as in the following examples:

Example 4.1 What is the upper $\frac{1}{2}$% point of χ_5^2? Call this $y_{.995}$. With $\nu = 5$ the approximating random variable X in (4.12) is distributed $N(1 - \frac{1}{45}, \sigma^2)$ where $\sigma = (90)^{-1/2} + (90)^{-3/2} = 0.1066$.

From Table A1, the upper $\frac{1}{2}$% point of $N(0, 1)$ is 2.5758, so the upper $\frac{1}{2}$% point of X is

$$x_{.995} = 1 - \frac{1}{45} + (2.4785 \times 0.1066) = 1.2523.$$

This is approximately equal to $1/\{1 - \frac{1}{6}\log(y_{.995}/5)\}$, so $1 - \frac{1}{6}\log(y_{.995}/5) \approx 1/1.2523 = 0.7985$, $\log(y_{.995}/5) \approx 1.2088$, and hence $y_{.995} \approx 16.748$. (The exact value to 3 decimal places is 16.750).

Example 4.2 In a goodness-of-fit test with 4 categories, the value of the goodness-of-fit statistic Y of (4.9),

$$Y = \sum_{j=1}^{4} (n_j - np_j)^2/(np_j),$$

comes out to 11.345. What is the significance probability?

On the null hypothesis that the model (p_1, p_2, p_3, p_4) is valid we can take Y to be distributed as χ_3^2. We have an observed value $y = 11.345$ from the Y-distribution, and we want the value of the significance probability $P = \text{Prob}(Y \geq y)$. The approximating normal random variable X for χ_3^2 in (4.12) is distributed $N(1 - \frac{1}{27}, \sigma^2)$ where $\sigma = (54)^{-1/2} + (54)^{-3/2} = 0.1386$. With $y = 11.345$,

$$x = 1/\{1 - \tfrac{1}{6}\log(11.345/3)\} = 1.2848$$

and the equivalent $N(0,1)$ value is

$$\left(1.2848 - \frac{26}{27}\right) \Big/ 0.1386 = 2.3221.$$

From Table A2, $\text{Prob}(N(0, 1) > 2.3221) = 0.0101$ or 1.01%, the required approximation to the significance probability. In fact 11.345 is the upper 1% point of χ_3^2, so the true P-value is 1% exactly.

Tables of percentage points of χ_ν^2, typically for $\nu = 1(1)30(10)100$ or a similar coverage, are commonly available in statistical texts and reference books. Tables of the distribution function of χ_ν^2 for $\nu = 1(1)30(2)70$ are given in Pearson and Hartley (1970, Table 7).

4.2.5 The F (or variance ratio) distribution

Associated with the χ^2-distribution is another important sampling distribution, the F-distribution, which is, like χ^2, an essential tool in statistical data analysis. Basically an F random variable is the ratio of two independent χ^2 random variables each scaled to have mean 1. More precisely, if Y_1 and Y_2 are independent χ^2 random variables with

respective degrees of freedom ν_1 and ν_2, then the ratio
$$F = (Y_1/\nu_1) \div (Y_2/\nu_2)$$
is a random variable whose distribution is called the *F-distribution with parameters* ν_1 *and* ν_2, or as statisticians commonly say 'with ν_1 and ν_2 degrees of freedom'.

It follows that, if the variance σ^2 of a normal distribution $N(\mu, \sigma^2)$ is estimated separately by the variances s_1^2 and s_2^2 of two independent samples of sizes n_1 and n_2, then the ratio of the estimates, s_1^2/s_2^2, is distributed as F with ν_1 and ν_2 degrees of freedom, where $\nu_1 = n_1 - 1$, $\nu_2 = n_2 - 1$; so the F-distribution can be used to compare the two estimates. This is a simple but important example of the wide variety of applications of the F-distribution in comparing quantities (parameters) on the basis of χ^2-distributed estimates, and hence assessing the acceptability of a model for describing a given set of data.

General form of the distribution If the random variable R is distributed as F with ν_1 and ν_2 degrees of freedom, its pdf is

$$f(r) = \begin{cases} Cr^{m_1-1}/(m_1 r + m_2)^{m_1+m_2} & \text{for } r > 0, \\ 0 & \text{otherwise,} \end{cases} \quad (4.13)$$

where $m_1 = \tfrac{1}{2}\nu_1$, $m_2 = \tfrac{1}{2}\nu_2$.

(The constant C is equal to
$$m_1^{m_1} m_2^{m_2} \Gamma(m_1 + m_2)/[\Gamma(m_1)\Gamma(m_2)]$$
where $\Gamma(m)$ is given by (4.11).)

There are *two parameters*, ν_1 and ν_2 (or equivalently m_1 and m_2).

The random variable takes positive values only. For $\nu_1 = 1$ and $\nu_1 = 2$ (not cases of great practical interest) the pdf decreases steadily as r increases, in the case $\nu_1 = 1$ from an infinite value at $r = 0$ and in the case $\nu_1 = 2$ from the value $f(0) = 1$. Otherwise the pdf is unimodal, with its mode near, but not actually at, $r = 1$ (modal r is $[(m_1 - 1)/m_1] \times [m_2/(m_2 + 1)]$).

Notation F_{ν_1, ν_2}

Moments The mean of F_{ν_1, ν_2} is $\nu_2/(\nu_2 - 2)$ provided $\nu_2 > 2$.

For $\nu_2 = 1$ or 2, the mean is infinite (reflecting the extreme skewness of the F-distributions in these cases).

Tables A table of upper percentage points for F_{ν_1, ν_2} is given in Appendix A5. Lower percentage points can be found from this table by the following formula:

The lower $100\alpha\%$-point of $F_{\nu_1, \nu_2} = 1/$(the upper $100\alpha\%$-point of F_{ν_2, ν_1})
(4.14)

4.3.2. The von Mises distribution

Similar tables of percentage points of F_{ν_1,ν_2} are commonly available in statistical texts and reference books; see, e.g. Owen (1962, Section 4), Pearson & Hartley (1970, Table 5), Fisher & Yates (1963, Table V), and, for a book of extensive tabulations of F, Mardia & Zemroch (1978).

4.3 Useful distributions on the circle

Distribution	Notation	Section
Uniform	U_C	4.3.1
von Mises	$VM(\alpha, \kappa)$	4.3.2
Wrapped Normal	$WN(\alpha, \sigma^2)$	4.3.3

4.3.1 The uniform distribution (on the circle)

In this distribution, all directions Θ from 0 to 2π are equally likely: the distribution is spread uniformly round the circle.

Form of the distribution The pdf is

$$f(\theta) = \text{constant} = 1/(2\pi) \quad (0 \leq \theta < 2\pi).$$

The distribution is unique; there are *no parameters*.

Notation U_C

Moments There is no mean direction. The resultant length is zero.

Applications In the analysis of spherical data, the particular importance of the circular uniform distribution is as the model for azimuth angles in any rotationally symmetric distribution referred to its axis of symmetry.

While the distribution is a valid model for some directional phenomena in the plane (a trivial example is the position in which a roulette wheel settles), its essential importance in the analysis of circular data is as a *null model*, in other words an over-simple model to be *disproved* by statistical analysis (i.e., shown to be *inconsistent with the data*). For example, to demonstrate the existence of a preferred direction of flight by birds of some particular species under given conditions, it may be necessary to test statistically that the observed directions of flight cannot reasonably be fitted by the circular uniform distribution.

4.3.2 The von Mises distribution

This is a symmetric unimodal distribution, serving as an all-purpose probability model for directions in the plane (much as the normal distribution for observations on the line).

Form of the distribution The pdf is

$$f(\theta) = C \exp[\kappa \cos(\theta - \alpha)], \quad 0 \leq \theta < 2\pi. \tag{4.15}$$

The constant C is equal to $1/[2\pi I_0(\kappa)]$, where $I_0(\kappa)$, known as a 'modified Bessel function', is a well-tabulated function given in series form by

$$I_0(\kappa) = 1 + \frac{(\frac{1}{2}\kappa)^2}{1!1!} + \frac{(\frac{1}{2}\kappa)^4}{2!2!} + \frac{(\frac{1}{2}\kappa)^6}{3!3!} + \ldots.$$

There are *two parameters* κ and α. α is a location parameter, the distribution having a mode at α and symmetry about the direction α. κ is a shape parameter called the *concentration parameter*, because the larger the value of κ the more the distribution is concentrated around α (see Figure 4.6, where the pdf is illustrated for $\kappa = 2$ and $\kappa = 20$ respectively).

Notation $\mathbf{VM}(\alpha, \kappa)$. The limiting form $\mathbf{VM}(\alpha, 0)$ as $\kappa \to 0$ is uniform $\mathbf{U_C}$.

Moments The mean direction is α. The resultant length, in other words the mean value of $\cos(\Theta - \alpha)$, is $I_1(\kappa)/I_0(\kappa)$, a ratio of modified Bessel functions; this is approximately equal to

$$1 - 1/(2\kappa) \tag{4.16}$$

the approximation being good for κ greater than, say, 3.

Tables The distribution function for a von Mises random variable is tabulated for $\kappa = 0.0(0.2)10.0$ in Table 3 of Batschelet (1981) and Appen-

Figure 4.6 Probability density function of the von Mises $\mathbf{VM}(\alpha, \kappa)$ distribution, for $\kappa = 2, 20$. The curves also represent the profile graph of $h(\theta, \phi)$ for the Fisher distribution for $\kappa = 2, 20$, the straight line representing the mean direction.

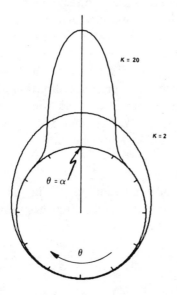

dix 2.1 of Mardia (1972). $\rho = I_1(\kappa)/I_0(\kappa)$ is tabulated as a function of κ for $\kappa = 0.0(0.1)10.0(1)20, 50, 100, 250, 500, \infty$ in Table C of Batschelet (1981) and for $\kappa = 0.0(0.1)9.0$ (0.2) 10.0, 12, 15, 20, 24, 30, 40, 60, 120, ∞ in Appendix 2.2 of Mardia (1972). Also to be found in Batschelet (1981, Table F) are tabulations of the modified Bessel functions I_0 and I_1. κ is tabulated as a function of ρ for $\rho = 0.00(0.01)1.00$ in Table B of Batschelet (1981) and Appendix 2.3 of Mardia (1972).

Reference A detailed account of the properties and applications of the von Mises distribution is given in Mardia (1972).

4.3.3 The wrapped normal distribution

Like the von Mises distribution this is a symmetric unimodal distribution, which could serve as a probability model for any of the directional data sets in the plane otherwise modelled by a von Mises distribution. For most purposes the calculations involved in statistical analyses based on this model are less convenient than those based on a von Mises model; but see below.

Form of the distribution The pdf is

$$f(\theta) = [\sigma\sqrt{(2\pi)}]^{-1} \sum_{r=-\infty}^{\infty} \exp[-\tfrac{1}{2}(\theta - \alpha - 2r\pi)^2/\sigma^2], \quad 0 \leq \theta < 2\pi. \tag{4.17}$$

There are *two parameters* σ^2 and α, α being a location parameter; the distribution has a mode at α and symmetry about the direction α. σ^2 is a shape parameter, reflecting the extent to which the distribution is concentrated around α; the smaller the value of σ^2 the greater the concentration, and *vice versa*.

Notation WN(α, σ^2). The limiting form WN(α, ∞) as $\sigma^2 \to \infty$ is uniform U_C.

Moments The mean direction is α. The resultant length ρ (the mean value of $\cos(\Theta - \alpha)$) is $\exp[-\tfrac{1}{2}\sigma^2]$; providing σ^2 is less than, say, 0.4 this is approximately equal to

$$1 - \tfrac{1}{2}\sigma^2. \tag{4.18}$$

Assuming $\kappa > 3$, it is clear from (4.16) and (4.18) that when $1 - 1/(2\kappa) = 1 - \tfrac{1}{2}\sigma^2$, i.e. when $\sigma^2 = 1/\kappa$, a wrapped normal distribution WN(α, σ^2) has approximately the same resultant length as the 'matching' von Mises distribution VM(α, κ) and so can be expected to have much the same dispersion. In fact, a stronger result turns out to hold – the two distributions WN($\alpha, 1/\kappa$) and VM(α, κ) agree very closely, and VM(α, κ) is approximated to very good accuracy by WN($\alpha, 1/\kappa$). For a detailed discussion with illustrative diagrams, see Kendall (1974); see also Lewis (1974), Collett & Lewis (1981) and Watson (1983a, Chapter 3). This result underlies the probability plotting procedure described in §**5.3.2(ii)** for investigating the fit of a Fisher distribution to a sample of spherical data.

4.4 Distributions for modelling spherical data

The distributions surveyed in this Section are summarised in Table 4.1.

4.4.1 The point "distribution"

Sometimes a distribution has a limiting form in which the angles Θ, Φ do not vary at all but have fixed values $\Theta = \alpha$, $\Phi = \beta$. In such a case we can regard (Θ, Φ) as a random variable having the *degenerate* distribution defined by

$$P(\Theta = \alpha, \Phi = \beta) = 1.$$

The whole of the probability is concentrated at the point (α, β).

Notation $\mathbf{D}(\alpha, \beta)$ or $\mathbf{D}(\lambda, \mu, \nu)$ or $\mathbf{D}(\boldsymbol{\lambda})$

4.4.2 The uniform distribution on the sphere

In this distribution, all directions (Θ, Φ) are equally likely: the distribution is spread uniformly over the surface of the sphere.

Form of the distribution The pde is constant, and so has the value $(1/4\pi)\,dS$. Equivalently, the pdf is

$$f(\theta, \phi) = \sin\theta/(4\pi) \quad (0 \leq \theta \leq \pi, 0 \leq \phi < 2\pi)$$

The distribution is unique; there are *no parameters* (in contrast to the 2-parameter family of uniform distributions on the line, §**4.2.1**).

Whatever directions are chosen for the coordinate axes, Θ and Φ are independent; the marginal distribution of Θ has pdf

$$g(\theta) = \tfrac{1}{2}\sin\theta \quad (0 \leq \theta \leq \pi)$$

and the marginal distribution of Φ is circular uniform $\mathbf{U}_C$.

Basic form of pde $C e^T \, dS \quad T = 0$

Notation $\mathbf{U}_S$

Moments There is no mean direction. The resultant length is zero, and equivalently the spherical variance is 1. All three moments of inertia are equal (to $\tfrac{1}{3}$).

Applications The most important role of the uniform distribution $\mathbf{U}_S$ in the analysis of spherical data is as a *null model*, to be set up as a possible

Table 4.1. *Summary of distributions on the sphere used as models for data in this book*

Distribution	Notation	Vectorial (V) or axial (A)	Characteristics	Section
Point	$\mathbf{D}\{\alpha, \beta\}$ or $\mathbf{D}\{\boldsymbol{\lambda}\}$			**4.4.1**
Spherical Uniform	U_S			**4.4.2**
Fisher	$\mathbf{F}\{(\alpha, \beta), \kappa\}$ or $\mathbf{F}\{(\lambda, \mu, \nu), \kappa\}$ or $\mathbf{F}\{\boldsymbol{\lambda}, \kappa\}$	V	unimodal; rotationally symmetric	**4.4.3**
Watson	$\mathbf{W}\{(\alpha, \beta), \kappa\}$ or $\mathbf{W}\{(\lambda, \mu, \nu), \kappa\}$ or $\mathbf{W}\{\boldsymbol{\lambda}, \kappa\}$	A	bipolar ($\kappa > 0$) or girdle ($\kappa < 0$); rotationally symmetric	**4.4.4**
Kent (Fisher–Bingham 5-parameter)	$\mathbf{K}\{\boldsymbol{\xi}_1, \boldsymbol{\xi}_2, \boldsymbol{\xi}_3\}, \kappa, \beta\}$ or $\mathbf{K}\{\mathbf{G}, \kappa, \beta\}$	V	unimodal ($\kappa \geq 2\beta$) or bimodal with equal modal strengths ($\kappa < 2\beta$); asymmetric	**4.4.5**
Wood (bimodal)	$\mathbf{WB}\{\gamma, \delta, \alpha, \beta, \kappa\}$ or $\mathbf{WB}\{(\boldsymbol{\mu}_1, \boldsymbol{\mu}_2, \boldsymbol{\mu}_3), \alpha, \beta, \kappa\}$	V	bimodal with equal modal strengths; symmetric about great-circle plane through modes	**4.4.6**

explanation for a set of data with a view to showing it unacceptable, thus providing evidence of the existence of non-isotropic features in the population from which the data are drawn; see Example 6.5, for instance. However, there do exist examples of a converse implication being sought: see Examples 5.5 and 5.9. Similar remarks apply to the uniform distribution on the circle; see §**4.3.1**.

4.4.3 The Fisher distribution

This important distribution is the basic model for directions distributed unimodally with rotational symmetry, and serves more generally as an all-purpose probability model for directions in space and directional measurement errors (much as the von Mises distribution for directions in the plane (§**4.3.2**), and the normal distribution for observations on the line, (§**4.2.2**)). It is the direct generalisation to the sphere of the von Mises distribution on the circle, and for this reason is sometimes called the *von Mises-Fisher distribution*. In fact, these two distributions are respectively the circular and spherical members ($p=2$ and $p=3$) of a family of distributions on the p-sphere called *Langevin distributions* by Watson (1983a). Another name given to the Fisher distribution is the *von Mises-Arnold-Fisher distribution* (Batschelet, 1981). See the historical account of this distribution in Chapter 1.

General form of the distribution The pde is

$$h(\theta, \phi)\, dS = C_F \exp[\kappa(\sin\theta \sin\alpha \cos(\phi-\beta) + \cos\theta \cos\alpha)]\, dS \quad (4.19)$$

where

$$C_F = \kappa/(4\pi \sinh\kappa) \quad (4.20)$$
$$= \kappa/[2\pi(\exp(\kappa)-\exp(-\kappa))] \quad (4.21)$$

Correspondingly, the pdf is

$$f(\theta, \phi) = C_F \exp[\kappa(\sin\theta \sin\alpha \cos(\phi-\beta) + \cos\theta \cos\alpha)]\sin\theta \quad (0\leq\theta\leq\pi, 0\leq\phi<2\pi) \quad (4.22)$$

From (4.22), the conditional distribution of Φ given $\Theta = \theta$ is **VM**$(\beta, \kappa \sin\theta \sin\alpha)$.

There are three *parameters*, κ, α and β. α and β (or the equivalent direction cosines (λ, μ, ν)) are location parameters, the distribution having rotational symmetry about the direction (α, β). κ is a shape parameter called the *concentration parameter*, since the larger the value of κ the more the distribution is concentrated towards the direction (α, β).

If we write $\alpha = 0$ in (4.19), so that Θ is measured with reference to the

4.4.3. The Fisher distribution

axis of symmetry, we get

$$h(\theta, \phi) = C_F \exp(\kappa \cos \theta), \tag{4.23}$$

the rotational symmetry being shown by the non-appearance of ϕ in the expression for $h(\theta, \phi)$ in (4.23). The polar graph of this function against θ shows the profile of the distribution in any plane through the axis of symmetry, and has already been illustrated in Figure 4.6 for the particular cases $\kappa = 2$ and $\kappa = 20$.

A typical data set that could reasonably be modelled by a Fisher distribution is illustrated in Figure 5.2.

Basic form of the pde $C e^T \, dS$

$$T = \kappa(\mathbf{x}'\boldsymbol{\lambda}) \tag{4.24}$$

Notation

$\mathbf{F}\{(\alpha, \beta), \kappa\}$ or $\mathbf{F}\{(\lambda, \mu, \nu), \kappa\}$ or $\mathbf{F}\{\boldsymbol{\lambda}, \kappa\}$ as convenient

Moments The mean direction is $(\alpha, \beta) = \boldsymbol{\lambda}$ say. If Ψ is the angle between a random direction $\mathbf{x}$ from the Fisher distribution and the mean direction $\boldsymbol{\lambda}$, so that $\cos \Psi = \mathbf{x}'\boldsymbol{\lambda}$, the mean value of $\cos \Psi$, in other words the resultant length ρ, is given in terms of the concentration parameter κ by

$$\rho = \frac{1+\exp(-2\kappa)}{1-\exp(-2\kappa)} - \frac{1}{\kappa} = \coth \kappa - \frac{1}{\kappa}. \tag{4.25}$$

The second trigonometric moment α_2, defined as the average value of $\cos(2\Psi)$, is given by

$$\alpha_2 = 1 - \frac{4}{\kappa} \coth \kappa + \frac{4}{\kappa^2} = 1 - \frac{4\rho}{\kappa}. \tag{4.26}$$

These expressions enter implicitly into the main methods used in the statistical analysis of Fisher samples, e.g. the calculation of a confidence cone for an unknown mean direction (**§5.3.2(iv)**).

Standardised form of the distribution This is the Fisher distribution referred to the direction $(0, 0, 1)$ as its axis of symmetry. For this direction $\alpha = 0$ and β is arbitrary; we therefore denote the distribution by $\mathbf{F}\{0, \kappa\}$ rather than $\mathbf{F}\{(0, \beta), \kappa\}$. Equivalently we can denote it by $\mathbf{F}\{(0, 0, 1), \kappa\}$. From (4.19) and (4.22), its pde is

$$h(\theta, \phi) \, dS = C_F \exp(\kappa \cos \theta) \, dS \tag{4.27}$$

and its pdf is

$$f(\theta, \phi) = C_F \exp(\kappa \cos \theta) \sin \theta \quad (0 \leq \theta \leq \pi, 0 \leq \phi < 2\pi) \tag{4.28}$$

with C_F given by (4.20), (4.21).

Limiting forms of the distribution When $\kappa = 0$, $F\{(\alpha, \beta), \kappa\}$ becomes the uniform distribution U_S.

When $\kappa \to \infty$, $F\{(\alpha, \beta), \kappa\}$ tends to total concentration at (α, β), i.e. the point distribution $D(\alpha, \beta)$.

Properties of the distribution The density of $F\{(\lambda, \mu, \nu), \kappa\}$ is greatest for the direction (λ, μ, ν), which is in other words the mode, and is least in the opposite direction $(-\lambda, -\mu, -\nu)$, which can be called the antimode; see Figure 4.6. For $F\{(\alpha, \beta), \kappa\}$ the mode is (α, β) and the antimode $(\pi - \alpha, \beta \pm \pi)$.

No generality is lost by assuming the concentration parameter κ to be positive; for $F\{(\lambda, \mu, \nu), -\kappa\}$ is clearly the same as $F\{(-\lambda, -\mu, -\nu), \kappa\}$, and so denotes a Fisher distribution with positive concentration parameter κ but with (λ, μ, ν) as *antimode* and $(-\lambda, -\mu, -\nu)$ as mode.

Marginal and conditional distributions For the general Fisher distribution $F\{(\alpha, \beta), \kappa\}$, where the z-direction $(0, 0, 1)$ is not the modal direction (the axis of symmetry), Θ and Φ are not independent. The conditional distribution of Φ given $\Theta = \theta$ depends on θ; from (4.22), it is von Mises $VM\{\beta, \kappa \sin \theta \sin \alpha\}$. Unless $\kappa \sin \theta \sin \alpha$ is small, say less than 3, this von Mises distribution can be approximated to good accuracy by the wrapped normal distribution $WN\{\beta, \sigma^2\}$ with $\sigma^2 = 1/(\kappa \sin \theta \sin \alpha)$; see §**4.3.3**.

For the standardised Fisher distribution $F\{0, \kappa\}$ referred to its mean direction as z-axis, it follows from (4.28) that Θ and Φ are independently distributed; the marginal distribution of Φ is uniform U_C, and the marginal distribution of Θ is given from (4.28) by the pdf

$$f(\theta) = \frac{\kappa}{2 \sinh \kappa} \exp(\kappa \cos \theta) \sin \theta \quad (0 \leq \theta \leq \pi). \quad (4.29)$$

This implies that, if $X = 1 - \cos \Theta$, the distribution of X is given by the pdf

$$g(x) = [\kappa/(1 - \exp(-2\kappa))] \exp(-\kappa x) \quad (0 \leq x \leq 2) \quad (4.30)$$

i.e. X has an exponential distribution $E(\kappa)$ truncated at $x = 2$. If $\exp(-2\kappa)$ is negligible, say if $\kappa > 2$, the truncation can be ignored, and X can be taken to have the exponential distribution $E(\kappa)$.

Tables A table of 10%, 5%, 1% and 0.1% points of the distribution (4.29) is given in Mardia (1972), pp 320-321, for $\kappa = 0.0$ (0.1) 5.0 (0.2) 10.0 (0.5) 13.0 (1) 15, 20 (10) 50, 100.

References Fisher (1953), Mardia (1972), Batschelet (1981).

4.4.4 The Watson distribution

As described in Chapter 1 this distribution is sometimes attributed, jointly with Watson, to Dimroth (1962) or to Scheidegger (1965), but we consider that the most appropriate title for it is the Watson distribution. It is the basic model for undirected lines (axes) distributed with rotational symmetry in either bipolar or girdle form.

General form of the distribution The pde is

$$h(\theta, \phi)\, dS = C_W \exp[\kappa(\sin\theta \sin\alpha \cos(\phi-\beta) + \cos\theta\cos\alpha)^2]\, dS \quad (4.31)$$

where

$$C_W = 1 \Big/ \left(4\pi \int_0^1 \exp(-\kappa u^2)\, du\right). \quad (4.32)$$

Correspondingly, the pdf is

$$f(\theta, \phi) = C_W \exp[\kappa(\sin\theta \sin\alpha \cos(\phi-\beta) - \cos\theta\cos\alpha)^2]\sin\theta \quad (0 \leq \theta \leq \pi, 0 \leq \phi < 2\pi). \quad (4.33)$$

There are three *parameters* κ, α and β, as with the Fisher distribution. α and β (or the equivalent direction cosines λ, μ, ν) are location parameters, the distribution having rotational symmetry about the axis $\pm(\lambda, \mu, \nu)$. In the bipolar case we call this the *principal axis*, and in the girdle case we call it the *polar axis*. κ is a shape parameter which may be either positive or negative; for positive values of κ the distribution is *bipolar*, and for negative values of κ it is *girdle* (equatorial). The larger the value of $|\kappa|$ the more the distribution is concentrated in the bipolar case ($\kappa > 0$) round the poles (α, β) and $(\pi - \alpha, \beta \pm \pi)$, and the more it is concentrated in the girdle case ($\kappa < 0$) round the great circle in the plane normal to (α, β); thus κ can be regarded as a *concentration parameter*.

If we set $\alpha = 0$ in (4.31), so that Θ is measured with respect to the axis of symmetry, we get

$$h(\theta, \phi) = C_W \exp(\kappa \cos^2\theta). \quad (4.34)$$

As with the Fisher distribution in equation (4.23), the rotational symmetry is shown by the non-appearance of ϕ in this expression for $h(\theta, \phi)$. The polar graph of the function against θ shows the profile of the distribution in any plane through the axis of symmetry; this is illustrated in Figure 4.7 for the bipolar cases $\kappa = +2$, $\kappa = +20$ and for the girdle cases $\kappa = -2$, $\kappa = -20$. In the bipolar case, the straight line in Figure 4.7 represents the principal axis of the distribution, and in the girdle case it represents the equatorial plane of the distribution.

Two typical data sets that could reasonably be modelled by a Watson bipolar and a Watson girdle distribution are illustrated in Figures 6.1, 6.2 respectively.

Basic form of the pde $Ce^T \, dS$

$$T = \kappa (\mathbf{x}'\boldsymbol{\lambda})^2 \tag{4.35}$$

Notation

$\mathbf{W}\{(\alpha, \beta), \kappa\}$ or $\mathbf{W}\{(\lambda, \mu, \nu), \kappa\}$ or $\mathbf{W}\{\boldsymbol{\lambda}, \kappa\}$ as convenient

Standardised form of the distribution This is the Watson distribution referred to the direction $(0, 0, 1)$ as its axis of rotational symmetry. We denote it by $\mathbf{W}\{\mathbf{0}, \kappa\}$ rather than $\mathbf{W}\{(0, \beta), \kappa\}$, β being arbitrary when $\alpha = 0$; equivalently it is $\mathbf{W}\{(0, 0, 1), \kappa\}$. From (4.31) and (4.33), its pde is

$$h(\theta, \phi) \, dS = C_W \exp(\kappa \cos^2 \theta) \, dS \tag{4.36}$$

Figure 4.7 Profile graphs of $h(\theta, \phi)$ for the Watson bipolar and girdle distributions. For the bipolar case, $\kappa = 2, 20$, and the straight line corresponds to the principal axis of the distribution. For the girdle case, $\kappa = -2, -20$, and the straight line corresponds to the equatorial plane of the distribution.

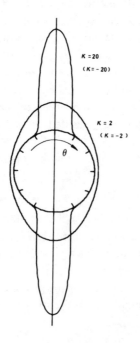

4.4.4. The Watson distribution

and its pdf is
$$f(\theta, \phi) = C_W \exp(\kappa \cos^2 \theta) \sin \theta$$
$$(0 \leq \theta \leq \pi, 0 \leq \phi < 2\pi) \tag{4.37}$$
with C_W given by (4.32).

Limiting forms of the distribution When $\kappa = 0$, $W\{(\alpha, \beta), \kappa\}$ becomes the uniform distribution of undirected lines, and so in effect the uniform distribution U_S.

When $\kappa \to \infty$, the Watson bipolar distribution $W\{(\alpha, \beta), \kappa\}$ tends to total concentration at the two ends (α, β), $(\pi - \alpha, \beta \pm \pi)$ of the principal axis.

When $\kappa \to -\infty$, the Watson girdle distribution tends to a distribution of type U_C totally concentrated on the great circle in the plane normal to the polar axis.

Marginal distributions For the standardised Watson distribution $W\{0, \kappa\}$, Θ and Φ are independently distributed; the marginal distribution of Φ is uniform U_C, and the marginal distribution of Θ is given by the pdf
$$f(\theta) = 2\pi C_W \exp(\kappa \cos^2 \theta) \sin \theta \quad (0 \leq \theta \leq \pi). \tag{4.38}$$

These results follow at once from (4.37).

Consider first the girdle case ($\kappa < 0$). Write $\kappa' = -\kappa$, so $\kappa' > 0$. The positive random variable X defined by $X = 2\kappa' \cos^2 \Theta$ will, from (4.38), be distributed with pdf proportional to
$$\frac{1}{\sqrt{x}} \exp(-\tfrac{1}{2}x) \quad (0 \leq x \leq 2\kappa'). \tag{4.39}$$

In other words, X has a χ^2 distribution on 1 degree of freedom truncated at $x = 2\kappa'$. The upper tail probability of χ_1^2 has values 2.5%, 1.4%, 0.8% when $x = 5, 6, 7$; so if $2\kappa' > 6$, say, i.e. $\kappa < -3$, the truncation can be ignored, and X can be taken to have the full χ_1^2 distribution.

Turning now to the bipolar case ($\kappa > 0$), define the positive random variable $Y = \kappa(1 - \cos^2 \Theta)$. From (4.38), this has pdf proportional to
$$[1 - (y/\kappa)]^{-\frac{1}{2}} \exp(-y) \quad (0 \leq y \leq \kappa). \tag{4.40}$$

Near the poles $\theta = 0$ and $\theta = \pi$, where the distribution tends to be concentrated, $\cos^2 \theta$ is near 1 and y is small; the factor $[(1 - (y/\kappa)]^{-\frac{1}{2}}$ in (4.40) can thus be approximated by 1 provided the concentration parameter κ is large enough, say $\kappa > 8$. The truncation $y \leq \kappa$ in (4.40) can then be ignored, and Y can be taken to have the standard exponential distribution $E(1)$. (This corrects equation (9.7.21), page 281 of Mardia (1972), where the degrees of freedom of the approximating χ^2 should be 2, corresponding to an exponential distribution, and not 1, and where correspondingly n, $n-2$, $n-4$ should be replaced as degrees of freedom by $2n$, $2n-2$, $2n-4$ in the next three equations and in (9.7.23)).

The χ_1^2 and $E(1)$ approximations to X and Y have useful applications in examining whether data can reasonably be modelled by Watson girdle or Watson bipolar distributions (see Chapter 6).

References Bingham (1964), Bingham (1974), Watson (1965), Mardia (1972), Batschelet (1981).

4.4.5 The Kent distribution

This was introduced by Kent (1982) as a particular model, specially suited for practical application, from an 8-parameter family of *Fisher-Bingham* distributions introduced by him in the same paper (see §4.5 below for a brief account), which he constructed as a generalisation of the Fisher distribution (§4.4.3) and the *Bingham* distribution (see §4.5). He called this particular model "the FB_5 distribution" (5-parameter Fisher-Bingham), though it is not the only usable 5-parameter member of his Fisher-Bingham family of models. It is a flexible model for vectors, with the important feature of *asymmetry*, and it is capable of representing both unimodal and bimodal data by appropriate choice of the parameter values. We confine its use in this book to modelling asymmetric *unimodal* data, and employ another distribution, the Wood distribution (§4.4.6) for modelling bimodal data, since it is computationally simpler than the Kent bimodal distribution for purposes of statistical analysis.

Basic form of the pde $C e^T dS$ This is given by

$$T = \kappa(\mathbf{x}'\boldsymbol{\xi}_1) + \beta[(\mathbf{x}'\boldsymbol{\xi}_2)^2 - (\mathbf{x}'\boldsymbol{\xi}_3)^2] \tag{4.41}$$

where $(\boldsymbol{\xi}_1, \boldsymbol{\xi}_2, \boldsymbol{\xi}_3)$ is an orthogonal triple of unit vectors; the 3×3 orthogonal matrix $(\boldsymbol{\xi}_1, \boldsymbol{\xi}_2, \boldsymbol{\xi}_3)$ can be regarded as a rotation matrix, and we denote it by $\mathbf{G}$. There are five *parameters*, three location parameters specifying the directions $(\boldsymbol{\xi}_1, \boldsymbol{\xi}_2, \boldsymbol{\xi}_3)$, and two non-negative shape parameters κ and β (we use the symbol β to keep in line with Kent's notation: it is not used here to denote longitude). The pde is

$$h(\theta, \phi) \, dS = C_K \exp(T) \, dS$$

where T is given by (4.41) with $\mathbf{x}' = (\sin\theta\cos\phi, \sin\theta\sin\phi, \cos\theta)$. The constant C_K is a function of κ and β, given when $2\beta < \kappa$ (which is, as we shall see, the unimodal case of practical interest in this book) by

$$C_K = 1 \Big/ \left\{ (2\pi)^{\frac{1}{2}} \kappa^{-\frac{1}{2}} \sum_{r=0}^{\infty} \frac{(2r)!}{r!r!} \left(\frac{\beta}{\kappa}\right)^{2r} I_{2r+\frac{1}{2}}(\kappa) \right\}$$

where the functions $I_{2r+\frac{1}{2}}(\kappa)$ are modified Bessel functions. When κ is

4.4.5. The Kent distribution

large, we have the simple approximate formula

$$C_K \approx \exp(-\kappa)(\kappa^2 - 4\beta^2)^{\frac{1}{2}}/(2\pi). \quad (4.42)$$

Interpreting first the location parameters in (4.41), ξ_1 is the *mean direction* or *pole* of the distribution. In any plane normal to ξ_1 the profile of the distribution is oval, the density (which is the same at opposite ends of any diameter of the small circle) being greatest in the directions $\pm\xi_2$ and least in the directions $\pm\xi_3$. Thus the distribution has rotational *asymmetry* about the mean direction, which gives it desirable flexibility as a model. The undirected lines $\pm\xi_2$ and $\pm\xi_3$ are called respectively the *major axis* and the *minor axis* of the distribution.

Turning to the two shape parameters, the greater the value of κ the more the distribution is concentrated towards the pole ξ_1; thus κ is a *concentration parameter*, as in the Fisher and Watson distributions. If now we look at the profile of the distribution in any plane normal to ξ_1, the greater the value of β the greater is the ratio of the maximum density h (in the direction of the major axis) to the minimum density (in the direction of the minor axis); in other words, the greater the value of β the more the profile departs from circular symmetry. β is accordingly termed the *ovalness parameter*. The distribution has two main forms, according to the value of κ/β; if $\kappa/\beta \geq 2$ it is unimodal and if $\kappa/\beta < 2$ it is bimodal. Further details are given below after the standardised form of the distribution has been introduced.

Notation

$\mathbf{K}\{(\xi_1, \xi_2, \xi_3), \kappa, \beta\}$ or $\mathbf{K}\{\mathbf{G}, \kappa, \beta\}$

Moments Kent (1982) gives formulae for the lower moments in terms of C_K.

Standardised form of the distribution Referring the distribution to $(0, 0, 1)$ as mean direction, $(1, 0, 0)$ as major axis and $(0, 1, 0)$ as minor axis, we get the standardised form $\mathbf{K}\{(\mathbf{z}, \mathbf{x}, \mathbf{y}), \kappa, \beta\}$ which we will write $\mathbf{K}\{\mathbf{0}, \kappa, \beta\}$. This has pde

$$h(\theta, \phi) \, dS = C_K \exp(\kappa \cos \theta + \beta \sin^2 \theta \cos 2\phi) \, dS \quad (4.43)$$

and equivalently it has pdf

$$f(\theta, \phi) = C_K \exp(\kappa \cos \theta + \beta \sin^2 \theta \cos 2\phi) \sin \theta$$
$$(0 \leq \theta \leq \pi, 0 \leq \phi < 2\pi). \quad (4.44)$$

Unimodal and bimodal forms If $\kappa/\beta \geq 2$, the distribution is *unimodal* with mode at the pole ξ_1 where (putting $\theta = 0$ in (4.43)) the density h has its maximum value $C_K \exp(\kappa)$. Away from the mode the density is everywhere less than $C_K \exp(\kappa)$ and attains its minimum value

$C_K \exp(-\kappa)$ at $-\boldsymbol{\xi}_1$, the diametrically opposite point to the mode ($\theta = \pi$ in (4.43)); this is thus the distribution's *antimode*. If on the other hand $\kappa/\beta < 2$, the distribution is *bimodal*. In terms of the standardised form (4.43), the density is still $C_K \exp(\kappa)$ at the pole $\theta = 0$, but takes greater values over a region of the spherical surface away from $\theta = 0$, so that this point is no longer the mode. It turns out that there are two modes, positioned at the points $\phi = 0$ and $\phi = \pi$ on the circle of latitude $\theta = \cos^{-1}(\kappa/2\beta)$, with equal strengths (densities) given by

$$h = C_K \exp[(\kappa^2 + 4\beta^2)/(4\beta)]. \tag{4.45}$$

There are also two equal antimodes, at $\phi = \pi/2$ and $\phi = 3\pi/2$ on the circle of latitude $\theta = \pi - \cos^{-1}(\kappa/2\beta)$, at which

$$h = C_K \exp[-(\kappa^2 + 4\beta^2)/(4\beta)].$$

The angular distance between the two modes, and likewise between the two antimodes, is $2\cos^{-1}(\kappa/2\beta)$.

Limiting forms of the distribution The distribution $\mathbf{K}\{(\boldsymbol{\xi}_1, \boldsymbol{\xi}_2, \boldsymbol{\xi}_3), \kappa, 0\}$ obtained when $\beta = 0$ is the Fisher distribution $\mathbf{F}\{\boldsymbol{\xi}_1, \kappa\}$. $\mathbf{K}\{(\boldsymbol{\xi}_1, \boldsymbol{\xi}_2, \boldsymbol{\xi}_3), 0, 0\}$ is the uniform distribution $\mathbf{U}_S$. When $\kappa \to \infty$ the distribution tends to the point "distribution" $\mathbf{D}\{\boldsymbol{\xi}_1\}$ whatever the value of β.

References See Kent (1982) for a detailed account of the properties of the model, and Wood (1982) for a comparison of the bimodal form of Kent's model ($\kappa/\beta < 2$) with Wood's bimodal model described in §**4.4.6**.

4.4.6 The Wood distribution

This distribution was introduced by Wood (1982) as a model for vectorial data with two modes of equal strengths (i.e., equal densities).

Basic form of the pde $Ce^T \, d\mathbf{S}$ This is given by

$$T = \kappa \cos \alpha (\mathbf{x}'\boldsymbol{\xi}_1)$$
$$+ \kappa \sin \alpha [(\mathbf{x}'\boldsymbol{\xi}_2) - (\mathbf{x}'\boldsymbol{\xi}_3)^2]/[1 - (\mathbf{x}'\boldsymbol{\xi}_1)^2]^{1/2} \tag{4.46}$$

where $(\boldsymbol{\xi}_1, \boldsymbol{\xi}_2, \boldsymbol{\xi}_3)$ is an arbitrary orthogonal triple of unit vectors. To keep to Wood's notation we rewrite this in terms of an orthogonal triple of unit vectors $(\boldsymbol{\mu}_1, \boldsymbol{\mu}_2, \boldsymbol{\mu}_3)$ of the special form

$$\boldsymbol{\mu}_1 = (\cos \gamma \cos \delta, \cos \gamma \sin \delta, -\sin \gamma)',$$
$$\boldsymbol{\mu}_2 = (-\sin \delta, \cos \delta, 0)', \qquad (0 \leq \gamma \leq \pi, 0 \leq \delta < 2\pi) \tag{4.47}$$
$$\boldsymbol{\mu}_3 = (\sin \gamma \cos \delta, \sin \gamma \sin \delta, \cos \gamma)'$$

with

$$\boldsymbol{\xi}_1 = \boldsymbol{\mu}_3,$$
$$\boldsymbol{\xi}_2 = \boldsymbol{\mu}_1 \cos(\tfrac{1}{2}\beta) + \boldsymbol{\mu}_2 \sin(\tfrac{1}{2}\beta), \tag{4.48}$$
$$\boldsymbol{\xi}_3 = -\boldsymbol{\mu}_1 \sin(\tfrac{1}{2}\beta) + \boldsymbol{\mu}_2 \cos(\tfrac{1}{2}\beta).$$

4.4.6. The Wood distribution

That is to say, $\frac{1}{2}\beta$ is the particular rotation of $\boldsymbol{\xi}_2, \boldsymbol{\xi}_3$ in their joint plane required to give the two-parameter form (4.47). The expression (4.46) can then be written

$$T = \kappa \cos \alpha (\mathbf{x}'\boldsymbol{\mu}_3) + \kappa \sin \alpha \cos \beta \, [(\mathbf{x}'\boldsymbol{\mu}_1)^2 - (\mathbf{x}'\boldsymbol{\mu}_2)^2]/[1-(\mathbf{x}'\boldsymbol{\mu}_3)^2]^{1/2}$$
$$+ \kappa \sin \alpha \sin \beta [2(\mathbf{x}'\boldsymbol{\mu}_1)(\mathbf{x}'\boldsymbol{\mu}_2)]/[1-(\mathbf{x}'\boldsymbol{\mu}_3)^2]^{1/2}. \quad (4.49)$$

As with the Kent distribution there are five *parameters*. Three of these are location parameters specifying the directions $(\boldsymbol{\xi}_1, \boldsymbol{\xi}_2, \boldsymbol{\xi}_3)$; in the form given by (4.49) and (4.47), which is the one we shall use, these parameters are γ, δ and β. As regards the other two parameters, α is an angular shape parameter between 0 and π, 2α or $2(\pi - \alpha)$ being the angular distance between the two modes according as α lies between 0 and $\frac{1}{2}\pi$ or between $\frac{1}{2}\pi$ and π; while κ, which can be assumed positive, is a concentration parameter (see below).

The pde is

$$h(\theta, \phi) \, dS = C_F \exp(T) \, dS$$

where T is given by (4.49) with $\mathbf{x}' = (\sin \theta \cos \phi, \sin \theta \sin \phi, \cos \theta)$.

The constant C_F is the same as the one defined in (4.21) for the Fisher distribution.

Notation

$$\mathbf{WB}\{\gamma, \delta, \alpha, \beta, \kappa\} \text{ or } \mathbf{WB}\{(\boldsymbol{\mu}_1, \boldsymbol{\mu}_2, \boldsymbol{\mu}_3), \alpha, \beta, \kappa\}$$

(the letters denoting 'Wood bimodal').

Moments The mean direction is $\boldsymbol{\mu}_3$. The resultant length ρ is given by

$$\rho = (\coth \kappa - (1/\kappa)) \cos \alpha,$$

i.e., from (4.25),

$$\rho = \cos \alpha \times \text{the resultant length of } \mathbf{F}\{\boldsymbol{\lambda}, \kappa\} \quad (4.50)$$

Standardised form of the distribution Referring the distribution to $(0, 0, 1)$ as mean direction (so that $\gamma = 0$), and taking the longitude-angle location parameter β as 0, we get the standardised form $\mathbf{WB}\{0, \delta, \alpha, 0, \kappa\}$ which can be written $\mathbf{WB}\{0, \alpha, 0, \kappa\}$ since it is the same for all δ. Its pde is

$$h(\theta, \phi) \, dS = C_F \exp(\kappa \cos \alpha \cos \theta$$
$$+ \kappa \sin \alpha \sin \theta \cos 2\phi) \, dS \quad (4.51)$$

and equivalently its pdf is

$$f(\theta, \phi) = C_F \exp(\kappa \cos \alpha \cos \theta + \kappa \sin \alpha \sin \theta \cos 2\phi) \sin \theta$$
$$(0 \leq \theta \leq \pi, 0 \leq \phi < 2\pi) \quad (4.52)$$

The Wood distribution in relation to the Kent bimodal distribution The similarity between equations (4.52) and (4.44) indicates that there are some resemblances between the Wood and Kent bimodal models, which we now describe. In any plane at right angles to the mean direction the profile of the Wood distribution has an oval shape similar to that for the Kent distribution. In fact, the profile for the standardised **WB**$\{0, \alpha, 0, \kappa\}$ in the plane $\theta = \theta_1$ (in other words, the conditional distribution of Φ given $\theta = \theta_1$) is the *same* as the profile for the standardised **K**$\{0, \kappa^*, \beta^*\}$ in the same plane $\theta = \theta_1$ when the ovalness β^* is equal to $\kappa \sin \alpha / \sin \theta_1$, whatever the value of the Kent concentration parameter κ^*. The Wood distribution has two modes, which in the standardised form (4.51) are positioned at the points $\phi = 0$ and $\phi = \pi$ on the circle of latitude $\theta = \alpha$, and which have the same maximum strength given by

$$h = C_F \exp(\kappa). \qquad (4.53)$$

There are also two antimodes, at the points $\phi = \pi/2$ and $\phi = 3\pi/2$ on the circle of latitude $\theta = \pi - \alpha$, with the same minimum strength given by $h = C_F \exp(-\kappa)$. The angular distance between the two modes, and likewise between the two antimodes, is 2α or $2(\pi - \alpha)$ whichever is smaller. This compares with the angular distance $2\cos^{-1}(\kappa/2\beta)$ between the modes of the Kent distribution.

Limiting forms of the distribution The distribution **WB**$\{(\mu_1, \mu_2, \mu_3), \alpha, \beta, 0\}$ obtained when $\kappa = 0$ is the uniform distribution U_S. The distribution **WB**$\{(\mu_1, \mu_2, \mu_3), 0, \beta, \kappa\}$ obtained when $\alpha = 0$ is the Fisher distribution **F**$\{\mu_3, \kappa\}$, the two modes at angular distance 2α coalescing into the single mode at μ_3 of the Fisher distribution. When $\kappa \to \infty$, **WB**$\{(\mu_1, \mu_2, \mu_3), \alpha, \beta, \kappa\}$ tends to total concentration at the two modes, each having probability $\frac{1}{2}$.

Negative values of κ The concentration parameter κ is essentially positive; the greater its value, the more the distribution is concentrated towards the two modal directions. A negative value of κ simply gives another Wood distribution, since

$$\mathbf{WB}\{(\mu_1, \mu_2, \mu_3), \alpha, \beta, -\kappa\} = \mathbf{WB}\{(\mu_1, \mu_2, \mu_3), \pi - \alpha, \beta \pm \pi, \kappa\}$$

which is the mirror image in the origin of

$$\mathbf{WB}\{(\mu_1, \mu_2, \mu_3), \alpha, \beta, \kappa\}.$$

Reference Wood (1982).

4.5 Other distributions on the sphere

The spherical distributions required for the various techniques covered in this book have been described in §4.4. To round off our account, we list in this final section some other distributions for modelling spherical data which have been proposed in the literature. Axial distribu-

4.5. Other distributions on the sphere

tions are considered first and then vectorial, for reasons of convenience (so that early reference can be made to the Bingham distribution).

A. Axial models – multipurpose

The Bingham Distribution. This is a 5-parameter axial model. According to the values of the parameters, it can represent distributions which are either symmetric or asymmetric, and of either girdle or bipolar form. The basic form of the pde is $C e^T \, dS$ with

$$T = \kappa_1(\mathbf{x}'\boldsymbol{\xi}_1)^2 - \kappa_2(\mathbf{x}'\boldsymbol{\xi}_2)^2 \qquad (4.54)$$

where $(\boldsymbol{\xi}_1, \boldsymbol{\xi}_2)$ is an orthogonal pair of unit vectors (defined by 3 parameters) and κ_1, κ_2 are positive shape parameters (making 5 parameters in all); as pointed out by Wood (1986), an equivalent form for T is

$$\kappa(\mathbf{x}'\boldsymbol{\lambda}_1)(\mathbf{x}'\boldsymbol{\lambda}_2) \qquad (4.55)$$

where $\boldsymbol{\lambda}_1, \boldsymbol{\lambda}_2$ are arbitrary unit vectors (defined by 4 parameters, including one shape parameter, the angle between the vectors), and the positive quantity κ is the other shape parameter.

References Bingham (1964, 1974); Mardia (1972, 1975a); Mardia & Zemroch (1977), who give a table for estimating the parameters κ_1 and κ_2 from a set of data.

B. Axial models – girdle

The Arnold Distribution. This is a girdle distribution with rotational symmetry. Its standardised pde is of the form

$$C \exp(-\kappa|\cos \theta|) \, dS, \quad (\kappa > 0),$$

with just one shape parameter κ measuring the extent of concentration about the great circle $\theta = \tfrac{1}{2}\pi$.

References Arnold (1941), Selby (1964). These authors do not discuss the case when κ is negative, which would give a bipolar axial distribution.

The Selby Distribution. This is, like the Arnold distribution, a girdle distribution with rotational symmetry. Its standardised pde is of the form

$$C \exp(\kappa \sin \theta) \, dS, \quad (\kappa > 0),$$

with a shape parameter measuring the extent of concentration about the great circle $\theta = \tfrac{1}{2}\pi$.

Reference Selby (1964). As with the Arnold distribution, negative values of κ give bipolar axial distributions, a case not discussed in the literature.

C. Axial models – small circle girdle

The Kelker–Langenberg Distribution. This distribution was proposed as a model for geological fold orientations. It models an elliptical cluster of points concentrated asymmetrically towards a *small* circle. The

standardised pde is of the form $Ce^T\,dS$ with
$$T = -\kappa_1 \sin^2(\theta - \gamma) + \kappa_2 \cos^2(\theta - \gamma)\sin^2\phi$$
and θ restricted to the range $0 \leq \theta \leq \frac{1}{2}\pi$; γ, lying in the range $0 < \gamma \leq \frac{1}{2}\pi$, is the colatitude of the small-circle girdle plane, and κ_1, κ_2 are positive shape parameters. The distribution is a generalisation of the Bingham distribution, reducing to it when $\gamma = \frac{1}{2}\pi$.

Reference Kelker & Langenberg (1982).

D. Vectorial models – multipurpose

The General Fisher–Bingham Distribution. This very flexible 8-parameter family of models was introduced by Kent (1982). He called it the *Fisher–Bingham distribution*, since the function h in its pde $h(\theta, \phi)\,dS$ is obtained by multiplying the h-functions for a Fisher and a Bingham distribution. This gives the general form of its pde as

$$C \exp\left[\kappa_0(\mathbf{x}'\boldsymbol{\xi}_0) + \kappa_1(\mathbf{x}'\boldsymbol{\xi}_1)^2 - \kappa_2(\mathbf{x}'\boldsymbol{\xi}_2)^2\right]dS \qquad (4.56)$$

where $\boldsymbol{\xi}_0, \boldsymbol{\xi}_1, \boldsymbol{\xi}_2$ are unit vectors with $\boldsymbol{\xi}_1, \boldsymbol{\xi}_2$ at right angles (thus involving 5 parameters), and $\kappa_0, \kappa_1, \kappa_2$ are shape parameters.

Depending on the values of $\kappa_0, \kappa_1, \kappa_2$ and the orientation of $\boldsymbol{\xi}_0$ with respect to $(\boldsymbol{\xi}_1, \boldsymbol{\xi}_2)$, the distribution (4.56) can be unimodal or bimodal or can be concentrated around a closed curve; it may be symmetrical or asymmetrical; and, while in general vectorial, it will be axial in the particular case $\kappa_0 = 0$.

References Kent (1982), Wood (1986). Beran (1979) has proposed a general class of exponential models, which also includes the Fisher and Bingham models.

Clearly the 8-parameter general form (4.56) includes a variety of particular models. One of these, the Kent distribution (involving 5 parameters) has been described in detail in §**4.4.5** above. Another particular model from the Fisher–Bingham family, the Fisher–Watson distribution, is briefly described next.

The Fisher–Watson Distribution. This is a 6-parameter distribution from the Fisher–Bingham family, introduced by Wood (1985) who calls it the 'Fisher–Dimroth–Watson distribution'. The general form of its pde is

$$C \exp\left[\kappa_0(\mathbf{x}'\boldsymbol{\xi}_0) + \kappa(\mathbf{x}'\boldsymbol{\xi})^2\right]dS \qquad (4.57)$$

where $\boldsymbol{\xi}_0, \boldsymbol{\xi}$ are arbitrary unit vectors, and where each of the shape parameters κ_0 and κ may be either positive or negative. Like the full 8-parameter Fisher–Bingham model, this 6-parameter model covers unimodal, bimodal and closed-curve distribution types, both symmetric and asymmetric. It appears to have convenient properties for purposes of practical application. For details, see Wood (1985).

4.5. Other distributions on the sphere

The Bingham–Mardia Distribution. This 4-parameter distribution is usually cited as a model for vectorial data concentrated towards a small circle, but for suitable values of the parameters it can also represent other forms (see below). The standardised form of the pde is

$$C \exp[\kappa(\cos\theta - \rho)^2] \, dS \tag{4.58}$$

with rotational symmetry about the z-axis as mean direction, and with two shape parameters κ and ρ; ρ can be taken as positive. The small-circle girdle form corresponds to the case $\kappa < 0$, $\rho < 1$, the small circle of maximum concentration lying in the plane of constant colatitude $\theta = \cos^{-1}\rho$. When $\rho > 1$, the distribution has a mode and an antimode respectively at $\theta = 0$ and $\theta = \pi$ (if $\kappa > 0$) or at $\theta = \pi$ and $\theta = 0$ (if $\kappa < 0$), and has the same general type of profile as a Fisher distribution. When $\kappa > 0$ and $\rho < 1$ we get a "squeezed belt" distribution, with a trough of minimum concentration around the small circle in the plane $\theta = \cos^{-1}\rho$, a mode of maximum strength at π, and a second mode of less strength at 0.

Reference Bingham & Mardia (1978).

E. Vectorial models – unimodal

The Saw Distribution. This is a unimodal distribution with rotational symmetry, proposed by Saw (1978). Like the Fisher distribution, it is fully specified by its mean direction and a concentration parameter κ. Its standardised pde is given as a power series in $\cos\theta$ by

$$h(\theta,\phi)\,dS = \left[(4\pi)^{-1} \exp(-\kappa^2) \sum_{r=0}^{\infty} \frac{2r+1}{r!} (\kappa\cos\theta)^{2r} \right.$$
$$\left. + \tfrac{1}{2}\pi^{-\frac{3}{2}} \exp(-\kappa^2) \sum_{r=0}^{\infty} \frac{(r+1)!}{(2r+1)!} (2\kappa\cos\theta)^{2r+1} \right] dS$$

Reference Saw (1978) discusses the properties of the distribution and applies it to a well-known set of data on the orbits of planets (Watson, 1970; Mardia, 1972).

F. Vectorial models – small circle

The Mardia–Gadsden Distribution. This is a 4-parameter model for vectorial data concentrated towards a small circle, and with rotational symmetry about the line perpendicular to the plane of the small circle, which is its mean direction. The standardised form of the pde is

$$C \exp[\kappa \cos(\theta - \alpha)] \, dS, \quad (\kappa > 0) \tag{4.59}$$

The Bingham-Mardia distribution with $\rho < 1$ is also a rotationally symmetric model for data concentrated towards a small circle; setting $\rho = \cos\alpha$ in (4.58), we can compare the two small-circle models (4.58) and (4.59). The expressions $(\cos\theta - \cos\alpha)^2$ and $\cos(\theta - \alpha)$ are respectively minimised and maximised on the small circle where $\theta = \alpha$, hence the condition that κ is negative for the Bingham-Mardia small-circle model and positive for the Mardia-Gadsden small-circle model.

Reference Mardia & Gadsden (1977) introduce the model, give an account of its properties, and apply it to data from Vulcanology and from Palaeomagnetism.

5
Analysis of a single sample of unit vectors

5.1 Introduction

The methods described in this chapter are concerned with the analysis of samples of independent observations from some common population of unit vectors. As illustrations, consider the following data sets.

Example **5.1** 50 pole positions determined from the palaeomagnetic study of New Caledonian laterites. A plot of the data is shown in Figure 5.1. (The data are listed in Appendix B1).

Example **5.2** 26 measurements of magnetic remanence made on samples collected from red-beds. A plot of the data is shown in Figure 5.2. (The data are listed in Appendix B2).

Figure 5.1. Equal-area projection of 50 virtual geomagnetic pole positions. The coordinate system is (Latitude, Longitude). See Example 5.1.

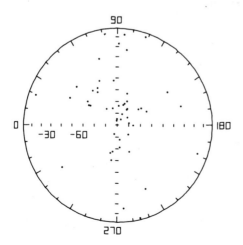

Example 5.3 148 measurements of arrival directions of low mu showers of cosmic rays. A plot of the data is shown in Figure 5.3. (The data are listed in Appendix B3).

Example 5.4 155 measurements of facing directions of conically folded bedding planes. A plot of the data is shown in Figure 5.4. (The data are listed in Appendix B4).

Example 5.5 52 measurements of magnetic remanence made on samples collected from red-beds. A plot of the data is shown in Figure 5.5. (The data are listed in Appendix B5).

Figure 5.2 Equal-area projection of 26 measurements of magnetic remanence (See §**3.3.1** for display conventions). The coordinate system is (Declination, Inclination). See Example 5.2.

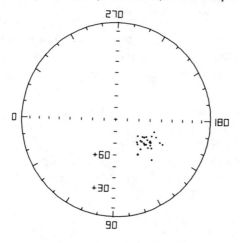

Figure 5.3 Equal-area projection of 148 arrival directions of low mu cosmic showers. The coordinate system is (Declination, Right Ascension). See Example 5.3.

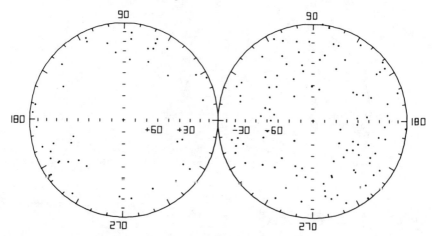

5.1. Introduction

***Example* 5.6** 107 measurements of magnetic remanence in samples of Precambrian volcanics. A plot of the data is shown in Figure 5.6. (The data are listed in Appendix B6).

The simplest questions one can ask about such data relate to the general shape of the underlying distribution, for example, whether it is unimodal (Example 5.1, Example 5.2) or multimodal (Example 5.6), whether it is rotationally symmetric about a reference direction (Example 5.2, Example 5.3), concentrated along an arc of a great or small circle (Example 5.4), or even whether the distribution is completely isotropic

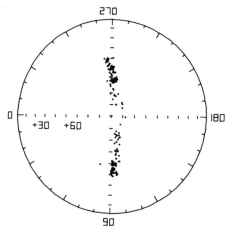

Figure 5.4 Equal-area projection of 155 facing directions of conically folded planes. The coordinate system is (Plunge, Plunge azimuth). See Example 5.4.

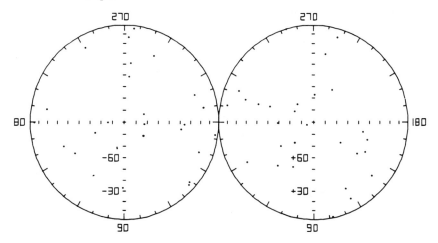

Figure 5.5 Equal-area projection of 52 measurements of magnetic remanence. The coordinate system is (Declination, Inclination). See Example 5.5.

(Example 5.5) and so exhibits no structural feature such as a mode. In many situations, the shape of the distribution will be known *a priori*. Thus, in Example 5.2, provided that there have not been any sampling problems, the scientist can be reasonably certain that the distribution is unimodal, because a single component of magnetisation is present. On the other hand, Example 5.6 contains an unknown number of modes, evidently caused by an unknown number of events relating to its magnetisation history. To a considerable extent, these questions can be resolved during the initial, *exploratory* phase of data analysis (§5.2), in which the data are displayed in one or more ways to highlight features of the types described above.

Having chosen a suitable model for the data (e.g. a unimodal model for the data of Example 5.1 or Example 5.2) we can examine a number of problems relating to this model (in Example 5.1 or Example 5.2, how to estimate a reference direction and some measure of the error of this estimate). A simplified diagrammatic summary of the treatment of a sample of data is shown in Figure 5.7.

Sections §**5.3**, §**5.4** and §**5.5** are concerned, respectively, with methods for analysing unimodal data, data distributed along an arc of a great or small circle, and multimodal (in particular, bimodal) data, and §**5.6** with general tests for isotropy (or uniformity) and rotational symmetry. In each of §**5.3** – §**5.5**, there is a dichotomy of methods into those based on only a few assumptions (which we call *simple* procedures) and those which involve fitting, or trying to fit, a particular probability distribution to the data (so-called *parametric* procedures). To clarify this distinction,

Figure 5.6 Equal-area projection of 107 measurements of magnetic remanence. The coordinate system is (Declination, Inclination). See Example 5.6.

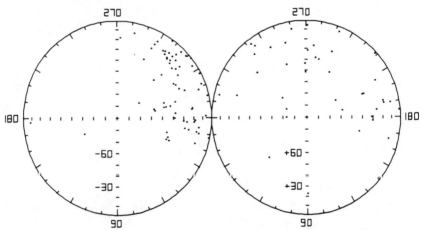

5.1. Introduction

consider one of the problems encountered in §5.3, namely, estimation of the mean direction of a unimodal distribution, together with some assessment of the error of this estimate. If we assume that the distribution is symmetrically distributed about its mean direction (i.e. that the azimuths, or longitudes, of the deviations from this mean direction are isotropic), the direction of the resultant of the sample unit vectors (see §3.2.1) is a natural estimate of the mean direction and, provided that the sample size is not too small, we can get an adequate estimate of the error of the sample mean direction by a simple procedure. On the other hand, if a probability distribution such as the Fisher distribution (§4.4.3) can be fitted satisfactorily to the data, we are in effect assuming additional information and so can get an improved estimate of the error in the sample mean direction.

Figure 5.7 Summary of development of methods for analysing a single sample of vectorial data.

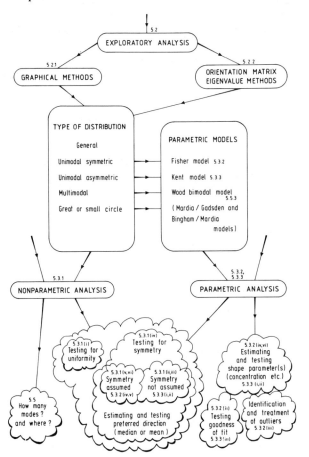

106   5. Analysis of a single sample of unit vectors

An example of how a complete analysis of a sample of vector data might be carried out is given in Chapter 1 (Example 1.1). It shows how the various stages of analysis described in this chapter fit together to give a logical statistical approach to one particular problem.

5.2   Exploratory analysis
5.2.1   Graphical methods

Apart from simple projections of the raw data, the main graphical method for initial study of a sample of data is to compute an estimate of the density of the distribution and to display contours of this estimate. A general description of this technique is given in §**3.3.3**, to which the reader is referred for notation. Here, we look at its application to some of our examples.

> *Example* **5.7** Figure 5.8 shows the contours of the estimated density for the data (set B1) in Example 5.1, using the smoothing value $C_n = 16.6$. The data are approximately centred without any rotation, yet the contours are clearly not circular, emphasizing the impression we gain from Figure 5.1 that the distribution is not symmetrically distributed about its mean direction.
>
> Figure 5.9 shows corresponding contours for the data (set B4) of Example 5.4, using the smoothing value $C_n = 4.5$. There appears to be a single mode; otherwise, the contours stray over most of the surface. In fact, a test of uniformity (see Examples 5.9 & 5.34) indicates that the underlying distribution is probably uniform.
>
> Figure 5.10 shows the contours for the data (set B6) in Example 5.6, using the smoothing value $C_n = 13.6$. The contours highlight two, possibly three distinct modes, whose identification is further facilitated by looking at the contours of the rotated data (see Figure 5.11; Figures 3.13 and 5.25 give alternative displays of these data).

Figure 5.8 Contour plot of the data in Figure 5.1.

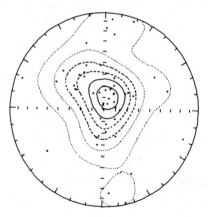

5.2.1. Exploratory analysis: graphical methods

Figure 5.9 Contour plot of the data in Figure 5.5.

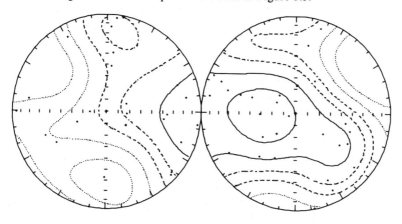

Figure 5.10 Contour plot of the data in Figure 5.6.

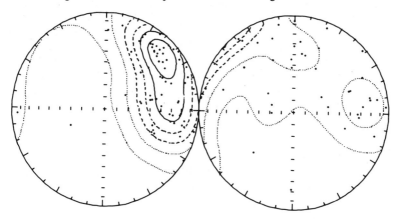

Figure 5.11 Contour plot of the data in Figure 5.6, rotated to the sample mean direction as reference direction.

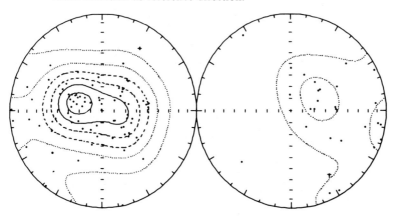

Another way of displaying the estimated density is the shade plot, described in §3.3.3. However, this is rather less useful as an interactive technique, unless sophisticated graphical hardware and software are available.

5.2.2 Quantitative methods

The principal exploratory tool is to look at the eigenvalues of the orientation matrix, as described in §3.4. Here, we apply the ideas described in that section to some of our data sets.

> **Example 5.8** The eigenvalues and associated ratios for the data sets B1–B6 in Examples 5.1, 5.2, 5.3, 5.4, 5.5 and 5.6 are given in Table 5.1. We may now consult Figure 3.15 to see what information is yielded.
>
> For Example 5.1, with $\hat{\gamma}$ between 1 and 2, we find a distribution of cluster type, with some tendency towards a girdle; this is verified by the contour plot in Figure 5.8. For Example 5.2, we find a strongly unimodal (large $\hat{\zeta}$, $\hat{\gamma}$) distribution with evidence of symmetry about the preferred direction. The data of Example 5.3 have all eigenvalues rather close to $\frac{1}{3}$, which would suggest a uniform distribution; however an inspection of the rotated data and contours in Figure 5.12 (based on a smoothing value $C_n = 11.4$) indicates a possible bimodal distribution. For Example 5.4, we find a girdle distribution, but the display in Figure 5.4 needs to be consulted to identify the distribution as being of small-circle type. For Example 5.5, the indications are that the distribution is close to uniform (low $\hat{\zeta}$), whereas for Example 5.6, it appears that the data are non-uniform but multimodal.

We can now proceed to statistical analysis based on the form of the data distribution, as described in the following sections.

Figure 5.12(a) Contour plot of the data in Figure 5.3

(a)

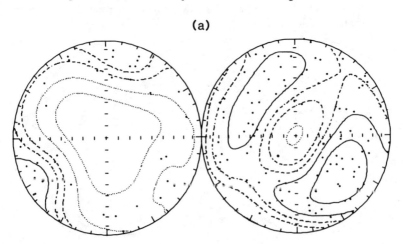

5.2.2. Exploratory analysis: quantitative methods

Table 5.1 *Normalised eigenvalues, log ratios* $r_1 = \log(\bar{\tau}_2/\bar{\tau}_1)$, $r_2 = \log(\bar{\tau}_3/\bar{\tau}_2)$ *and estimated strength and shape indicators* $\hat{\zeta} = \log(\bar{\tau}_3/\bar{\tau}_1)$ *and* $\hat{\gamma} = r_2/r_1$ *for data in Examples 5.1–5.6.*

Example number	$\bar{\tau}_1$	$\bar{\tau}_2$	$\bar{\tau}_3$	r_1	r_2	$\hat{\gamma}$	$\hat{\zeta}$
5.1	0.1160	0.2088	0.6752	0.59	1.17	2.00	1.76
5.2	0.0071	0.0104	0.9825	0.38	4.55	4.93	11.92
5.3	0.2554	0.3187	0.4258	0.22	0.29	0.51	1.31
5.4	0.0022	0.3347	0.6631	5.02	0.68	5.71	0.14
5.5	0.2569	0.3499	0.3933	0.31	0.12	0.43	0.38
5.6	0.1206	0.2731	0.6063	0.82	0.80	1.61	0.98

Figure 5.12 (*cont.*)
(b) Data in Figure 5.3 rotated to the sample mean direction as reference direction
(c) Contour plot of data in Figure 5.12(b).

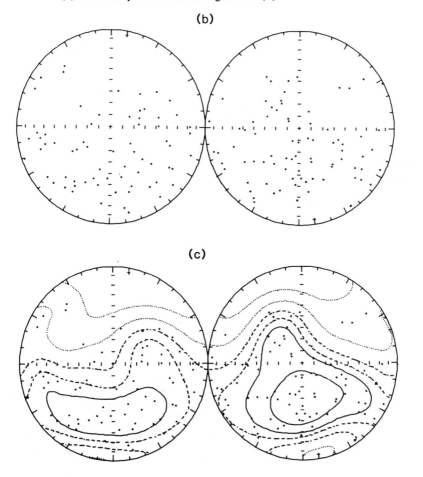

5.3 Analysis of a sample of unit vectors from a unimodal distribution

5.3.1 Simple procedures

Given the sample of observations $(\theta_1, \phi_1), \ldots, (\theta_n, \phi_n)$ (with corresponding direction cosines $(x_1, y_1, z_1), \ldots, (x_n, y_n, z_n)$) from a unimodal spherical distribution, we consider various simple procedures for the following problems:

(i) Testing the sample to see if it has been drawn from a uniform distribution, against the alternative that it has been drawn from a unimodal distribution.

(ii) Estimating the median direction and assigning an error to this estimate (distribution not assumed symmetric about median direction).

(iii) Testing the sample to see if the median direction takes a specified value (distribution not assumed symmetric about median direction).

(iv) Testing the sample to see if it is drawn from a rotationally symmetric distribution.

(v) Estimating the mean direction of a rotationally symmetric distribution, and assigning a cone of confidence to this estimate.

(vi) Testing the sample to see if the mean direction of a rotationally symmetric distribution takes a specified value.

Except for the test in §5.3.1(i), the tests and confidence cones given below in §5.3.1 are based on asymptotic statistical theory, and caution should be exercised in interpreting the results from samples of size less than, say, 25.

5.3.1(i) *Test for uniformity against a unimodal alternative.* Calculate the squared resultant length of the data (cf. (3.3)):

$$R^2 = \left(\sum x_i\right)^2 + \left(\sum y_i\right)^2 + \left(\sum z_i\right)^2 \tag{5.1}$$

The hypothesis of uniformity is rejected if R^2 is too large.

Critical values

sample size $n < 10$ use critical values for R in Appendix A7.
sample size $n \geq 10$ compute $3R^2/n$, and test as a value of the χ_3^2 distribution in Appendix A3.

> **Example 5.9** The data (set B5) described in Example 5.3 (see Figure 5.5), by their nature, may well be randomly distributed on the surface of the unit sphere. Suppose we wish to test this hypothesis at the 95% level against the alternative that there remains some residual component

5.3.1(ii). Estimation of median direction

of magnetisation. We have $n = 52$, $R^2 = 59.3$, $3R^2/n = 3.42$; since $3R^2/n < \chi_3^2(0.05) = 7.81$, we conclude that it is reasonable to assume that the directions of magnetisation are isotropic.

References and footnotes The test is due to Rayleigh (1919). Watson (1956b) made a tabulation of some percentage points for small n; Stephens (1964) extended these tabulations, and also gave a table of sample sizes $n(\kappa, \alpha, \lambda)$ required to detect a Fisher distribution (with specified concentration parameter κ) with a probability λ, when the test of randomness has significance level α. Diggle, Fisher & Lee (1985) have made a study of Rayleigh's test and compared it with other tests of randomness.

5.3.1(ii) Estimation of the median direction of a unimodal distribution. Suppose our data appear to be drawn from a unimodal distribution, but that the pattern of points does not look isotropic, or rotationally symmetric, about some preferred direction. For example, Figure 5.1 shows some data (set B6) somewhat streaked along part of a great circle, and exhibiting other asymmetry. In such a case, we may prefer a more general idea of a reference direction than the mean direction, such as a quantity analogous to the median of a sample of linear data. A suitable measure for this purpose is the sample *spherical median* $\tilde{P} = (\hat{\gamma}, \hat{\delta})$, which is defined as the direction for which the sum of arc lengths from $\tilde{P}$ to each $P_i = (\theta_i, \phi_i)$ is minimised. So if $(\tilde{x}, \tilde{y}, \tilde{z})$ are the direction cosines of $(\tilde{\theta}, \tilde{\phi})$, then they minimise the quantity

$$\sum_{i=1}^{n} \arccos(\tilde{x}x_i + \tilde{y}y_i + \tilde{z}z_i) \tag{5.2}$$

The minimisation requires the use of an optimisation routine such as that of Nelder and Mead (1965), E04JAF in the NAG library, or ZXMIN in the IMSL package.

[The sample spherical median is an estimate of the population spherical median. Let **X** be a random vector from the population under study, and **a** an arbitrary direction, and let $\theta_a = \arccos(\mathbf{X}'\mathbf{a})$, the angular deviation of **X** from **a**. The spherical median is the direction **a** for which the average value of θ_a is minimum. By way of comparison, the mean direction is the direction for which the average value of $\cos \theta_a$ is minimum.]

For small sample size n, there is no direct way of obtaining a confidence region for the true population median direction (γ, δ) other than by a bootstrap technique (see §3.6.1). For $n \leq 25$, an approximate elliptical confidence cone can be calculated, based on the fact that $(\tilde{x}, \tilde{y})$ has an approximate bivariate normal distribution. Rotate the sample so that it is measured relative to $(\hat{\gamma}, \hat{\delta})$ using (3.8) with $\mathbf{A}(\hat{\gamma}, \hat{\delta}, 0)$, to obtain

$(\theta_1', \phi_1'), \ldots, (\theta_n', \phi_n')$. Calculate

$$\left. \begin{array}{l} C_{11} = (1/n) \sum_{i=1}^{n} \cot \theta_i'(1-\cos(2\phi_i'))/2 \\[4pt] C_{22} = (1/n) \sum_{i=1}^{n} \cot \theta_i'(1+\cos(2\phi_i'))/2 \\[4pt] C_{12} = C_{21} = -(1/n) \sum_{i=1}^{n} \cot \theta_i' \sin(2\phi_i')/2 \end{array} \right\} \quad (5.3)$$

and form the matrix

$$\mathbf{C} = \begin{pmatrix} C_{11} & C_{12} \\ C_{21} & C_{22} \end{pmatrix}. \quad (5.4)$$

Next, calculate

$$\left. \begin{array}{l} \sigma_{11} = 1 + (1/n) \sum_{i=1}^{n} \cos(2\phi_i') \\[4pt] \sigma_{22} = 1 - (1/n) \sum_{i=1}^{n} \cos(2\phi_i') \\[4pt] \sigma_{12} = \sigma_{21} = (1/n) \sum_{i=1}^{n} \sin(2\phi_i') \end{array} \right\} \quad (5.5)$$

and form the matrix

$$\Sigma = \frac{1}{2} \begin{pmatrix} \sigma_{11} & \sigma_{12} \\ \sigma_{21} & \sigma_{22} \end{pmatrix}. \quad (5.6)$$

Then let

$$\mathbf{W} = \mathbf{C} \Sigma^{-1} \mathbf{C} = \begin{pmatrix} w_{11} & w_{12} \\ w_{21} & w_{22} \end{pmatrix}. \quad (5.7)$$

An approximate $100(1-\alpha)\%$ elliptical confidence cone centred on $(\tilde{\gamma}, \tilde{\delta})$ has as its boundary the points (x,y,z) which satisfy

$$w_{11}x^2 + 2w_{12}xy + w_{22}y^2 = -2\log(\alpha)/n \quad (5.8)$$

(and $x^2 + y^2 + z^2 = 1$, $z \geq 0$). These can be determined by the method in §3.2.5: (5.8) corresponds to (3.20).

Example 5.10 For the data (set B1) of Example 5.1, the spherical median direction is (lat. 78.9°, long. 98.4°). To obtain a 95% elliptical confidence cone ($\alpha = 0.05$) we require $\mathbf{W}$ (cf. (5.7)), which is calculated to be

$$\mathbf{W} = \begin{pmatrix} 4.839 & -2.939 \\ -2.939 & 9.918 \end{pmatrix}$$

with eigenvalues $\lambda_1 = 3.49$ and $\lambda_2 = 11.26$, and eigenvectors

$$\mathbf{u} = \begin{pmatrix} 0.9094 \\ -0.4160 \end{pmatrix} \text{ and } \mathbf{v} = \begin{pmatrix} 0.4160 \\ 0.9094 \end{pmatrix}.$$

Finally, the semi-vertical angles of the elliptical cone are 10.7° and 5.9°. The confidence region is shown in Figure 5.13.

5.3.1(iii). Test for specified median direction

References and footnotes The spherical median enjoys robustness properties analogous to those of the linear median. See Fisher (1985) for further details. No useful large-sample theory is currently available for the mean direction of a sample from a unimodal distribution which is not rotationally symmetric about its mean direction.

5.3.1(iii) *Test for a specified value of the population median direction.* Suppose we wish to test whether the population median has the value (γ_0, δ_0). Let ( ) be the sample median direction, and calculate $(\theta_1', \phi_1'), \ldots, (\theta_n', \phi_n')$ and hence Σ (see (5.6)) as in (ii) above. Now calculate the deviations of the sample from the hypothesized median (γ_0, δ_0) by using (3.8) with $\mathbf{A}(\gamma_0, \delta_0, 0)$, to obtain $(\theta_1^0, \phi_1^0), \ldots, (\theta_n^0, \phi_n^0)$, and let

$$\mathbf{U} = n^{-1/2} \begin{pmatrix} \Sigma \cos \phi_i^0 \\ \Sigma \sin \phi_i^0 \end{pmatrix}. \tag{5.9}$$

The test statistic is

$$X^2 = \mathbf{U}' \Sigma^{-1} \mathbf{U} \tag{5.10}$$

and the hypothesis that the spherical median direction is $(\hat{\gamma}_0, \hat{\delta}_0)$ is rejected if X^2 is too large.

Critical values
sample size $n < 25$ Use bootstrap methods (cf. §3.6.1)
sample size $n \geq 25$ Compare with $-2 \log \alpha$, the upper $100\alpha\%$ point of χ_2^2.

Figure 5.13 Approximate 99% elliptical confidence cone for the spherical median direction, for the data in Figure 5.1. See Example 5.10.

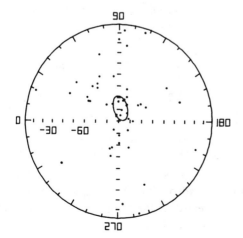

Example 5.11 Continuing the analysis of data set B1 from Example 5.10, suppose we wish to test at the 99% level the hypothesis that the true median direction is at latitude $-90°$ (the South Pole). Since the raw data are measured relative to $(0°, 0°)$ (transformed to polar coordinates) we can use the raw longitudes $\phi_1, \ldots, \phi_n$ for $\phi_1^0, \ldots, \phi_n^0$. Then from (5.9)

$$\mathbf{U} = 50^{-1/2}\begin{pmatrix} 2.4 \\ 16.4 \end{pmatrix}; \text{ also } \Sigma^{-1} = \begin{pmatrix} 1.828 & -0.330 \\ -0.330 & 2.339 \end{pmatrix}.$$

Hence, from (5.10), $X^2 = 12.32$, and the hypothesis is rejected, since X^2 exceeds $\chi_2^2(0.01) = 9.21$. The significance probability is $\exp(-12.32/2) \simeq 0.002$.

References and footnotes See Fisher (1985).

5.3.1(iv) *Test for rotational symmetry about the mean direction.* Suppose first that the true mean direction is known, and has polar coordinates (α_0, β_0). We shall describe both graphical and formal methods. For each of these, we require the data to be rotated so that they are measured relative to (α_0, β_0), using (3.8) with $\mathbf{A}(\alpha_0, \beta_0, 0)$, to obtain $(\theta_1', \phi_1'), \ldots, (\theta_n', \phi_n')$. Under the symmetry assumption, the longitudes $\phi_1', \ldots, \phi_n'$ should be uniformly distributed on $[0, 2\pi]$.

Graphical method Perform a Q-Q plot of the data

$$X_1 = \phi_1'/2\pi, \ldots, X_n = \phi_n'/2\pi$$

as described in §3.5.1; the assumption of symmetry is implausible if the resulting plot is not approximately linear, passing through the origin, with approximate slope 45°.

Formal test Calculate Kuiper's V_n-statistic (3.39) described in §3.5.2 using $F(x) = x$; the hypothesis of symmetry is rejected if V_n is too large.

Critical values sample size $n \geq 9$. Compute V_n^* (3.41) and compare with the appropriate critical value in Appendix A6.

Example 5.12 Suppose we have reason to believe that the data (set B2) in Example 5.2 are drawn from a population symmetric about the direction (Dec. 150°, Inc. 60°) and wish to test this hypothesis. For the graphical method, we obtain Figure 5.14, in which the departure from the 45° line passing through the origin is evident. For the formal test, $D_n^+ = 0.036$, $D_n^- = 0.316$, so that $V_n^* = 1.870$, which exceeds the upper 5% critical value of 1.747 listed in Appendix A6. So we reject the hypothesis of symmetry about the pole direction (Dec. 150°, Inc. 60°).

Now suppose that (α, β) is *not* known. The graphical procedure described above can be used, except that the sample mean direction $(\hat{\alpha}, \hat{\beta})$ replaces (α_0, β_0) as the reference direction. For a formal test, the procedure given in §**6.2.1(iv)** should be employed.

5.3.1(v). Estimation of mean direction

5.3.1(v) *Estimation of the mean direction of a symmetric unimodal distribution.* Suppose we can assume that our sample is drawn from a distribution which is rotationally symmetric about its (unknown) mean direction (α, β). The sample estimate $(\hat{\alpha}, \hat{\beta})$ of the mean direction is $\hat{\alpha} = \hat{\theta}, \hat{\beta} = \hat{\phi}$, where $(\hat{\theta}, \hat{\phi})$ are defined in §3.2.1. As in §5.3.1(ii) above, there is no direct way of obtaining a confidence cone for (α, β), and a bootstrap technique (§3.6.1) is the only method of assessing the variability of our estimate. For $n \geq 25$, an approximate confidence cone can be calculated, based on the fact that $\hat{x}$ and $\hat{y}$ (cf. (3.2)-(3.4)) are approximately independently distributed as normal random variables. Compute $\bar{R} = R/n$ (cf. (3.3)),

$$d = 1 - (1/n) \sum_{i=1}^{n} (x_i \hat{x} + y_i \hat{y} + z_i \hat{z})^2 \tag{5.11}$$

and

$$\hat{\sigma} = [d/(n\bar{R}^2)]^{1/2} \tag{5.12}$$

$\hat{\sigma}$ is the (estimated) *spherical standard error* of the sample mean direction

Figure 5.14 Uniform Q-Q plot for the data in Figure 5.2, to test for rotational symmetry about a specified direction. See Example 5.12.

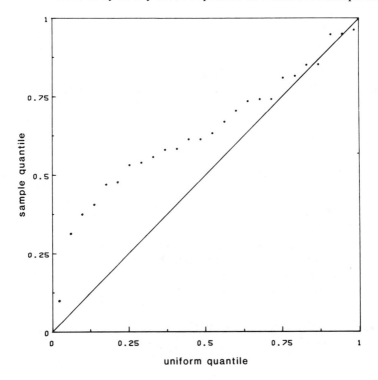

$(\hat{\alpha}, \hat{\beta})$. An approximate $100(1-\alpha)\%$ confidence cone for (α, β) is centred on $(\hat{\alpha}, \hat{\beta})$ with semi-vertical angle

$$q = \arcsin(e_\alpha^{1/2} \hat{\sigma}) \qquad (5.13)$$

where $e_\alpha = -\log \alpha$. Thus, for a 95% confidence cone, $\alpha = 0.05$, $e_\alpha = -\log 0.05 = 2.996$, and (5.13) becomes $q = \arcsin(1.731\hat{\sigma})$.

> **Example 5.13** Under the assumption that the data (set B2) of Example 5.2 are sampled from a distribution symmetric about its mean direction (which can be checked as described in §**5.3.1(iv)**), we obtain a large-sample confidence cone. The estimated mean direction is $(\hat{\alpha}, \hat{\beta}) = $ (Dec. 144.2°, Inc. 57.2°), $R = 25.8$ ($n = 26$), and from (5.12), $\hat{\sigma} = 0.0262$. Hence $q = \arcsin(1.731 \times 0.0262) = 2.6°$.
>
> So an approximate 95% confidence cone for the unknown mean direction is centred on $(\hat{\alpha}, \hat{\beta})$ with semi-vertical angle $q = 2.6°$.

References and footnotes See Fisher & Lewis (1983) and Watson (1983a, Chapter 4).

5.3.1(vi) *Test for a specified mean direction of a symmetric distribution.* Suppose we have a random sample of data from a distribution which is assumed to be rotationally symmetric about its unknown mean direction (α, β), and that we wish to test the hypothesis that $(\alpha, \beta) = (\alpha_0, \beta_0)$. This can be done graphically, using the method in §**5.3.1(iv)** above. For a formal test, let $(\hat{x}, \hat{y}, \hat{z})$ be the direction cosines of the sample mean direction (cf. (3.2)-(3.4)) and $(\lambda_0, \mu_0, \nu_0)$ those of (α_0, β_0). Calculate the spherical standard error $\hat{\sigma}$ from (5.12) and then the test statistic

$$h_n = [1 - (\lambda_0 \hat{x} + \mu_0 \hat{y} + \nu_0 \hat{z})^2]/\hat{\sigma}^2. \qquad (5.14)$$

The hypothesis that (α_0, β_0) is the true mean direction is rejected if h_n is too large.

Critical values sample size $n \geq 25$. Reject the hypothesis at level $100\alpha\%$ if h_n exceeds $-\log \alpha$.

Note that if we have already calculated a $100(1-\alpha)\%$ confidence cone for the unknown mean direction, we have an equivalent test (at the $100\alpha\%$ level) by checking whether the hypothesized value (α_0, β_0) lies inside this cone.

> **Example 5.14** For the data (set B2) of Example 5.13, to see whether the direction (Dec. 150°, Inc. 60°) is acceptable as the true mean direction, we see immediately from Example 5.13 that it is not, as it differs from the sample mean direction (Dec. 144.2°, Inc. 57.2°) by far more than the semi-angle $q = 2.6°$ of the 95% confidence cone.

References and footnotes The test is due to Watson (1983a, Chapter 4). For small sample sizes, use the test given in §**5.3.1(iv)**.

5.3.2 Parametric models: unimodal distributions with rotational symmetry

The only distribution in common use as a model for examples of unit vectors from a unimodal distribution symmetric about its polar direction is the Fisher distribution, as defined in §**4.4.3**. We shall consider the following problems in relation to a sample of data supposedly drawn from a Fisher distribution.

(i) Testing the sample for uniformity against the alternative that it was drawn from a Fisher distribution.
(ii) Testing to see whether the Fisher distribution is a suitable model for the data.
(iii) Testing individual observations which are unusually far from the main data mass to see if they are to be judged discordant (statistically inconsistent with the main data mass), assuming that the underlying distribution is Fisherian.
(iv) Estimating the parameters of the underlying Fisher distribution, including confidence regions for the mean direction and concentration parameter.
(v) Testing whether the mean direction of the distribution has a specified value.
(vi) Testing whether the concentration parameter of the distribution has a specified value.
(vii) Other problems relating to the Fisher distribution.

5.3.2(*i*) *Test for uniformity against a Fisher* $F\{(\alpha, \beta), \kappa\}$ *alternative.* Since uniformity corresponds to $\kappa = 0$, this is equivalent to the procedures given in §**5.3.1(i)**.

5.3.2(*ii*) *Goodness-of-fit of the Fisher* $F\{(\alpha, \beta), \kappa\}$ *model.* We present two methods of assessing the adequacy of fit of this model to a set of data. The first comprises a set of graphical displays which permit a general assessment of the plausibility of the model, and ready identification of any isolated points subject to whose omission or adjustment the model would appear reasonable. The second comprises a set of formal significance tests companion to the graphical procedures. Both methods are based on marginal distribution properties (see §**4.4.3**) of the Fisher distribution. A restriction on the use of these methods is that they are based on approximations which are only valid if κ exceeds say, 5. However, in many data contexts, κ will not have such a low value.

Graphical procedures Calculate the sample mean direction $(\hat{\alpha}, \hat{\beta}) = (\hat{\theta}, \hat{\phi})$ as in §3.2.1, and rotate the data to the pole $(\hat{\alpha}, \hat{\beta})$ using (3.8) with $A(\hat{\alpha}, \hat{\beta}, 0)$, to obtain $(\theta'_1, \phi'_1), \ldots, (\theta'_n, \phi'_n)$.

(a) *Colatitude plot* Let $X_i = 1 - \cos\theta'_i$, $i = 1, \ldots, n$, and re-arrange the X_i's in increasing order, obtaining $X_{(1)}, \ldots, X_{(n)}$. Perform a Q-Q plot using the exponential model, as described in §3.5.1. The plot should be approximately linear, passing through the origin. The slope of the plot gives an estimate of $1/\kappa$.

(b) *Longitude plot* Let $X_i = \phi'_i/2\pi$, $i = 1, \ldots, n$, and re-arrange the X_i's in increasing order, obtaining $X_{(1)}, \ldots, X_{(n)}$. Perform a Q-Q plot using the uniform model, as described in §3.5.1. The plot should be approximately linear, passing through the origin, with slope 45°.

(c) *Two-variable plot* Rotate the original data $(\theta_1, \phi_1), \ldots, (\theta_n, \phi_n)$ to the pole $((3\pi/2 - \hat{\alpha}, \hat{\beta} - \pi)$ using §3.2.1 with $A(3\pi/2 - \hat{\alpha}, \hat{\beta} - \pi, 0)$, to obtain $(\theta''_1, \phi''_1), \ldots, (\theta''_n, \phi''_n)$, set $X_i = (\phi''_i - \pi)\sqrt{\sin\theta''_i}$, $i = 1, \ldots, n$, and re-arrange the X_i's in increasing order to obtain $X_{(1)}, \ldots, X_{(n)}$. Perform a Q-Q plot using the normal distribution, as described in §3.5.1. The plot should be approximately linear, passing through the origin with slope approximately $1/\sqrt{\kappa}$.

The first and second plots are specifically concerned with the distribution of the deviations from the pole, and with the distribution of the longitudes or azimuths of these deviations. The third plot is a check on the independence or otherwise of the deviation and the longitude of this deviation. Any outliers in the data set will tend to show up in the colatitude plot, as will be seen in Example 5.15 below. The first and third plots provide separate graphical estimates of κ, which should be in reasonable agreement if the underlying distribution is Fisherian.

> **Example 5.15** The three plots for the data (set B2) of Example 5.2 are shown in Figure 5.15. No plot suggests any significant departure from the hypothesis that the underlying distribution is Fisherian. The two graphical estimates of κ, from Figures 5.15(a) and 5.15(c) are $1/\hat{\kappa}_a = 0.0090$, $\hat{\kappa}_a \approx 111$ and $1/\sqrt{\hat{\kappa}_c} = 0.095$, $\hat{\kappa}_c \approx 110$, which are in excellent agreement.

> **Example 5.16** Another set of measurements of magnetic remanence in red-beds is shown in Figure 5.16 (see Appendix B7). The corresponding probability plots for goodness-of-fit of Fisher's distribution are given in Figure 5.17. The colatitude plot in Figure 5.17(a) reveals two points, $P = $ (Dec. 25°, Inc. −14°) and $Q = $ (Dec. 122°, Inc. −50°), which are outliers with respect to the Fisherian model. The two-variable plot also reveals two points, R and S, which appear aberrant: these are also the points P and Q, respectively. So, either the Fisher distribution can be reasonably assumed as a model for the remaining 15 observations, with P and Q being discordant, or a Fisher distribution is not the appropriate model for the 15 observations, and we must seek another model to fit

5.3.2(ii). Goodness-of-fit of the Fisher model

Figure 5.15 Probability plots for checking goodness-of-fit of the Fisher distribution to the data in Figure 5.2. (a) colatitude plot (b) longitude plot (c) two-variable plot. See Example 5.15.

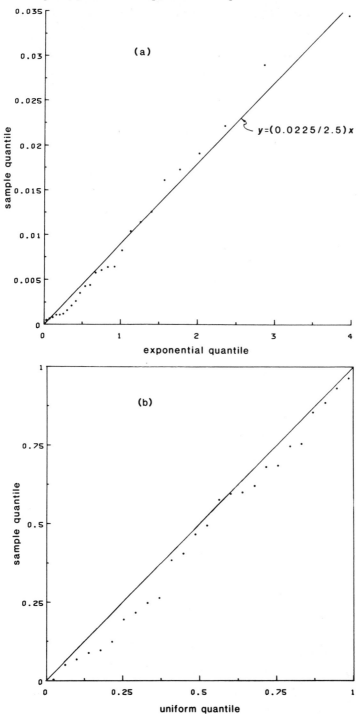

120   5. Analysis of a single sample of unit vectors

them. This new model may well fit the points *P* and *Q* as well, in which case they would no longer be outliers.

As a first step, then, we consider analysing the reduced sample of 15 observations. The probability plots for these are shown in Figure 5.18. These now appear reasonable, with the possible exception of the point

Figure 5.15 (*cont.*)

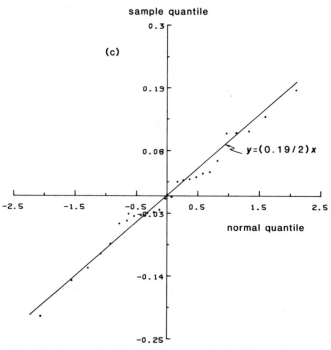

Figure 5.16 Equal-area projection of 17 measurements of magnetic remanence. The coordinate system is (Declination, Inclination). See Example 5.16.

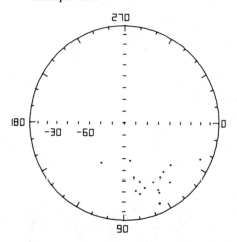

5.3.2(ii). Goodness-of-fit of the Fisher model 121

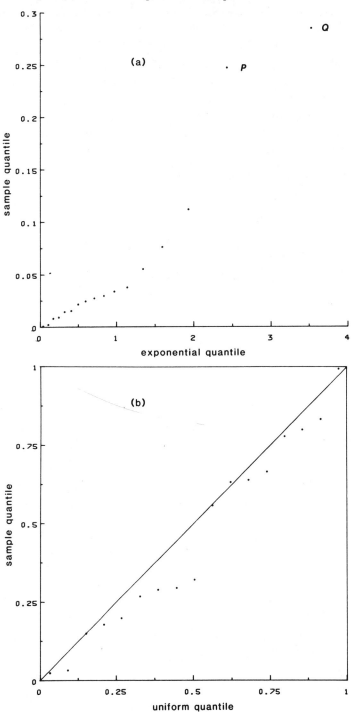

Figure 5.17 Probability plots for checking goodness-of-fit of the Fisher distribution to the data in Figure 5.16. (a) colatitude plot (b) longitude plot (c) two-variable plot. See Example 5.16.

$U = $ (Dec. 81°, Inc. −30°) in Figure 5.18(c). However, this point corresponds to T in Figure 5.17(a), which is clearly not an outlying point in that plot. The two graphical estimates of κ are $\hat{\kappa}_a \approx 30$ and $\hat{\kappa}_c \approx 28$, which agree well. So we accept the Fisher distribution as a model for the reduced data set. The question of how to deal with the outlying points is discussed further in the next Section.

Generally speaking, we need to take care in using probability plots when the sample size is small, say $n \leq 25$. They can be subject to quite substantial fluctuations due solely to sampling variability. In the last example, this did not appear to be a problem. However, a graphical method is only an informal guide in model-fitting, and will in general be followed by formal testing.

Formal testing procedures Corresponding to each of (a), (b) and (c) above, we have a formal test, as follows:

(a) Colatitude test Calculate the Kolmogorov-Smirnov statistic' D_n defined by (3.38), using $X'_1, \ldots, X'_n$ and $F(x) = 1 - \exp(-\hat{\kappa}x)$ with $\hat{\kappa} = (n-1)/\Sigma X_i$. The test statistic is

$$M_E(D_n) = (D_n - 0.2/n)(\sqrt{n} + 0.26 + 0.5/\sqrt{n}) \tag{5.15}$$

Figure 5.17 (*cont.*)

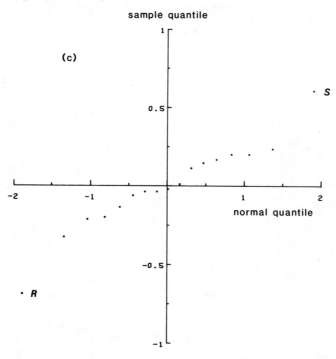

5.3.2(ii). Goodness-of-fit of the Fisher model

The hypothesis that the distribution is Fisherian is rejected if $M_E(D_n)$ is too large.

Critical values Use critical values in Appendix A8.

(b) Longitude test Calculate Kuiper's V_n defined by (3.39), using $X_{(1)}, \ldots, X_{(n)}$ as calculated for the longitude plot and $F(x) = x$. The test statistic is

$$M_U(V_n) = V_n(\sqrt{n} - 0.567 + 1.623/\sqrt{n}) \qquad (5.16)$$

The hypothesis that the distribution is Fisherian is rejected if $M_U(V_n)$ is too large.

Critical values Use critical values in Appendix A8.

Note: a test specifically designed to detect an alternative distribution with elliptical contours is given in §**5.3.3(iii)**.

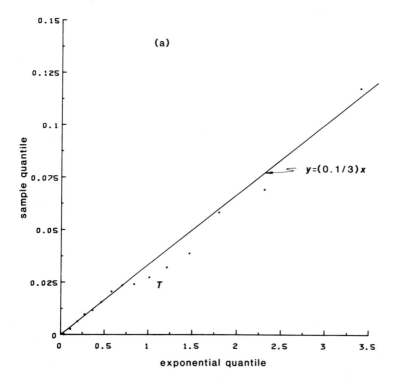

Figure 5.18 Probability plot for checking goodness-of-fit of the Fisher distribution to the data in Figure 5.16 with (25°, −14°) and (122°, −50°) omitted. (a) colatitude plot (b) longitude plot (c) two-variable plot. See Example 5.16.

124   5. Analysis of a single sample of unit vectors

Figure 5.18 (cont.)

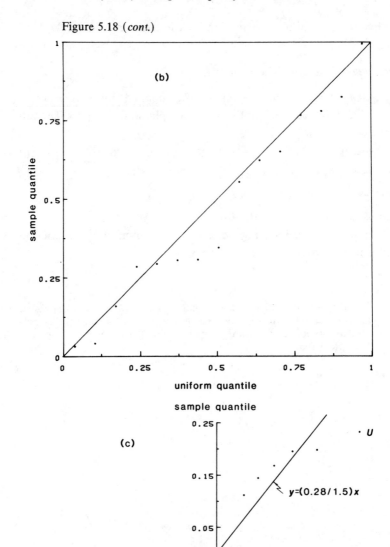

5.3.2(iii). Outlier test for discordancy

(c) **Two-variable test** Calculate $s^2 = (1/n)\Sigma X_i^2$, $x_i = X_i/s$, $i = 1, \ldots, n$, and re-order the x_i's to obtain $x_{(1)} \leq \ldots \leq x_{(n)}$. Calculate the Kolmogorov-Smirnov statistic D_n in (3.38) using $x_{(1)}, \ldots, x_{(n)}$ and the normal distribution function $F(x)$ in Table 3.2. The test statistic is

$$M_N(D_n) = D_n(\sqrt{n} - 0.01 + 0.85/\sqrt{n}) \tag{5.17}$$

The hypothesis that the distribution is Fisherian is rejected if M_N is too large.

Critical values Use critical values in Appendix A8.

> **Example 5.17** For the data (set B2) of Example 5.2, we obtain the following results for the three tests:
>
> Colatitude test: $D_n^+ = 0.1107$, $D_n^- = 0.0$, $D_n = 0.1107$,
> $\quad M_E = 0.5619 < 0.926$, the 15% point
> Longitude test: $D_n^+ = 0.1205$, $D_n^- = 0.0370$, $V_n = 0.1575$,
> $\quad M_U = 0.7798$
> Two-variate test: $D_n^+ = 0.0895$, $D_n^- = 0.1038$, $D_n = 0.1038$,
> $\quad M_N = 0.5457 < 0.775$, the 15% point
>
> This confirms the conclusions reached from the graphical analysis in Example 5.15.
>
> **Example 5.18** For the data (set B7) of Example 5.16, we obtain the statistics
>
> $M_E = 1.1010$, $M_U = 1.0558$, $M_N = 0.5645$
>
> The significance probability exceeds 10% for both the second and third statistics, but the colatitude test statistic lies between the 5% and 1% points, suggesting some departure from the Fisher model.

References and footnotes See Lewis & Fisher (1982) for the graphical methods, Fisher & Best (1984) for the formal tests, and Fisher (1982a) for a brief summary of these and outlier detection methods; these matters are also discussed briefly in Watson (1983a, Chapter 1) and Watson (1983b); Mardia, Holmes & Kent (1984) give an omnibus goodness-of-fit test, designed to detect an axially asymmetric alternative model.

5.3.2(iii) Outlier test for discordancy. Generally speaking, whereas the formal goodness-of-fit tests described above are satisfactory in detecting general departures in shape for the marginal colatitude and longitude distributions, they may not be particularly sensitive to the presence of one or two outlying points. On the other hand, the graphical procedures are very useful for identifying isolated points worthy of further investigation. An appropriate method for doing this formally is a test of discordancy.

Suppose, then, that one of the observations in the sample, say (θ_n, ϕ_n), is an outlier, that is, it appears to be suspiciously far from the main mass of data as judged by a plot of the data, or by one of the graphical procedures in §5.3.2(ii) above. We may test to see whether (θ_n, ϕ_n) is significantly far away, i.e. whether it should be judged *discordant*, as follows.

We shall write, for emphasis, R_n for the resultant length of the whole sample (cf. (3.3)) and define R_{n-1} by

$$R_{n-1} = \left\{ \left(\sum_{i=1}^{n-1} x_i \right)^2 + \left(\sum_{i=1}^{n-1} y_i \right)^2 + \left(\sum_{i=1}^{n-1} z_i \right)^2 \right\}^{1/2} \tag{5.18}$$

the resultant length of the sample after omitting the outlier (θ_n, ϕ_n). The test statistic is

$$E_n = (n-2)(1 + R_{n-1} - R_n)/(n - 1 - R_{n-1}) \tag{5.19}$$

and the outlying point (θ_n, ϕ_n) is classified as discordant if E_n is too large.

Critical values Suppose the calculated value of E_n for the given data is x. Compute

$$p = \begin{cases} n[(n-2)/(n-2+x)]^{n-2}, & \text{if } x \geq n-2, \\ n[(n-2)/(n-2+x)]^{n-2} - (n-1)[(n-2-x)/(n-2+x)]^{n-2}, & \text{if } x < n-2. \end{cases} \tag{5.20}$$

Then $\text{Prob}(E_n > x) = p$, so if p is less than, say, 0.05, (θ_n, ϕ_n) is significantly far away from the sample mean direction. This approximation is adequate except for very dispersed distributions ($\kappa < 2.5$). No critical values are available for $\kappa < 2.5$.

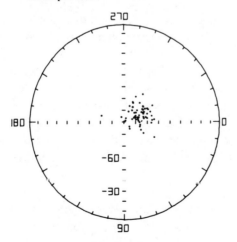

Figure 5.19 Equal area projection of 62 measurements of magnetic remanence. The coordinate system is (Declination, Inclination). See Example 5.19.

5.3.2(iii). Outlier test for discordancy

Example 5.19 Data set A in Appendix B8 constitutes measurements of magnetic remanence in 62 rock specimens, after each specimen had been partially thermally demagnetised to the same stage. (See Example 8.1 for other details about both sets). In this set, observation number 16 was entered incorrectly as (Dec. 194°, Inc. −69°) (shown parenthetically in the data listing). A plot of these data with the incorrect measurement is shown in Figure 5.19, in which the incorrect point appears as an outlier. It also appears as an outlier in the colatitude plot (cf. §**5.3.2(ii)** above) in Figure 5.20. If the point is omitted, the methods of the previous subsection indicate that the sample is well-modelled by a Fisher distribution. Accordingly, we test the outlying point for discordancy. We obtain $E_n = 11.79$, corresponding to a significance probability of 0.0013. So the point is classified as discordant.

A subsequent check of the data entry enabled this point to be corrected. Other ways of dealing with discordant observations are discussed later in this subsection.

Sometimes there will seem to be more than one outlying value in the sample. If the outliers are clustered together, it may be appropriate to test whether there are in fact two populations present. A suitable procedure for this is described in §**7.2.3(ii)**. On the other hand, the outliers

Figure 5.20 Colatitude plot for the data in Figure 5.19, checking on the goodness-of-fit of the Fisher distribution. See Example 5.19.

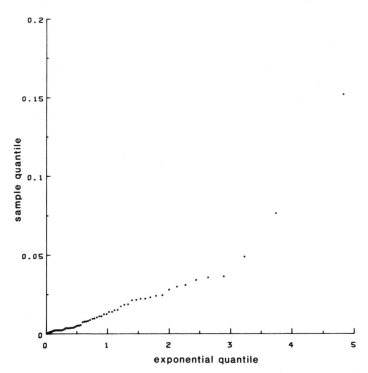

may be scattered, in which case two possible methods suggest themselves:

(a) test the outliers one at a time, starting with the "most extreme" on some suitable definition (note that it may not be intuitively obvious which of a sample of observations on the sphere *is* the most aberrant),

(b) test them *en bloc*

Each of these approaches has advantages and disadvantages (see, for example, Barnett & Lewis 1984, Chapter 5 for a general discussion). Suppose the t outliers are $(\theta_{n-t+1}, \phi_{n-t+1}), \ldots, (\theta_n, \phi_n)$.

As a generalisation of (5.18), define R_{n-t} by

$$R_{n-t} = \left[\left(\sum_{i=1}^{n-t} x_i \right)^2 + \left(\sum_{i=1}^{n-t} y_i \right)^2 + \left(\sum_{i=1}^{n-t} z_i \right)^2 \right]^{1/2}$$

the resultant length of the sample when the t suspect observations are omitted. Calculate

$$E_n^{(t)} = [(n-t-1)/t](t + R_{n-t} - R_n)/(n-t-R_{n-t}). \quad (5.21)$$

The t outliers are jointly classified as discordant if $E_n^{(t)}$ is too large.

Critical values Use critical values in Appendix A9.

> **Example 5.20** The analysis of the data (set B7) in Example 5.16 suggests that there may be two discordant data points. For these data, we obtain from (5.21), $E_n^{(2)} = 8.11$.
>
> By interpolating in the table in Appendix A19, we find that this value of $E_n^{(2)}$ corresponds to a significance probability of approximately 0.025. So there is quite strong evidence for taking the two suspect points to be discordant.

References and footnotes See Fisher, Lewis & Willcox (1981). Note that their Table 5, $t = 3$ entries are incorrect, and should all be multiplied by $\frac{2}{3}$, as noted by Best and Fisher (1986). Appendix A9 contains corrected values.

Having decided that one or more points in a sample are discordant, what do we do about it? To answer this, we have to consider

(a) how were the data obtained

and

(b) how do we wish to use them.

Two monographs have been written on the subject of outliers, by Barnett & Lewis (1984), and Hawkins (1980), in which these questions are considered in some depth. Here, we confine ourselves to a few remarks.

5.3.2(iv). Parameter estimation for Fisher distribution

In connection with (a), one obvious way in which outliers occur is through mis-measurement or mis-recording (see, for example, Example 5.19). If this can be established as the cause, perhaps the adjustment required is clear; otherwise, that observation must be deleted. Another possibility is that there is actually more than one population represented. This circumstance can arise due to incorrect sampling, or in preparation of a specimen for measurement (for example, incomplete purging of overprint components of remanent magnetisation prior to isolating the primary or original component). One may then wish to re-model the data as a mixture of distributions, or, if interested in just one population, delete those which are classified as not being from that population. A third possibility is that the outlying data do come from the same population as the others, but that the vagaries of sampling have produced wayward points (as must happen from time to time). The question of how to deal with the outliers leads us to the question (b) above.

A principal use of a single sample of data relates to the mean direction – what is it? is it the same as another specified mean direction? how precisely has it been measured? – so we consider how to handle outliers in this light. In brief, the presence of one or two outliers in a sample does not have an important effect on estimation of the mean direction, but it may be important when calculating confidence cones, or testing hypotheses about specified mean directions, or in the comparison of several samples. Accordingly, in §**5.3.2(iv)** below we shall see how outlying values can be *accommodated* when estimating κ.

5.3.2(iv) *Parameter estimation for the Fisher distribution.* The usual (so-called maximum likelihood) estimate of the mean direction (α, β) of the Fisher distribution is the sample mean direction $(\hat{\alpha}, \hat{\beta}) = (\hat{\theta}, \hat{\phi})$ as defined by (3.4)-(3.6). Let $(\hat{\lambda}, \hat{\mu}, \hat{\nu})$ be the direction cosines of $(\hat{\alpha}, \hat{\beta})((\hat{x}, \hat{y}, \hat{z})$ in (3.4)). The maximum likelihood estimate of κ is the solution of the equation

$$\coth(\kappa) - 1/\kappa = R/n \tag{5.22}$$

If the mean direction is known, and has direction cosines (λ, μ, ν), R in (5.22) is replaced by R^*, where

$$R^* = RC \text{ and } C = \lambda\hat{\lambda} + \mu\hat{\mu} + \nu\hat{\nu} \tag{5.23}$$

Tabulated values of the solution of equation (5.22) are given in Appendix A10. For the range of values $0.95 \leq R/n \leq 1$, the estimate of κ exceeds 2.5, a common occurrence with palaeomagnetic data. The solution of (5.22) is then well-approximated by

$$\hat{\kappa}_{MLE} = n/(n-R) \tag{5.24}$$

An approximately unbiased estimate of κ which is usually used in practice when $\bar{R} = R/n \geq 0.95$ is

$$\hat{\kappa} = (1 - 1/n)\hat{\kappa}_{MLE} = (n-1)/(n-R) \qquad (R/n \geq 0.95) \qquad (5.25)$$

and, unless otherwise stated, it is the one we shall use in this book. However, for small sample sizes, this estimate can be substantially biased upward, resulting in a deceptively small confidence cone for the mean direction. For sample sizes less than 16, the estimate

$$\hat{\kappa}_1 = (1 - 1/n)^2 \kappa_{MLE} \qquad (5.26)$$

is to be preferred.

As discussed at the end of the last subsection on testing outliers, one may wish to estimate κ from a sample with a few outliers present. In such a case, we would like to downweight, rather than discount completely, the contribution of the outliers. To do this, rotate the sample $(\theta_1, \phi_1), \ldots, (\theta_n, \phi_n)$, to the pole $(\hat{\alpha}, \hat{\beta})$ using (3.8) with $\mathbf{A}(\hat{\alpha}, \hat{\beta}, 0)$, to obtain $(\theta'_1, \phi'_1), \ldots, (\theta'_n, \phi'_n)$. Let $X_i = 1 - \cos\theta'_i$, $i = 1, \ldots, n$, and rearrange the X_i's into increasing order, obtaining $X_{(1)} \leq \ldots \leq X_{(n)}$.

Define the weights

$$c_{(i)} = 1/n - 2i/[n^2(n+1)], \qquad i = 1, \ldots, n. \qquad (5.27)$$

A *robust* estimate of κ is then given by

$$\hat{\kappa}_R = 1 \Big/ \sum_{i=1}^{n} c_{(i)} X_{(i)}. \qquad (5.28)$$

Since $X_{(1)}$ corresponds to the smallest deviation from the sample mean direction, and since the $c_{(i)}$'s are decreasing, we see that the contribution of each data point in (5.28) is increasingly down-weighted the greater its deviation from $(\hat{\alpha}, \hat{\beta})$.

Example 5.21 The analysis in Examples 5.15 and 5.18 indicate that the Fisherian model is acceptable for the data (set B2) in Example 5.2. The parameter estimates are

$(\hat{\alpha}, \hat{\beta}) = $ (Dec. 144.2°, Inc. 57.2°)

$\hat{\kappa} = (26-1)/(26-25.77) \approx 109$

(Note the good agreement with the graphical estimates in Example 5.15).

Example 5.22 Examples 5.16 and 5.19 contain data sets (B7 and B8(A)) in which outliers are present. It is interesting to look at the two estimates of κ using (5.25) and (5.28). We obtain

(*Example* 5.16 $n = 17$) $\hat{\kappa} = 40.1$, $\hat{\kappa}_R = 47.7$

(*Example* 5.19 $n = 62$) $\hat{\kappa} = 64.8$, $\hat{\kappa}_R = 67.6$

The improved estimates will, of course, result in smaller confidence cones for the mean direction. In the second case, the improvement is only marginal, because of the sample size.

5.3.2(iv). Parameter estimation for Fisher distribution

We now turn to the question of finding a confidence cone for the unknown mean direction (α, β). Such a cone will be centred on $(\hat{\alpha}, \hat{\beta})$, with semi-vertical angle θ_α say corresponding to some specified $100(1-\alpha)\%$ confidence. For emphasis, we recall the notation $\chi_\nu^2(\beta)$ for the upper $100\beta\%$ point of the χ_ν^2 distribution, i.e.

$$\text{Prob } (\chi_\nu^2 \geq \chi_\nu^2(\beta)) = \beta.$$

Suppose first that κ is known ($\kappa = \kappa_0$ say)†

$\kappa_0 \leq 5$ Find the tabulated value, $\bar{R}_{1-\alpha}^*$ say, corresponding to κ_0 and $1-\alpha$ in Appendix A11. Then

$$\theta_\alpha = \text{arc cos } (R_{1-\alpha}^*/R), \quad R_{1-\alpha}^* = n\bar{R}_{1-\alpha}^*, \tag{5.29}$$

R being the sample resultant length.

$\kappa_0 > 5$

$$\theta_\alpha = \text{arc cos } \{(n - \chi_{2n}^2(\alpha))/R\}. \tag{5.30}$$

Now suppose that κ is unknown Depending on the sample size n and the estimated value $\hat{\kappa}$ of κ, we use either an exact procedure or a simpler approximate procedure. For $n < 30$ and $\hat{\kappa} < 5$, the exact procedure should be followed. Let $R_z = R\hat{\nu}$ ($\hat{\lambda}, \hat{\mu}, \hat{\nu}$ being the direction cosines of the estimated mean direction).

$n \leq 8$ Find the tabulated value, $R_{1-\alpha}^0$ say, corresponding to $1-\alpha$, n and R_z, in Appendix A12. Then use (5.29).

$8 < n < 30$

$0 < R_z < n/4$. The exact procedure is as for $n \leq 8$.
An approximate procedure is to calculate

$$R_{\alpha,1} = \{R_z^2 - 2/3n \log (\alpha)\}^{1/2} \tag{5.31}$$

and then use (5.29) with $R_{1-\alpha}^* = R_{\alpha,1}$.

$n/4 \leq R_z \leq 3n/5$. The exact procedure is as for $n \leq 8$.
An approximate procedure is to calculate

$$R_\alpha = \tfrac{1}{2}(R_{\alpha,1} + R_{\alpha,2}) \tag{5.32}$$

where $R_{\alpha,1}$ is defined by (5.31) and

$$R_{\alpha,2} = n(1-\alpha^*) + R_z\alpha^*, \quad \alpha^* = \alpha^{1/(n-1)} \tag{5.33}$$

and then use (5.29) with $R_{1-\alpha}^* = R_\alpha$.

$R_z > 3n/5$.

Calculate $R_{\alpha,2}$ as in (5.33) and then use (5.29) with $R_{1-\alpha}^* = R_{\alpha,2}$.

If $\hat{\kappa} \geq 5$, the following approximate procedure is adequate for all sample sizes:

$$\theta_\alpha = \text{arc cos } \{1 - [(n-R)/R][(1/\alpha)^{1/(n-1)} - 1]\}. \tag{5.34}$$

† An improved method for this case is given by Stephens (1975).

If $n \geq 30$ and $\hat{\kappa} < 5$, the approximation

$$\theta_\alpha = \arccos\{1 + \log(\alpha)/(\hat{\kappa}R)\} \tag{5.35}$$

can be used.

Example 5.23 For the data (set B2) analysed in Example 5.21, a 95% confidence cone for the unknown mean direction can be computed from (5.34), since $\hat{\kappa} = 109$. We have

$$\theta_\alpha = \arccos\{1 - [(26 - 25.77)/25.77][(1/0.05)^{1/25} - 1]\} = 2.7°.$$

Note that this is in fact fractionally larger than the result in Example 5.13, even though we have used a parametric model for the data. The explanation is that the calculation in Example 5.13 is based on an approximate (large-sample) result, which has been applied for n close to its minimum acceptable value, so the resulting semi-angle is not as reliable as that computed here.

Calculation of a $100(1-\alpha)\%$ confidence interval for an unknown concentration parameter κ depends on whether or not the mean direction is known.

Mean direction known Let the mean direction have direction cosines (λ, μ, ν), and calculate R^* from (5.23). Enter Appendix A11 with α and $\bar{R}^* = R^*/n$, and locate a lower value κ_L for which the tabulated $\frac{1}{2}\alpha$ value is $\bar{R}^*$ and an upper value κ_U for which the tabulated $1 - \frac{1}{2}\alpha$ value is $\bar{R}^*$. If the $\frac{1}{2}\alpha$ entry for $\kappa = 0$ exceeds 0, set $\kappa_L = 0$. If the $1 - \frac{1}{2}\alpha$ entry for $\kappa = 5$ is less than $\bar{R}^*$, or if both the $\frac{1}{2}\alpha$ and $1 - \frac{1}{2}\alpha$ entries for $\kappa = 5$ are less than $\bar{R}^*$, use instead

and
$$\kappa_L = \tfrac{1}{2}\chi^2_{2n}(1-\tfrac{1}{2}\alpha)/(n - R^*)$$
$$\kappa_U = \tfrac{1}{2}\chi^2_{2n}(\tfrac{1}{2}\alpha)/(n - R^*) \tag{5.36}$$

as appropriate.

So, unless the distribution is quite disperse, the limits given in (5.36) will commonly be used.

Mean direction unknown Use the above procedure with R and $\bar{R}$ replacing R^* and $\bar{R}^*$, referring to Appendix A13 instead of A11, and using

and
$$\kappa_L = \tfrac{1}{2}\chi^2_{2n-2}(1-\tfrac{1}{2}\alpha)/(n - R)$$
$$\kappa_U = \tfrac{1}{2}\chi^2_{2n-2}(\tfrac{1}{2}\alpha)/(n - R) \tag{5.37}$$

instead of (5.36).

Example 5.24 Continuing Example 5.23, a 95% confidence interval for κ is, from (5.34),

$$\kappa_L = \tfrac{1}{2}(32.36)/(26-25.77) = 70.3,$$
$$\kappa_U = \tfrac{1}{2}(71.42)/(26-25.77) = 155.3.$$

References and footnotes Point estimation of the mean direction (α, β) and concentration parameter κ was considered initially by Fisher (1953);

5.3.2(v). Test for specified mean direction

see also Mardia (1972, Section 9.2) and Watson (1983a, Chapter 4). Estimation of κ when (α, β) is known was also studied by Fisher (1953); see also Watson (1956a) and Mardia (1972, page 251). Other estimates of κ have been proposed by Schou (1978), Watson (1983a, Appendix A) and Watson (1983b); by Best & Fisher (1981) (correction for bias in small samples); and by Fisher (1982b) and Watson (1983a) in the context of small samples. For discussion of robustness aspects, see also Watson (1967, 1970). A confidence cone for (α, β) when κ is unknown but assumed to be sufficiently large that $\exp(-2\kappa)$ is negligible was obtained by Fisher (1953). Watson & Williams (1956) derived an exact procedure, applicable for all values of κ. Watson (1956a) derived confidence intervals for the unknown value of κ. Stephens (1962, 1967) provided tables enabling the Watson & Williams procedure to be implemented. Stephens (1967) also gives the theory and tables for exact interval estimation of κ. General summaries of these procedures are given by Mardia (1972); see also Watson (1983a) and Mardia (1975a,b).

5.3.2(v) *Test for a specified mean direction.* The method of testing whether the mean direction has a specified value, with direction cosines $(\lambda_0, \mu_0, \nu_0)$ say, is complementary to that used to obtain a confidence cone for (λ, μ, ν). If a $100(1-\alpha)\%$ cone has already been computed, as in §**5.3.2(iv)** above, the hypothesis is rejected at the $100\alpha\%$ level if $(\lambda_0, \mu_0, \nu_0)$ does not fall inside the cone. Alternatively, we can test it directly. The procedures used follow the format in §**5.3.2(iv)** for calculating a confidence cone. Suppose first that $\kappa = \kappa_0$, a known value, and calculate

$$C_0 = \lambda_0 \hat{\lambda} + \mu_0 \hat{\mu} + \nu_0 \hat{\nu}, \tag{5.38}$$

the cosine of the angle between the hypothesised mean direction and the sample mean direction $(\hat{\lambda}, \hat{\mu}, \hat{\nu})$. Also calculate

$$R_z = RC_0. \tag{5.39}$$

$\kappa_0 < 5$ Reject the hypothesis at the $100\alpha\%$ level if $R_z < n\bar{R}^*_{1-\alpha}$ where $\bar{R}^*_{1-\alpha}$ is obtained from Appendix A11.

$\kappa_0 \geq 5$ Reject the hypothesis at the $100\alpha\%$ level if

$$2\kappa_0(R - R_z) > \chi^2_{2n}(\alpha). \tag{5.40}$$

If κ is unknown, we use either an exact or an approximate procedure, depending on the values of n and $\hat{\kappa}$. If $n < 30$ and $\hat{\kappa} < 5$, the exact procedures should be followed.

$n \leq 8$ Reject the hypothesis at the $100\alpha\%$ level if $R > R^*_{1-\alpha}$, where $R^*_{1-\alpha}$ is obtained from Appendix A12.

$8 \leq n < 30$, $0 < R_z < \frac{1}{4}n$. The exact procedure is as for $n \leq 8$.

An approximate procedure is to calculate $R_{\alpha,1}$ from (5.31) and to reject the hypothesis at the $100\alpha\%$ level if $R > R_{\alpha,1}$

$\frac{1}{4} \leq R_z \leq \frac{3}{5}n$. The exact procedure is as for $n \geq 8$.

An approximate procedure is to calculate R_α from (5.32) and to reject the hypothesis at the $100\alpha\%$ level if $R > R_\alpha$.

$R_z > \frac{3}{5}n$. Calculate $R_{\alpha,2}$ as in (5.33) and reject the hypothesis at the $100\alpha\%$ level if $R > R_{\alpha,2}$.

If $\hat{\kappa} \geq 5$, the following approximation can be used for all sample sizes. Calculate

$$R_\alpha = n - (n - R_z)\alpha^{1/(n-1)} \tag{5.41}$$

and reject the hypothesis at level $100\alpha\%$ if $R > R_\alpha$.

Example 5.25 In Example 5.12 we tested the data (set B2) of Example 5.2 for symmetry about the direction (Dec. 150°, Inc. 60°). Subsequent analysis suggested that the Fisher distribution was an appropriate model. Suppose we now test the hypothesis that the distribution has this mean. The inference will be different from the inference in Example 5.12 because it will use the extra assumption that the distribution is Fisherian. From (5.38), (5.41) and the results in Example 5.21, we have

$R_z = 25.77 \times 0.9974 = 25.70$

and from (5.41), $R_\alpha = 25.74$, whereas $R = 25.77$. So the significance probability is about 5% and there are some grounds for rejecting the null hypothesis.

References and footnotes See corresponding notes in §5.3.2(iv) above.

5.3.2(vi) Test for specified concentration parameter. As in §5.3.2(v), a test of the hypothesis $\kappa = \kappa_0$ at level $100\alpha\%$ can be performed by checking whether κ_0 lies within the $100(1-\alpha)\%$ confidence interval (κ_L, κ_U) computed in §5.3.2(iv). Alternatively, we may perform the test directly; as in §5.3.2(v), we follow the format in §5.3.2(iv) for calculating a confidence interval. Suppose first that the mean direction is known, with direction cosines (λ, μ, ν) and calculate R^* from (5.23), and then $\bar{R}^* = R^*/n$.

Mean direction known

$\kappa_0 < 5$ For given n, β, κ_0, let $\bar{R}^*_\beta$ be the corresponding entry in Appendix A11.

Test of $\kappa = \kappa_0$ against $\kappa > \kappa_0$: reject if $R^* > \bar{R}^*_\alpha$
Test of $\kappa = \kappa_0$ against $\kappa < \kappa_0$: reject if $R^* < \bar{R}^*_{1-\alpha}$
Test of $\kappa = \kappa_0$ against $\kappa \neq \kappa_0$: reject if either $R^* < \bar{R}^*_{\alpha/2}$
or $\bar{R}^* > \bar{R}^*_{1-\alpha/2}$

$\kappa_0 \geq 5$

Test of $\kappa = \kappa_0$ against $\kappa > \kappa_0$: reject if $R^* > n - \chi^2_{2n}(1-\alpha)/2\kappa_0$
Test of $\kappa = \kappa_0$ against $\kappa < \kappa_0$: reject if $R^* < n - \chi^2_{2n}(\alpha)/2\kappa_0$
Test of $\kappa = \kappa_0$ against $\kappa \neq \kappa_0$: reject if either

$$R^* > n - \chi^2_{2n}(\tfrac{1}{2}\alpha)/2\kappa_0$$

or

$$R^* < n - \chi^2_{2n}(\tfrac{1}{2}\alpha)/2\kappa_0$$

Mean direction unknown

$\kappa_0 < 5$

Use the procedure for (λ, μ, ν) known, with $\bar{R}$ replacing $\bar{R}^*$ and using Appendix A13 instead of A11.

$\kappa_0 \geq 5$

Test of $\kappa = \kappa_0$ against $\kappa > \kappa_0$: reject if $R > n - \chi^2_{2n-2}(1-\alpha)/2\kappa_0$
Test of $\kappa = \kappa_0$ against $\kappa < \kappa_0$: reject if $R < n - \chi^2_{2n-2}(\alpha)/2\kappa_0$
Test of $\kappa = \kappa_0$ against $\kappa \neq \kappa_0$: reject if either

$$R > n - \chi^2_{2n}(1-\tfrac{1}{2}\alpha)/2\kappa_0$$

or

$$R < n - \chi^2_{2n}(\tfrac{1}{2}\alpha)/2\kappa_0.$$

Example 5.26 Continuing Example 5.24, suppose we wish to test the hypothesis that $\kappa = 100$ against the alternative that $\kappa > 100$, at the 95% level. Using the procedure for (λ, μ, ν) unknown, $\kappa_0 \geq 5$, we have
$n - \chi^2_{2n-2}(1-\alpha)/2\kappa_0 = 26 - 34.76/200 = 25.83$.
Since $R = 25.77$, we accept the hypothesis that $\kappa = 100$.

References and footnotes See corresponding notes in §**5.3.2(iv)** above.

5.3.2(vii) *Other problems relating to the analysis of a single sample of data from the Fisher distribution.*

(a) Analysis of colatitudes. In some situations, only the colatitudes $\theta_1, \ldots, \theta_n$ (or equivalently, the Inclinations) are available for study, due to sampling problems (see e.g. Briden and Ward (1966) for a discussion of such data).

Parameter estimation is then considerably more complicated. A detailed discussion is given by Clark (1983).

(b) Approximate test that the mean direction lies in a specified plane Suppose that the plane is specified by an axis *normal* to it, with direction cosines $(\lambda_0, \mu_0, \nu_0)$, and define

$$R^* = R(\lambda_0 \hat{\lambda} + \mu_0 \hat{\mu} + \nu_0 \hat{\nu}) \tag{5.42}$$

where $(\hat{\lambda}, \hat{\mu}, \hat{\nu})$ are the direction cosines of the sample mean direction.

We consider two cases.

n large, $\hat{\kappa}$ small For a test at the $100\alpha\%$ level, calculate $R_\alpha = \{\frac{1}{3}n\chi_1^2(\alpha)\}^{1/2}$, and reject the hypothesis if $R^* > R_\alpha$

n large, $\hat{\kappa}$ large For a test at the $100\alpha\%$ level, calculate $R_\alpha = \{(n/\hat{\kappa})t_{n-1}(\frac{1}{2}\alpha)\}^{1/2}$, where $\hat{\kappa}$ is computed from (5.24) and $t_{n-1}(\frac{1}{2}\alpha)$ is the upper $100(\frac{1}{2}\alpha)\%$ point of the t-distribution with $n-1$ degrees of freedom and may be calculated as $\sqrt{F_{1,n-1}(\frac{1}{2}\alpha)}$ using Appendix A5. Reject the hypothesis if $R^* > R_\alpha$.

These approximate tests are due to Stephens (1962).

(c) Choice of sample size for estimating the mean direction with given precision This problem is discussed by Watson and Irving (1957) in the context of deciding on the number of sites to be sampled, and the number of specimens to be collected at each site, for the purpose of estimating a mean palaeomagnetic direction.

5.3.3 Parametric models: unimodal distributions without rotational symmetry

A generalisation of the Fisher distribution is the Kent distribution (§4.4.5), in which deviations from the preferred direction do not have to be isotropic. Detailed statistical analysis for this model is given by Kent (1982), and we confine ourselves here to a brief discussion of some problems of estimation and testing. No goodness-of-fit or outlier detection procedures are available at present.

5.3.3(i) Estimation of the parameters of the Kent distribution. Kent (1982) describes maximum likelihood estimation of the parameters, and then shows that estimation by moments is more straightforward and reasonably efficient. Accordingly, we describe the latter method. Let $(\theta_1, \phi_1), \ldots, (\theta_n, \phi_n)$ be a random sample of data from the Kent distribution, with corresponding direction cosines $(x_1, y_1, z_1), \ldots, (x_n, y_n, z_n)$.

Step 1 Calculate the sample mean direction as in §3.2.1 and thence $(\hat{\theta}, \hat{\phi})$ as in (3.4)-(3.6), and $\bar{R}$, the mean resultant length; also calculate the matrix $S = T/n$, where T is given by (3.15).

Step 2 Compute the matrix
$$H = \begin{pmatrix} \cos\hat{\theta} & -\sin\hat{\theta} & 0 \\ \sin\hat{\theta}\cos\hat{\phi} & \cos\hat{\theta}\cos\hat{\phi} & -\sin\hat{\phi} \\ \sin\hat{\theta}\sin\hat{\phi} & \cos\hat{\theta}\sin\hat{\phi} & \cos\hat{\phi} \end{pmatrix} \quad (5.43)$$

and then

$$B = H'SH$$

5.3.3(i). Parameter estimation for Kent distribution

with elements

$$\begin{pmatrix} b_{11} & b_{12} & b_{13} \\ b_{21} & b_{22} & b_{23} \\ b_{31} & b_{32} & b_{33} \end{pmatrix} \tag{5.44}$$

Define

$$\hat{\psi} = \tfrac{1}{2} \arctan \{2b_{23}/(b_{22} - b_{33})\} \tag{5.45}$$

Step 3 Compute the matrix

$$\mathbf{K} = \begin{pmatrix} 1 & 0 & 0 \\ 0 & \cos\hat{\psi} & -\sin\hat{\psi} \\ 0 & \sin\hat{\psi} & \cos\hat{\psi} \end{pmatrix} \tag{5.46}$$

and then

$$\hat{\mathbf{G}} = \mathbf{HK} = (\hat{\xi}_1, \hat{\xi}_2, \hat{\xi}_3) \tag{5.47}$$

where $\hat{\xi}_1, \hat{\xi}_2, \hat{\xi}_3$ are each 3×1 column vectors.

Finally, calculate

$$\mathbf{V} = \hat{\mathbf{G}}'\mathbf{S}\hat{\mathbf{G}} \tag{5.48}$$

with elements

$$\begin{pmatrix} v_{11} & v_{12} & v_{13} \\ v_{21} & v_{22} & v_{23} \\ v_{31} & v_{32} & v_{33} \end{pmatrix}$$

and let

$$Q = v_{22} - v_{33} \tag{5.49}$$

Step 4 For large κ, we have the approximate parameter estimates

$$\hat{\kappa} = (2 - 2\bar{R} - Q)^{-1} + (2 - 2\bar{R} + Q)^{-1} \tag{5.50}$$

$$\hat{\beta} = \tfrac{1}{2}\{(2 - 2\bar{R} - Q)^{-1} - (2 - 2\bar{R} + Q)^{-1}\}. \tag{5.51}$$

$\hat{\xi}_1'$ is just $(\hat{x}, \hat{y}, \hat{z})$, the sample mean direction.

Example 5.27 Kent (1982) has fitted this model to the data displayed in Figure 5.21 comprising 34 measurements of the direction of magnetisation for samples from the Great Whin Sill (Appendix B9). The estimate of the mean direction is (Dec. 187.8°, Inc. −4.9°). As intermediate stages of the calculation, we obtain

$$\mathbf{B} = \begin{pmatrix} 0.9443 & 0.0005 & 0.0009 \\ 0.0005 & 0.0385 & 0.0042 \\ 0.0009 & 0.0042 & 0.0172 \end{pmatrix},$$

$$\hat{\mathbf{G}} = \begin{pmatrix} 0.0848 & -0.9791 & 0.1846 \\ -0.9873 & -0.1076 & -0.1170 \\ 0.1345 & -0.1723 & -0.9758 \end{pmatrix},$$

138   5. Analysis of a single sample of unit vectors

$$V = \begin{pmatrix} 0.9443 & 0.0007 & 0.0008 \\ 0.0007 & 0.0393 & 0.0000 \\ 0.0008 & 0.0000 & 0.0164 \end{pmatrix}$$

and so

$\hat{\kappa} = 41.8, \hat{\beta} = 8.37$.

5.3.3(ii) *Elliptical confidence cone for the mean direction.*
Step 1 Rotate (θ_i, ϕ_i) to (θ'_i, ϕ'_i), using (3.8) with $\mathbf{A} = \hat{\mathbf{G}}$, and then relations of the form (3.5) and (3.6). Let $x_i^* = \cos \theta'_i$, $y_i^* = \sin \theta'_i \cos \phi'_i$, $z_i^* = \sin \theta'_i \sin \phi'_i$, $i = 1, \ldots, n$, and calculate

$$\tilde{\mu} = \sum x_i^*/n, \quad \tilde{\sigma}_2^2 = \sum y_i^{*2}/n, \quad \tilde{\sigma}_3^2 = \sum z_i^{*2}/n \quad (5.52)$$

Step 2 For a $100(1-\alpha)\%$ elliptical confidence cone, let

$$g = -2 \log(\alpha)/(n\tilde{\mu}^2) \quad (5.53)$$

and

$$s_2 = \tilde{\sigma}_2/g, \quad s_3 = \tilde{\sigma}_3/g \quad (5.54)$$

Then arcsin (s_2) and arcsin (s_3) are the major and minor semi-axes (in radian measure) of the $100(1-\alpha)\%$ elliptical confidence cone centred on the sample mean direction.

Step 3 To obtain a set of values on the perimeter of this cone, proceed as follows.

3.1 Let v_0 be any number between $-s_2$ and s_2 (from 5.54).

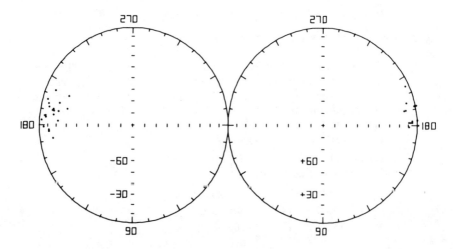

Figure 5.21 Equal-area projection of 34 measurements of magnetic remanence. The coordinate system is (Declination, Inclination). See Example 5.27.

5.3.3(iii). Test of Fisher model against Kent model

3.2 Let
$$w_0 = [s_3(g - v_0^2/s_2^2)]^{1/2}, \quad u_0 = (1 - v_0^2 - w_0^2)^{1/2}. \quad (5.55)$$

3.3 To obtain points on the cone rotated so that the z-axis is aligned with the sample mean direction, let $x_0 = v_0$, $y_0 = w_0$, $z_0 = u_0$, and $x_0' = v_0$, $y_0' = -w_0$, $z_0' = u_0$. (x_0, y_0, z_0) and (x_0', y_0', z_0') are then the direction cosines of two points (θ_0, ϕ_0) and (θ_0', ϕ_0') (obtainable using (3.4)-(3.6)) on the cone.

3.4 To obtain points on the cone in the original configuration, make the transformations

$$\begin{pmatrix} u_0^* \\ v_0^* \\ w_0^* \end{pmatrix} = \hat{G} \begin{pmatrix} u_0 \\ v_0 \\ w_0 \end{pmatrix}, \quad \begin{pmatrix} u_0^{**} \\ v_0^{**} \\ w_0^{**} \end{pmatrix} = \hat{G} \begin{pmatrix} u_0 \\ v_0 \\ -w_0 \end{pmatrix} \quad (5.56)$$

and then define $(x_0^*, y_0^*, z_0^*) = (v_0^*, w_0^*, u_0^*)$, $(x_0^{**}, y_0^{**}, z_0^{**}) = (v_0^{**}, w_0^{**}, u_0^{**})$. Convert (x_0^*, y_0^*, z_0^*) and $(x_0^{**}, y_0^{**}, z_0^{**})$ to (θ_0^*, ϕ_0^*) and $(\theta_0^{**}, \phi_0^{**})$ using (3.4)-(3.6).

3.5 Repeat Steps 3.1-3.4 for a range of values of u_0 between $-s_2$ and s_2.

Example 5.28 Continuing Example 5.27, a 95% elliptical confidence cone for the mean direction has semi-axes 4.9° and 3.2°; the cone is displayed in Figure 5.22, rotated to centralise the data

5.3.3(iii) *Test of whether a sample comes from a Fisher distribution, against the alternative that it comes from a Kent distribution.* The nature of the hypothesis being tested is whether the underlying distribution is rotationally symmetric about its mean direction, against the *specific* alternative

Figure 5.22 95% elliptical confidence cone for the mean direction of the data in Figure 5.21, using Kent's distribution as a model. See Example 5.28.

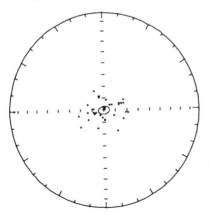

that the distribution is elliptically shaped (that is, its probability contours when projected using a Wulff or equal-angle projection will be elliptical). A general test of rotational symmetry for supposedly Fisherian data was given in §**5.3.2(ii)**. The test below should be more sensitive to this specific alternative. It is applicable for large samples.

Calculate the quantities Q and $\hat{\kappa}$ from (5.49) and (5.50), and obtain $G(\hat{\kappa})$ from the table in Appendix A14.

The test statistic is

$$K = n(\tfrac{1}{2}\hat{\kappa})^2 G(\hat{\kappa}) Q^2 \qquad (5.57)$$

and the hypothesis that the underlying distribution is Fisherian is rejected in favour of the Kent distribution if K is too large.

Critical values sample size $n \geq 30$ Reject the hypothesis at the $100\alpha\%$ level if K exceeds $-2 \log \alpha$. (The significance probability of an observed K value is $\exp(-\tfrac{1}{2}K)$).

> **Example 5.29** Continuing Example 5.27, we have $Q = 0.0229$, $\kappa = 41.76$, and $G(\hat{\kappa}) = 1.075$ by interpolating in Appendix A14. Hence $K = 8.36$, corresponding to a significance probability of about 0.015, so there is some evidence of departure from the Fisherian model. Using the goodness-of-fit procedures in §**5.3.2(ii)**, there is some suggestion of this from the Q-Q plot of longitudes, but the general test using $M_U(V_n)$ (see (5.16)) is not sensitive enough to detect it. The value $K = 8.36$ corrects the value 5.96 obtained by Kent (1982).

References and footnotes See Kent (1982).

5.4 Analysis of a sample of unit vectors from a great circle or small circle distribution

5.4.1 Introduction

There do not appear to be many physical mechanisms giving rise to such data. The principal mechanism appears to be related to cylindrical or conical folding in Structural Geology: if one measures the directed normals to bedding planes, with the direction being given by the facing, data distributions on a great circle will result if the planes are cylindrically folded, and on a small circle if the folding is conical. Data arising from other mechanisms (e.g. a hot-spot phenomenon) suggested in the literature are often better, or more appropriately, modelled as time series (§8). For the bedding planes example, the data may be distributed reasonably uniformly around the great or small circle (or even if distributed uniformly over a small arc, may sometimes be taken to be from a distribution uniform around the complete circle, if sampling problems have precluded sampling the complete range of azimuths). On

5.4.2. Parameter estimation

the other hand, there may be good (non-sampling) reasons why the data should not be isotropic with respect to the pole to the bedding planes. Such is the case with the data of Example 5.4 (see Figure 5.4).

In §**5.4.2**, we consider simple methods of estimating the polar axis to a great- or small-circle distribution. In §**5.4.3** there is a brief discussion of parametric models.

For a great circle distribution, the great circle C can be defined simply in terms of the pole (λ, μ, ν) to the plane containing C: C is the set of points (x_0, y_0, z_0) such that

$$(x_0, y_0, z_0)\begin{pmatrix}\lambda\\\mu\\\nu\end{pmatrix} = 0. \tag{5.58}$$

For a small-circle distribution, we need to specify not only the pole (λ, μ, ν) but the angular displacement ψ of a point on the small circle C_ψ from the nearer of the two ends of the polar axis (λ, μ, ν). So, C_ψ is defined as the set of points (x_0, y_0, z_0) such that

$$(x_0, y_0, z_0)\begin{pmatrix}\lambda\\\mu\\\nu\end{pmatrix} = \cos\psi. \tag{5.59}$$

Thus, when $\psi = 90°$, (5.59) becomes the great circle relationship (5.58).

5.4.2 Simple estimation of the pole to a great or small circle distribution

For great-circle distributions, it is possible to obtain an estimate of the polar axis and a large-sample confidence region for this estimate using the methods of §**6.3.1(ii)** for axial data: the polar axis is estimated by the eigenvector $\hat{\mathbf{u}}_1$ of the matrix $\mathbf{T}$ (see (3.15)) corresponding to the smallest eigenvalue $\hat{\tau}_1$ of $\mathbf{T}$. The reader is referred to §**6.3.1(ii)** for further details. For small-circle distributions, we proceed as follows. Given a random sample $\mathbf{X}'_1 = (x_1, y_1, z_1), \ldots, \mathbf{X}'_n = (x_n, y_n, z_n)$ (in direction cosines) from a small-circle distribution along the small circle C_ψ with pole $\boldsymbol{\lambda}' = (\lambda, \mu, \nu)$, the following iterative procedure can be used to estimate ψ and (λ, μ, ν).

Step 0 As an initial estimate of (λ, μ, ν) use $\hat{\boldsymbol{\lambda}}'_0 = (\hat{x}, \hat{y}, \hat{z})$, the direction cosines of the sample mean direction (cf. §**3.2.1**). Set $j = 1$.

Step 1 Calculate the jth estimate $\hat{\psi}_j$ of ψ from

$$\tan\hat{\psi}_j = \left[\sum_{i=1}^n \{1 - (\mathbf{X}'_i\hat{\boldsymbol{\lambda}}_{j-1})^2\}^{1/2}\right] / \sum_{i=1}^n \mathbf{X}'_i\hat{\boldsymbol{\lambda}}_{j-1}. \tag{5.60}$$

142   5. Analysis of a single sample of unit vectors

Step 2 Calculate the n vectors
$$\hat{\mathbf{X}}_i = \{(\mathbf{X}_i'\hat{\boldsymbol{\lambda}}_{j-1})\mathbf{X}_i - \hat{\boldsymbol{\lambda}}_{j-1}\}/\{1-(\mathbf{X}_i'\hat{\boldsymbol{\lambda}}_{j-1})^2\}^{1/2}, i=1,\ldots,n \quad (5.61)$$
and then
$$\mathbf{Y} = \cos\hat{\psi}_j \sum_{i=1}^n \mathbf{X}_i - \sin\hat{\psi}_j \sum_{i=1}^n \hat{\mathbf{X}}_i. \quad (5.62)$$
Now calculate the jth estimate $\hat{\boldsymbol{\lambda}}_j$ of $\boldsymbol{\lambda}$ as
$$\hat{\boldsymbol{\lambda}}_j = \mathbf{Y}/(\mathbf{Y}'\mathbf{Y})^{1/2}, \quad (5.63)$$

Step 3 Unless the new values $\hat{\boldsymbol{\lambda}}$ and $\hat{\psi}_j$ are acceptably close to the previous iterates $\hat{\boldsymbol{\lambda}}_{j-1}$ and $\hat{\psi}_{j-1}$, increase j by 1 and go to *Step 1*.

When the procedure has converged satisfactorily, denote the resulting estimates by $\hat{\boldsymbol{\lambda}}$ and $\hat{\psi}$. To obtain a confidence region for $\boldsymbol{\lambda}$ and a confidence interval for ψ, or equivalently, to obtain a confidence region for C_ψ, we can use bootstrap methods (cf. §3.6.1), as follows.

Step B1 Choose a number at random from the set $\{1,\ldots,n\}$, obtaining k_1 say. Selected the data point $\mathbf{X}_{k_1}$. Choose another number at random from $\{1,\ldots,n\}$ obtaining k_2 say, and select the data point $\mathbf{X}_{k_2}$. (There is probability $1/n$ that the same point will have been selected).

Continue, until n numbers have been selected, and a sample $\mathbf{X}_{k_1},\ldots,\mathbf{X}_{k_n}$ obtained.

Step B2 Calculate the estimates $\hat{\psi}^{(1)}$ and $\hat{\boldsymbol{\lambda}}^{(1)}$ say, for this sample.

Step B3 Obtain a total of say 200 "bootstrap" estimates
$$(\hat{\psi}^{(1)},\hat{\boldsymbol{\lambda}}^{(1)}),\ldots,(\hat{\psi}^{(200)},\hat{\boldsymbol{\lambda}}^{(200)})$$
by repeating *Steps B1* and *B2* a further 199 times.

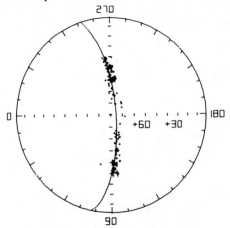

Figure 5.23 Best-fitting small circle for the data in Figure 5.4. See Example 5.30.

5.4.3. Parametric models

These 200 small-circle estimates give an indication of the variability of our sample estimate $(\hat{\psi}, \hat{\boldsymbol{\lambda}})$. This can be assessed visually by plotting the $\hat{\psi}^{(i)}$ values separately and the $\hat{\boldsymbol{\lambda}}^{(i)}$ values separately, or by plotting the 200 small circles.

Example **5.30** For the data (set B4) of Example 5.4, illustrated in Figure 5.4, we obtain the mean direction as (Dec. 244.0°, Inc. 87.6°), the pole to the fitted small circle as (Dec. 3.2°, Inc. 18.8°), and $\hat{\psi} = 75.9°$ (cf. 5.59). The best-fitting small circle is shown in Figure 5.23, and bootstrap estimates of $\boldsymbol{\lambda}$ and ψ separately, in Figure 5.24.

References and footnotes See Mardia & Gadsden (1977). (Note added in proof: An improved algorithm for least-squares fitting is given by Gray, Geiser & Geiser (1980)).

5.4.3 *Parametric models for great-circle or small-circle distributions, which are rotationally symmetric about the polar axis to the circle*

The two models proposed in the literature are due to Mardia & Gadsden (1977) and Bingham & Mardia (1978). Definitions and properties of these models are given in §**4.5**. The latter model is more tractable than the former; nevertheless, estimating the parameters is a somewhat complex procedure, and we refer the reader to these papers for further information.

Figure 5.24 Assessment of the variability of the estimated small circle in Figure 5.23, using 200 bootstrap samples. (a) variability of the estimate of the pole to the plane (b) variability of the estimate of the angular displacement of a point on the plane from the polar axis. See Example 5.30.

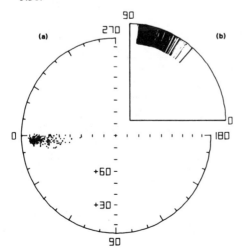

144   5. Analysis of a single sample of unit vectors

5.5 Analysis of a sample of unit vectors from a multimodal distribution

5.5.1 Introductory remarks.

The usual questions of interest with multimodal data are, how many modes are there, and where are they? The data of Example 5.6, pictured in Figure 5.6, indicate the sort of problem that this can present to the data analyst. Three lines of attack present themselves:

(i) Partition the data into groups visually, with the aid of a contour plot.
(ii) Use some form of clustering method to find the modes.
(iii) Use some probability model, such as a mixture of Fisher distributions, for the data, and estimate the number of modes, the parameters (mean direction and concentration) of each mode, and the properties of each component.

In the next subsection we indicate how (i) might be done for the data in Figure 5.6. We shall not pursue (ii) here, as there is little currently available in the literature which seems to be of practical use†. As for (iii), Stephens (1969b) has studied a distribution comprising a mixture of Fisher distributions and found that estimation is very tedious; for example, the number of parameters increases by four with each mode. (However, it may be reasonable in some cases to use the procedure in §**5.5.2** to separate the data into modal groups, and then to fit a Fisher distribution or some other unimodal distribution to each group). The only parametric model which is practicable to implement is a particular bimodal distribution with equal components, the Wood distribution. This is described in §**4.4.6** and applied below, in §**5.5.3**.

5.5.2 Simple modal analysis

Example 5.31 Continuing the remarks in §**5.5.1** on method (i), we investigate the modes of the data (set B6) of Example 5.6, illustrated in Figure 5.6. A rotated version of the contours is given in Figure 5.11, from which it appears that there are certainly three modes, and possibly four, if the elongated peak actually contains two modes. To investigate this point, we can use more contours, giving Figure 5.25. This shows the contours of the group in question seemingly shrinking towards a single peak. As it happened, there were scientific grounds for believing that four groups are present, and so the kidney-shaped mode was partitioned into two groups. (See Figure 3.9 for a summary picture). We can then calculate the mean or median direction of the data in each group.

† In a recent paper, Schaeben (1984) has put forward a nonparametric density approach to this problem.

5.5.3(i). Parameter estimation for Wood distribution

Where appropriate, we might then proceed further, to fit a Fisher distribution to each group. As it is, for these data one would probably go no further than computing a confidence region for the modal direction of each group using §5.3.1(ii) or §5.3.1(v) as appropriate.

5.5.3   A parametric model for a bimodal distribution of unit vectors

The Wood distribution (§4.4.6) is designed to model data concentrated about two modal directions with angular separation 2α, in approximately equal proportions, each mode having the same concentration. Let $(x_1, y_1, z_1), \ldots, (x_n, y_n, z_n)$ be the direction cosines of a sample of data from the Wood distribution; for convenience, write $\mathbf{X}'_i = (x_i, y_i, z_i)$, $i = 1, \ldots, n$.

5.5.3(i) *Estimation of the parameters.* This is a two-stage procedure, in the first stage the parameters γ and δ being estimated iteratively, and in the second stage α, β and κ being found. Define the vectors

$$\left.\begin{array}{l}\boldsymbol{\mu}'_1 = (\cos \gamma \cos \delta, \cos \gamma \sin \delta, -\sin \gamma) \\ \boldsymbol{\mu}'_2 = (-\sin \delta, \cos \delta, 0) \\ \boldsymbol{\mu}'_3 = (\sin \gamma \cos \delta, \sin \gamma \sin \delta, \cos \gamma)\end{array}\right\} \quad (5.64)$$

and then the quantities (functions of γ and δ)

$$\left.\begin{array}{l}U = \sum_{i=1}^{n} \mathbf{X}'_i \boldsymbol{\mu}_3 \\ V = \sum_{i=1}^{n} \{(\mathbf{X}'_i \boldsymbol{\mu}_1)^2 - (\mathbf{X}'_i \boldsymbol{\mu}_2)^2\}/\{1 - (\mathbf{X}'_i \boldsymbol{\mu}_3)^2\}^{1/2} \\ W = \sum_{i=1}^{n} 2(\mathbf{X}'_i \boldsymbol{\mu}_1) \cdot (\mathbf{X}'_i \boldsymbol{\mu}_2)/\{1 - (\mathbf{X}'_i \boldsymbol{\mu}_3)^2\}^{1/2}\end{array}\right\} \quad (5.65)$$

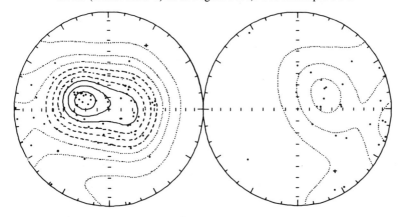

Figure 5.25 A contour plot of the data in Figure 5.6 using 9 contour levels (rather than 6, as in Figure 5.10). See Example 5.31.

Thus, for example,

$$\mathbf{X}'_i \boldsymbol{\mu}_3 = (x_i \ y_i \ z_i) \begin{pmatrix} \sin \gamma \cos \delta \\ \sin \gamma \sin \delta \\ \cos \gamma \end{pmatrix}$$

$$= x_i \sin \gamma \cos \delta + y_i \sin \gamma \sin \delta + z_i \cos \gamma$$

Finally, define the function

$$S^2(\gamma, \delta) = U^2 + V^2 + W^2 \tag{5.66}$$

The estimates of γ and δ are the values which maximise S^2. This will require an optimisation routine, such as Nelder and Mead (1965), subroutine EO4JAF in the NAG library, or ZXMIN in the IMSL package. So, the first stage is accomplished as follows:

Step 1 Calculate an initial estimate of γ and δ, $(\hat{\gamma}_0, \hat{\delta}_0)$ say, as described below.

Step 2 Use a numerical optimisation procedure to maximise S^2 as defined in (5.66), (5.65) and (5.64). This yields a final estimate $(\hat{\gamma}, \hat{\delta})$.

Initial estimation of (γ, δ) If the modes appear close together, use the sample mean direction (§3.2.1). If the modes are so widely separated that the mean resultant length $\bar{R}$ (cf. §3.2.1) is nearly zero, calculate the eigenvector $\hat{\mathbf{u}}_2$ corresponding to the second eigenvalue (§3.2.3) of the matrix $\mathbf{T}$ in (3.15), and perform two separate maximisations, one with $\hat{\mathbf{u}}_2$ as the initial direction, and one with $-\hat{\mathbf{u}}_2$ as the initial direction. The one yielding the larger value of S^2 should then be taken to be the correct initial vector.

At the end of the first stage, we have the estimates $\hat{\gamma}, \hat{\delta}$ of γ, δ. Substitute these into the set of equations (5.65) to obtain U, V and W. Then estimate α and β by $\hat{\theta}$ and $\hat{\phi}$, calculated from (3.4) - (3.6) with $\hat{x} = U$, $\hat{y} = V$, and $\hat{z} = W$.

Finally, let

$$R^* = (U^2 + V^2 + W^2)^{1/2} \tag{5.67}$$

κ is then estimated from (5.25) using $R = R^*$. If $\bar{R}^* = R^*/n < 0.95$, the estimate $\hat{\kappa}$ can be found from Appendix A10.

This gives us a set of parameter estimates $\hat{\alpha}, \hat{\beta}, \hat{\gamma}, \hat{\delta}, \hat{\kappa}$. The two modes of the fitted distribution are found as follows:

Step 1 Calculate the estimates $(\hat{\alpha}_1, \hat{\beta}_1) = (\hat{\alpha}, \frac{1}{2}\hat{\beta})$ and $(\hat{\alpha}_2, \hat{\beta}_2) = (\hat{\alpha}, \frac{1}{2}\hat{\beta} + \pi)$.

Step 2 Calculate $\hat{\boldsymbol{\mu}}_1, \hat{\boldsymbol{\mu}}_2, \hat{\boldsymbol{\mu}}_3$ from (5.64) using $\hat{\gamma}, \hat{\delta}$, and then form the 3×3 matrix

$$\mathbf{A} = (\hat{\boldsymbol{\mu}}_1, \hat{\boldsymbol{\mu}}_2, \hat{\boldsymbol{\mu}}_3)$$

5.5.3(i). Parameter estimation for Wood distribution

Step 3 Transform the estimates $(\hat{\alpha}_1, \hat{\beta}_1)$, $(\hat{\alpha}_2, \hat{\beta}_2)$ to $(\hat{\alpha}_1^*, \hat{\beta}_1^*)$, $(\hat{\alpha}_2^*, \hat{\beta}_2^*)$ respectively using the analogue of (3.8), e.g.

$$\begin{pmatrix} \sin \hat{\alpha}_1^* \cos \hat{\beta}_1^* \\ \sin \hat{\alpha}_1^* \sin \hat{\beta}_1^* \\ \cos \hat{\alpha}_1^* \end{pmatrix} = \mathbf{A} \begin{pmatrix} \sin \hat{\alpha}_1 \cos \hat{\beta}_1 \\ \sin \hat{\alpha}_1 \sin \hat{\beta}_1 \\ \cos \hat{\alpha}_1 \end{pmatrix} \quad (5.68)$$

$(\hat{\alpha}_1^*, \hat{\beta}_1^*)$ and $(\hat{\alpha}_2^*, \hat{\beta}_2^*)$ are then the estimates of the two modal directions.

This estimation assumes that $\hat{\alpha}$ is significantly greater than zero: recall that if $\alpha = 0$, the two modes coalesce and the Wood distribution reduces to a Fisher distribution. This can be checked using the methods in §**5.3.2**, although Wood (1982) gives a test of the hypothesis that the distribution is Fisherian against the specific alternative that it is the more general Wood distribution.

Wood also gives large sample confidence intervals for α, β, γ, δ and κ, but confidence cones for the two unknown modal directions are not available. These can be obtained by a form of bootstrap technique (see §**3.6.1**).

Example **5.32** Figure 5.26 illustrates the data in Appendix B10, comprising 33 estimates of the palaeomagnetic pole position from Tasmanian Dolerites obtained by Schmidt (1976). It is evident from the illustration that the Fisher distribution is not a good model. We describe Wood's (1982) analysis of the data, in this and the following example. (Note that, because the latitudes obtained by Schmidt (1976) are in the southern hemisphere, the values for colatitude in Wood (1982, Table 1) should be subtracted from 180° to obtain the correct conversion to polar coordinates. This has no effect on the analysis; however the results given below have been adjusted for this.)

Figure 5.26 Equal-area projection of 33 virtual geomagnetic pole positions. The co-ordinate system is (Latitude, Longitude). See Example 5.32.

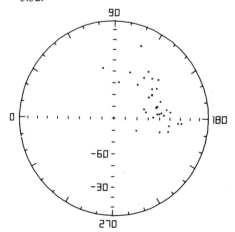

The mean pole position, in polar coordinates (θ, ϕ), is $(142.3°, 155.7°)$, with mean resultant length 0.932. Using the NAG routine referred to above yielded the values $(\hat{\gamma}, \hat{\delta}) = (142.3°, 152.5°)$ as maximising S^2 in (5.66), whence the estimates $\hat{\alpha} = 10.7°$, $\hat{\beta} = 178.0°$, $\hat{\kappa} = 19.3$. Finally, from (5.68), $(\hat{\alpha}_1^*, \hat{\beta}_1^*) = $ (Lat. $-51.0°$, Long. 135:7°), $(\hat{\alpha}_2^*, \hat{\beta}_2^*) = $ (Lat. $-50.5°$, Long. 170.0°).

References and footnotes See Wood (1982).

5.5.3(ii) Test for specified modal directions. Suppose we wish to test whether the true modal directions are (α_1^*, β_1^*) and (α_2^*, β_2^*). The following steps are needed.

Step 1 Rotate the data to the pole $(\frac{1}{2}(\alpha_1^* + \alpha_2^*), \frac{1}{2}(\beta_1^* + \beta_2^*))$ using (3.8); with this rotation, (α_1^*, β_1^*) and (α_2^*, β_2^*) transform to (α_1, β_1) and (α_2, β_2) say.

Step 2 Set $\alpha = \alpha_1$, $\beta = 2\beta_1$; put $\gamma = 0$, $\delta = 0$.

Step 3 Calculate

$$\left. \begin{array}{l} U_0 = \sum_{i=1}^{n} x_i \\[4pt] V_0 = \sum_{i=1}^{n} (x_i^2 - y_i^2)/(x_i^2 + y_i^2)^{1/2} \\[4pt] W_0 = \sum_{i=1}^{n} 2x_i y_i /(x_i^2 + y_i^2)^{1/2} \\[4pt] R_0 = (U_0^2 + V_0^2 + W_0^2)^{1/2} \end{array} \right\} \quad (5.69)$$

Estimate κ by $\hat{\kappa}_0$ as described in §5.5.3(i) above, using $R = R_0$.

Step 4 Compute

$$L_0 = -n \log (2\pi) - nf(\hat{\kappa}_0) + \hat{\kappa}_0 U_0 \cos \alpha$$
$$+ (V_0 \cos \beta + W_0 \sin \beta) \sin \alpha \quad (5.70)$$

where the function f is defined by

$$f(x) = \log\{(\exp(x) - \exp(-x))/x\}. \quad (5.71)$$

Step 5 Calculate the estimates $\hat{\alpha}, \hat{\beta}, \hat{\gamma}, \hat{\delta}$ and $\hat{\kappa}$ as in §5.5.3(i) above, then U, V and W as in §5.5.3(i), and finally

$$L_1 = -n \log(2\pi) - nf(\hat{\kappa}) +$$
$$+ \hat{\kappa}\{\hat{U} \cos \hat{\alpha} + (\hat{V} \cos \hat{\beta}) + \hat{W} \sin \hat{\beta}) \sin \hat{\alpha}\},$$

$f(\hat{\kappa})$ being calculated from (5.71).

Step 6 The test statistic is

$$\lambda = -2(L_0 - L_1). \quad (5.72)$$

5.6.1. General tests of uniformity

Critical values To test at the $100\alpha\%$ level, reject the hypothesis if λ exceeds the upper $100\alpha\%$ point of the χ_4^2 distribution.

Wood (1982) also gives a test for a specified value of the angle α, and for an interval range of values $0 < \alpha < \alpha_0$ (or $\alpha_0 \leq \alpha < \tfrac{1}{2}\pi$). ($\alpha$ is being used here to denote both the angular semi-distance between the modes and the significance level of the test, but there is no risk of confusion).

> **Example 5.33** Continuing the previous example, Schmidt and McDougall (1977) had suggested the values (Lat. $-37.7°$, Long. $123.5°$) and (Lat. $-50.7°$, Long. $174.5°$) for the two modes. If we carry out the test procedure above, we obtain $\lambda = 8.8$. The upper 5% point for χ_4^2 is 9.49, so the hypothesis that the data are from a distribution with these modes can be accepted at this level.

References and footnotes See Wood (1982).

5.6 General tests of uniformity and rotational symmetry
5.6.1 General tests of uniformity

In some situations, the data are extremely well dispersed over the unit sphere, and the alternative model of interest is that there are some (unspecified) preferred directions, compared with the null model that the underlying distribution is uniform (i.e. isotropic). Let the data have direction cosines $(x_1, y_1, z_1), \ldots, (x_n, y_n, z_n)$, and define ψ_{ij} as the angle between the ith and jth directions (cf. (3.1)). Calculate

$$A_n = n - (4/n\pi) \sum_{i=1}^{n-1} \sum_{j=i+1}^{n} \psi_{ij}, \tag{5.73}$$

$$G_n = \tfrac{1}{2}n - (4/n\pi) \sum_{i=1}^{n-1} \sum_{j=i+1}^{n} \sin \psi_{ij} \tag{5.74}$$

and

$$F_n = A_n + G_n, \tag{5.75}$$

A_n is Beran's (1968) statistic for testing uniformity against alternative models which are not symmetric with respect to the centre of the sphere. G_n is Gine's (1975) statistic for testing uniformity against alternative models which *are* symmetric with respect to the centre of the sphere. F_n is a test statistic to be used for testing against all alternative models. The hypothesis of uniformity is rejected in favour of the appropriate alternative if the corresponding statistic (e.g. F_n, against all alternative models) is too large.

Critical values sample size $n \geq 10$. Selected percentiles of A_n, G_n and F_n are given in Appendix A15. More detailed tables are given by Keilson et al. (1983).

Example 5.34 For the data (set B5) of Example 5.5, we can test whether all components of magnetisation have been purged. (This was tested, against a unimodal alternative, in Example 5.9). We obtain $F_n = 1.516$. Since the upper 20% point is 1.948, there is effectively no evidence that the underlying distribution is not uniform.

References and footnotes Beran (1968), Giné (1975), Prentice (1978), Watson (1983a, Chapter 2), Diggle, Fisher & Lee (1985). Another test of uniformity was proposed by Bingham (1964, 1974); see also Mardia (1972, Chapter 9) and Diggle *et al.* (1985).

5.6.2 General tests of rotational symmetry

It is sometimes of interest to test the hypothesis that a given sample of data has been drawn from a distribution rotationally symmetric about a specified or unspecified axis. (This was considered in §5.3.1(iv) in the particular context of unimodal distributions). In the general context, it can be done using the methods in §6.3.1(iv), based on eigenvalues of the orientation matrix **T** (see (3.15)), when the supposed axis of symmetry is unknown. If it is known, the graphical and formal methods in §5.3.1(iv) can be used.

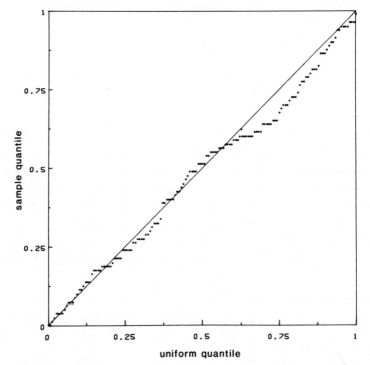

Figure 5.27 Uniform Q-Q plot of Right Ascension measurements for the data in Figure 5.3. See Example 5.34.

5.6.2. General tests of rotational symmetry

***Example* 5.35** For the data (set B3) of Example 5.3, which are illustrated in Figure 5.3 and Figure 5.12, consider testing the hypothesis that the Right Ascensions are distributed uniformly between 0° and 360° (that is, that the distribution is rotationally symmetric about an axis through Declination 90°). The uniform probability plot of Right Ascensions is shown in Figure 5.27, and appears reasonably linear. For a formal test, we obtain $D_n^+ = 0.0932$, $D_n^- = 0.0362$, and so from (3.39) and (3.41), $V_n^* = 1.597$. From Table A6, this is not an extreme value of V_n^*, and the hypothesis of rotational symmetry seems reasonable.

References and footnotes Another test of rotational symmetry, analogous to the tests of independence described in §**5.6.2**, is given by Jupp & Spurr (1983).

6

Analysis of a single sample of undirected lines

6.1 Introduction

In this chapter, we shall be concerned with techniques for statistical analysis of a sample of independent measurements from some common population of undirected lines, or axes. For example, consider the following data sets.

> ***Example* 6.1** Measurements of 75 poles to axial-plane cleavage surfaces of F_1 folds in Ordovician turbidites. A plot of the data is shown in Figure 6.1. (The data are listed as orientations of axial planes in Appendix B11).
>
> ***Example* 6.2** Measurements of 72 poles to axial-plane cleavage surfaces of F_1 folds in Ordovician turbidites. A plot of the data is shown in Figure 6.2. (The data are listed as orientations of axial planes in Appendix B12).

Figure 6.1 Equal-area projection of 75 poles to axial-plane cleavage surfaces. The coordinate system is (Plunge, Plunge azimuth). See Example 6.1.

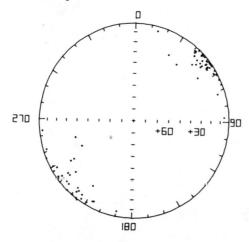

6.1. Introduction

***Example* 6.3** Measurements of the maximum susceptibility axis of magnetic fabric in 18 samples from ash strata. A plot of the data is shown in Figure 6.3. (The data are listed in Appendix B13).

***Example* 6.4** 221 poles to joint planes collected from Wanganderry Lookout, NSW, measured in Triassic sandstone. A plot of the data is shown in Figure 6.4. (The data are listed as orientations of joint planes in Appendix B14).

***Example* 6.5** Measurements of orientations of the dendritic fields at 94 sites in the retinas of six cats, in response to stimulation by various forms of polarised light. A plot of the data is shown in Figure 6.5. (The data are listed in Appendix B15).

Figure 6.2 Equal-area projection of 72 poles to axial-plane cleavage surfaces. The coordinate system is (Plunge, Plunge azimuth). See Example 6.2.

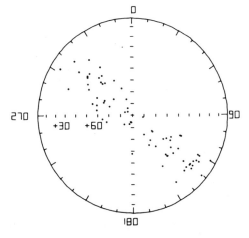

Figure 6.3 Equal-area projection of 18 maximum susceptibility axes. The coordinate system is (Plunge, Plunge azimuth). See Example 6.3.

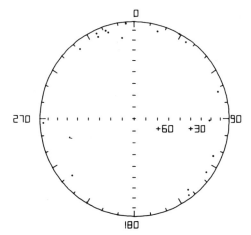

154   6. *Analysis of a single sample of undirected lines*

Our discussion of the analysis of such data will parallel, to some extent, that given in Chapter 5 for unit vectors, *and the reader is referred specifically to* §**5.1**, §**5.2**, §**5.4** and §**5.5**, although that material will be covered lightly here.

Axial data are sometimes measured directly as lines, but not infrequently are obtained as polar axes, or normals, to measured planes. Two important types of distribution for axial data are bipolar distributions, as in Example 6.1, and girdle distributions, as in Example 6.2;

Figure 6.4 Equal-area projection of 221 poles to joint planes. The coordinate system is (Plunge, Plunge azimuth). See Example 6.4.

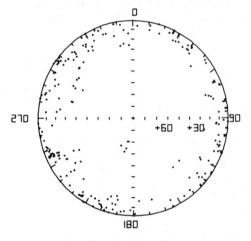

Figure 6.5 Equal-area projection of the orientations of the dendritic field at 94 sites in the retinas of six cats. The coordinate system is (Latitude, Longitude). See Example 6.5.

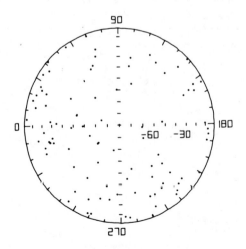

6.1. Introduction

distributions transitional between these two "extremes" also occur (Example 6.3 – see the analysis in Example 6.6), as do multimodal distributions (Example 6.4) and small-circle distributions (analogues of Example 5.4) and occasionally, seemingly isotropic distributions (Example 6.5). Exploratory analysis is usually concerned with trying to identify the type of distribution, and possibly with more detailed questions of shape, such as whether the distribution for the data in Example 6.1 is rotationally invariant, or isotropic, about the principal axis. Techniques for performing exploratory analysis are described in §**6.2**.

Having decided on the form of the distribution, we may then proceed to various problems of statistical inference concerning the data; for example, what is the "best" estimate of the principal or polar axis (Examples 6.1 and 6.2) and what error can we assign to this estimate? or, are the data really isotropic (Example 6.5) or is there some preferred axis? We may be prepared to make only very few assumptions about the nature of the distribution (e.g. that it is bipolar, but not necessarily rotationally symmetric, for the data in Example 6.1) in which case only simple methods are available. (In this context, we use the term "simple procedures" to refer to methods based on few assumptions, and hence of wide applicability). On the other hand, we may feel justified in using a parametric model such as the Watson distribution for bipolar data, enabling us to use more powerful techniques. Where possible, when discussing a given distribution-type, we consider simple methods first, and then methods based on parametric models. §**6.3** is concerned with bipolar data, §**6.4** with girdle, or great-circle data, §**6.5** with small-circle data and §**6.6** with multimodal data. §**6.7** is concerned with a general test of uniformity when no alternative model is proposed. A simplified diagrammatic summary of the treatment of a sample of axes is given in Figure 6.6.

The various facets of statistical analysis of a single sample of data can be seen as a whole, in the prototype analysis of a sample of vector data given in Chapter 1 (Example 1.1).

One omission from the discussion of parametric models is the Bingham distribution (§**4.5**) in its general form (the Watson bipolar and girdle distributions being special forms of this). There are two reasons for this: firstly, the difficulty in checking on the goodness-of-fit of the Bingham distribution to a sample of data, and secondly, that many of the problems for which one might wish to fit the Bingham distribution can be treated adequately using the simple methods described in this chapter. Specifically, in relation to this latter point, estimates of the three axes of the distribution are just the eigenvectors of the orientation matrix as used in this book, and the confidence regions we recommend for the individual

156   6. Analysis of a single sample of undirected lines

axes are, in many practical cases, essentially the same as those based on the Bingham theory. Estimation of confidence regions for all three axes is considered in §6.6.

6.2   Exploratory analysis
6.2.1   *Graphical methods*

The main technique for graphical analysis is the contour plot, as described in some detail in §3.3.3. Consider the application of this technique to the data of Examples 6.1-6.5.

Example **6.6** Figures 6.7-6.11 are the contour plots corresponding to the data in Figures 6.1-6.5 respectively. Figure 6.7 was obtained using a smoothing value $C_n = 53.5$. The data and contours have been rotated so that the principal axis plots in the middle of the figure. The high-level

Figure 6.6 Summary of development of methods for analysing a single sample of axial data.

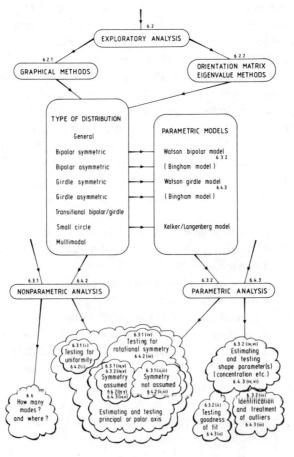

6.2.1. Exploratory analysis: graphical methods

Figure 6.7 Contour plot of the data in Figure 6.1, rotated to centralise the sample principal axis. See Example 6.6.

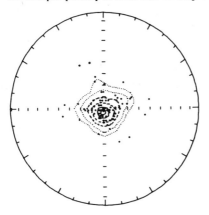

Figure 6.8 Contour plot of the data in Figure 6.2. See Example 6.6.

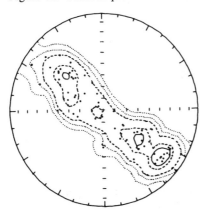

Figure 6.9 Contour plot of the data in Figure 6.3, rotated to centralise the sample principal axis. See Example 6.6.

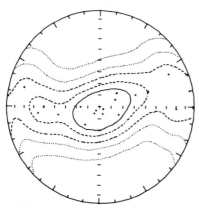

158   6. Analysis of a single sample of undirected lines

(interior) contours are reasonably circular, indicating a bipolar distribution with rotational symmetry about its principal axis. Figure 6.8 (obtained using $C_n = 31.7$) shows contours stretching around an equator like a belt, with local clustering within the belt, corresponding to a girdle distribution which is possibly rotationally symmetric with respect to its polar axis. Figure 6.9 (obtained using $C_n = 10.6$) suggests a distribution transitional between bipolar and girdle, although the small sample size prevents us from making a firm inference. Figure 6.10(a) (obtained

Figure 6.10 Contour plots of the data in Figure 6.4: (a) unrotated, (b) rotated to centralise the girdle. See Example 6.6.

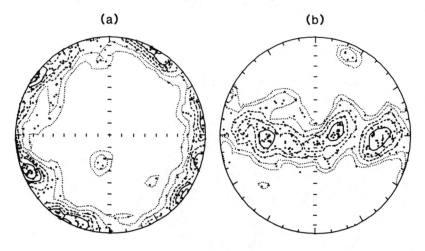

Figure 6.11 Contour plots of the data in Figure 6.5: (a) unrotated, (b) rotated to centralise the contours. See Example 6.6.

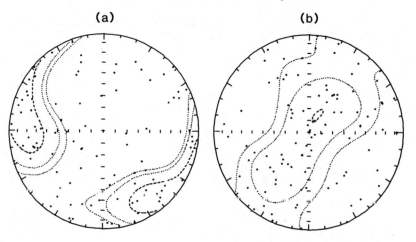

6.2.2. Exploratory analysis: quantitative methods

using $C_n = 63.7$) shows a girdle distribution with several local clusters, suggesting a mixture of distributions. The structure of the distribution is better appreciated by looking at a rotation of this figure, as shown in Figure 6.10(b). Figure 6.11(a) (obtained using $C_n = 4.9$) suggests the possibility of a girdle distribution, although the indication of the raw data plot (and the hypothesis which the data were collected to test) is that of an isotropic model. This is also made more useful by a rotation (Figure 6.11(b)).

6.2.2 Quantitative methods

The reader is referred to §3.4 for a description of the use of eigenvalue analysis in studying the shape of a data set. Here, we consider application of this form of analysis to our five examples discussed in §6.1 and §6.2.1.

Example **6.7** The eigenvalues and associated ratios for the data in Examples 6.1–6.5 are given in Table 6.1. We now check these various sets of numbers with Figure 3.15 to see what information is provided about the distributions.

For the data (set B11) in Example 6.1, we obtain a large value, $\hat{\gamma} = 9.53$, for shape and a moderate value, $\hat{\zeta} = 3.26$ for strength, corresponding to a cluster-type distribution; these indicate a well-defined bipolar distribution. By contrast, the data of Example 6.2 (set B12) have a near-zero value, $\hat{\gamma} = 0.07$, for shape, with moderate strength $\hat{\zeta} = 3.35$, indicating a well-defined girdle distribution. For the data (set B13) of Example 6.3, the shape $\hat{\gamma}$ is 0.22, still of girdle type but tending towards a bipolar distribution with a concentration of the data along part of the girdle. Example 6.4 (set B14) is also of girdle type, but with the girdle not so strongly defined (smaller value of $\hat{\zeta}$) due to the presence of several clusters in the data (note the contour plots in Figure 6.10). Finally, the data (set B15) in Figure 6.5 have a strength parameter close to zero ($\hat{\zeta} = 0.68$), suggesting a distribution close to uniform.

Table 6.1 *Normalised eigenvalues*, log *ratios* $r_1 = \log(\bar{\tau}_2/\bar{\tau}_1)$, $r_2 = \log(\bar{\tau}_3/\bar{\tau}_2)$, *and estimated shape and strength indicators* $\hat{\gamma} = r_2/r_1$ *and* $\hat{\zeta} = \log(\bar{\tau}_3/\bar{\tau}_1)$ *for the data in Examples 6.1-6.5*

Example number	$\bar{\tau}_1$	$\bar{\tau}_2$	$\bar{\tau}_3$	r_1	r_2	$\hat{\gamma}$	$\hat{\zeta}$
6.1	0.0353	0.0481	0.9165	0.31	2.95	9.53	3.26
6.2	0.0192	0.4354	0.5454	3.12	0.23	0.07	3.35
6.3	0.0218	0.3439	0.6343	2.76	0.61	0.22	3.37
6.4	0.0631	0.4196	0.5173	1.89	0.21	0.11	2.10
6.5	0.2162	0.3590	0.4248	0.51	0.17	0.33	0.68

6.3 Analysis of a sample of undirected lines from a bipolar distribution

6.3.1 Simple procedures

Given the sample of axes $\mathbf{X}_1, \ldots, \mathbf{X}_n$ with direction cosines (cf. §3.2)

$$\mathbf{X}_1 = \begin{pmatrix} x_1 \\ y_1 \\ z_1 \end{pmatrix}, \ldots, \mathbf{X}_n = \begin{pmatrix} x_n \\ y_n \\ z_n \end{pmatrix} \qquad (6.1)$$

from a bipolar spherical distribution, we consider various simple procedures for the following problems:

(i) Testing the sample to see whether it has been drawn from a uniform distribution, against the alternative that it has been drawn from a bipolar distribution (with specified or unspecified principal axis).
(ii) Estimating the principal axis and assigning an error to this estimate.
(iii) Testing whether the principal axis takes a specified value.
(iv) Testing the sample to see if it is drawn from a symmetric bipolar distribution.
(v) Estimating the principal axis of a symmetric bipolar distribution and assigning an error to this estimate.
(vi) Testing whether the principal axis of a symmetric bipolar distribution takes a specified value.

Except for the test in §6.3.1(i), the tests and confidence cones given below in §6.3.1 are based on asymptotic statistical theory, and caution should be exercised in interpreting the results from samples of size less than, say, 25.

6.3.1(i) Test for uniformity against a bipolar alternative. Suppose first that the principal axis under the alternative model is not specified. Calculate the normalised eigenvalues $\bar{\tau}_1$, $\bar{\tau}_2$ and $\bar{\tau}_3$ of the orientation matrix as defined in §3.2.3: the hypothesis of uniformity is rejected if τ_3 is too large.

Critical values See Appendix A16.

> **Example 6.8** The data in Figure 6.12 are 240 normals to cometary orbital planes. (The data are listed in Appendix B16). We may be interested in whether there is any evidence that these are not distributed randomly on the surface of the unit hemisphere, but have a tendency to be bipolar or girdle. The contour plot in Figure 6.13 does in fact suggest the existence of a weakly defined principal axis. For these data, $\bar{\tau}_3 = 0.4194$.

6.3.1(i). Uniformity test against bipolar alternative

From Appendix A16, the upper 1% point is 0.4003, so the significance probability is less than 1%, indicating substantial departure from uniformity. The strong evidence against the model of uniformity in favour of a seemingly weak mode is due to the large sample size.

Now suppose that the alternative model specifies the principal axis as having direction cosines $(\lambda_0, \mu_0, \nu_0)$. Calculate the statistic

$$S_n = (1/n) \sum_{i=1}^{n} (\lambda_0 x_i + \mu_0 y_i + \nu_0 z_i)^2. \tag{6.2}$$

The hypothesis of uniformity is rejected if S_n is too large.

Figure 6.12 Equal-area projection of the normals to 240 cometary orbits. The coordinate system is (Colatitude, Longitude). See Example 6.8.

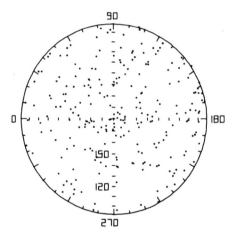

Figure 6.13 Contour plot of the data in Figure 6.12. See Example 6.8.

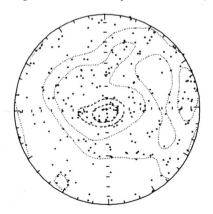

Critical values

Sample size $n \leq 100$ use critical values in Appendix A17
Sample size $n > 100$ compute $S = n^{\frac{1}{2}}(S_n - \frac{1}{3})/(4/45)^{\frac{1}{2}}$ and compare with appropriate percentile of $N(0,1)$ distribution in Appendix A1.

References and footnotes The test based on $\bar{\tau}_3$ is due to Anderson & Stephens (1972). The distribution of S_n was tabulated by Stephens (1966). See also Stephens (1965, 1972), Watson (1983a, Chapter 2).

6.3.1(ii) Estimation of the principal axis of a bipolar distribution. The usual estimate of the principal axis is the axis $\hat{\mathbf{u}}_3$ corresponding to the largest eigenvalue $\hat{\tau}_3$ of the orientation matrix **T**, as described in §3.2.4. A large-sample elliptical confidence cone can be calculated for the unknown principal axis $\mathbf{u}_3$, using the fact that $\hat{\mathbf{u}}_3$ has an asymptotic bivariate normal distribution. Define

$$\mathbf{H} = (\hat{\mathbf{u}}_1, \hat{\mathbf{u}}_2, \hat{\mathbf{u}}_3) \tag{6.3}$$

and

$$\mathbf{E} = \begin{pmatrix} e_{11} & e_{12} \\ e_{21} & e_{22} \end{pmatrix}. \tag{6.4}$$

The elements of **E** are given by

$$\left. \begin{aligned} e_{11} &= [n(\bar{\tau}_1 - \bar{\tau}_3)^2]^{-1} \sum_{i=1}^{n} (\hat{\mathbf{u}}_1' \mathbf{X}_i)^2 (\hat{\mathbf{u}}_3' \mathbf{X}_i)^2 \\ e_{22} &= [n(\bar{\tau}_2 - \bar{\tau}_3)^2]^{-1} \sum_{i=1}^{n} (\hat{\mathbf{u}}_2' \mathbf{X}_i)^2 (\hat{\mathbf{u}}_3' \mathbf{X}_i)^2 \\ e_{12} &= e_{21} = [n(\bar{\tau}_1 - \bar{\tau}_3)(\bar{\tau}_2 - \bar{\tau}_3)]^{-1} \sum_{i=1}^{n} (\hat{\mathbf{u}}_1' \mathbf{X}_i)(\hat{\mathbf{u}}_2' \mathbf{X}_i)(\hat{\mathbf{u}}_3' \mathbf{X}_i)^2 \end{aligned} \right\} \tag{6.5}$$

(cf. §3.2.3 and (6.1)). Let

$$\mathbf{F} = \mathbf{E}^{-1} \tag{6.6}$$

where **F** has elements

$$\begin{pmatrix} f_{11} & f_{12} \\ f_{21} & f_{22} \end{pmatrix} \tag{6.7}$$

A set of points on a $100(1-\alpha)\%$ elliptical confidence cone for $\mathbf{u}_3$ can now be obtained as described in §3.2.5, where in (3.20), $A = f_{11}$, $B = f_{12}$, $C = f_{22}$ and $D = -2 \log(\alpha)/n$. (If $n < 25$, use bootstrap methods - e.g. §3.6.1.)

> **Example 6.9** Figures 6.14 and 6.15 show the raw data and a rotated contour plot of 65 poles to axial-plane cleavage surfaces of F_1 folds. The data are listed as orientations of axial planes in Appendix B17.

6.3.1(iii). Test for a specified principal axis

The estimate of the principal axis u_3 has direction cosines (0.4072, 0.8687, −0.2819), corresponding to a plunge of 16.4° in the direction 295.1°. The major and minor semi-axes of an approximate 95% elliptical confidence cone for the true polar axis u_3 are 7.1° and 3.8° respectively: this confidence region is plotted in Figure 6.16.

References and footnotes See Watson (1983a, Chapter 5), Watson (1984a) and Prentice (1984).

6.3.1(*iii*) *Test for a specified value of the principal axis.* Suppose we wish to test whether the principal axis u_3 of the underlying distribution takes the specific value u_0, with direction cosines $(\lambda_0, \mu_0, \nu_0)$. Calculate the

Figure 6.14 Equal-area projection of 65 poles to cleavage-plane surfaces. The coordinate system is (Plunge, Plunge azimuth). See Example 6.9.

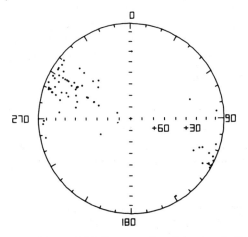

Figure 6.15 Contour plot of the data in Figure 6.14, rotated to centralise the sample principal axis. See Example 6.9.

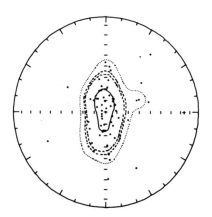

matrices **H**, **E** and **F** defined by (6.3), (6.4), (6.5) and (6.6), and form

$$\mathbf{F}^* = \begin{pmatrix} 0 & 0 & 0 \\ 0 & f_{11} & f_{12} \\ 0 & f_{21} & f_{22} \end{pmatrix} \tag{6.8}$$

The test statistic is

$$X^2 = n\mathbf{u}_0'\mathbf{H}\mathbf{F}^*\mathbf{H}'\mathbf{u}_0 \tag{6.9}$$

and the hypothesis that $\mathbf{u}_0$ is the principal axis is rejected if X^2 is too large. This is a large-sample test.

Critical values sample size $n \geq 25$ For a test at level $100\alpha\%$, reject the hypothesis if $X^2 > -2 \log(\alpha)$. The significance probability of an observed value, $X^2 = a$ say, is $\exp(-\tfrac{1}{2}a)$.

> **Example 6.10** Suppose that, for the data (set B17) of Example 6.9, we are interested in the hypothesis that the true polar axis has a plunge of 10° in the direction 300°. From (6.9) the value of X^2 is 15.3, with significance probability 0.0005, so it is unlikely that this is the true axis. An alternative method of testing, at the 95% level, would have been to check whether the hypothesized pole fell within the 95% elliptical confidence cone calculated in Example 6.9 and displayed in Figure 6.16.

References and footnotes See **6.3.1(ii)** above.

6.3.1(iv) *Test for rotational symmetry about the principal axis.* Suppose first that the true principal axis is known, and has direction cosines $(\lambda_0, \mu_0, \nu_0)$. Select one end of the axis (e.g. choose $\nu_0 > 0$) and convert $(\lambda_0, \mu_0, \nu_0)$ to polar coordinates (θ_0, ϕ_0) (§**2.2**). Rotate the data so that

Figure 6.16 Approximate 95% elliptical confidence cone for the principal axis, using the data in Figure 6.14. See Example 6.9.

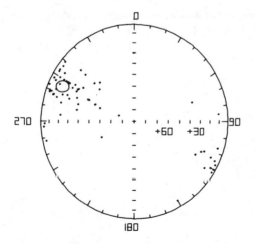

6.3.1(v). Estimation of principal axis (symmetric case)

they are measured relative to the principal axis through (θ_0, ϕ_0), as described in §3.2.2 (cf. equation (3.10)), obtaining $(x'_1, y'_1, z'_1), \ldots, (x'_n, y'_n, z'_n)$ say, and calculate the longitudes $\phi'_1, \ldots, \phi'_n$ relative to (θ_0, ϕ_0) using (3.4)-(3.6). Graphical and formal tests of the hypothesis of symmetry may now be performed as in §5.3.1(iv).

Now suppose that the principal axis is not known. The following procedure is approximate, but should be adequate for $n \geq 25$. Estimate the principal axis by $\hat{\mathbf{u}}_3$ as defined in §3.2.4, with direction cosines $(\hat{\lambda}, \hat{\mu}, \hat{\nu})$ say, and rotate the data as described above, using $(\hat{\lambda}, \hat{\mu}, \hat{\nu})$ instead of $(\lambda_0, \mu_0, \nu_0)$. The graphical procedure referred to above for $\phi'_1, \ldots, \phi'_n$ can then be performed. For formal testing, calculate the standardized eigenvalues $\bar{\tau}_1, \bar{\tau}_2$ and $\bar{\tau}_3$ as described in §3.2.4, and the quantity Γ defined below in (6.11). The test statistic is

$$P_n = 2n(\bar{\tau}_2 - \bar{\tau}_1)^2 / (1 - \bar{\tau}_3 + \Gamma). \tag{6.10}$$

The hypothesis of symmetry is rejected if P_n is too large.

Critical values sample size $n \geq 25$ For a test at the $100\alpha\%$ level reject the hypothesis if $P_n > -2 \log \alpha$. The significance probability of an observed value, $P_n = p_n$, say, is $\exp(-\frac{1}{2}p_n)$.

> **Example 6.11** For the data (set B11) of Example 6.1, a Q-Q plot of the quantities $\phi'_1, \ldots, \phi'_n$ is shown in Figure 6.17. The hypothesis of rotational symmetry seems reasonable. For a formal test, we obtain from (6.10) and the results in Table 6.1, $P_n = 1.81$. The 95% point is 5.99, so the conclusion from the graphical analysis is confirmed.

References and footnotes See Watson (1983a, Chapter 5), Prentice (1984).

6.3.1(v) *Estimation of the principal axis of a symmetric bipolar distribution.* As in **6.3.1(ii)** above, the usual estimate of the principal axis $\mathbf{u}_3$ is the eigenvector $\hat{\mathbf{u}}'_3 = (\hat{\lambda}, \hat{\mu}, \hat{\nu})$ corresponding to the largest eigenvalue $\hat{\tau}_3$ of the orientation matrix (see §3.2.4). However, the assumption of symmetry allows us to calculate a large-sample $100(1-\alpha)\%$ *circular* confidence cone for $\mathbf{u}_3$, instead of an *elliptical* confidence cone.

Calculate

$$\Gamma = (1/n) \sum_{i=1}^{n} (\hat{\lambda} x_i + \hat{\mu} y_i + \hat{\nu} z_i)^4 \tag{6.11}$$

and

$$\hat{\sigma} = [(\bar{\tau}_3 - \Gamma)/n]^{1/2} / (\bar{\tau}_3 - \bar{\tau}_2) \tag{6.12}$$

where $\bar{\tau}_2$ and $\bar{\tau}_3$ are the normalised eigenvalues (§3.2.4). It is natural to call $\hat{\sigma}$ the (estimated) *spherical standard error* of the sample principal axis $\hat{\mathbf{u}}_3$. An approximate $100(1-\alpha)\%$ confidence cone for $\mathbf{u}_3$ is centred on $\hat{\mathbf{u}}_3$ with semi-vertical angle

$$q = \arcsin(e_\alpha^{1/2}\hat{\sigma}) \qquad (6.13)$$

where $e_\alpha = -\log \alpha$. Thus, for a 95% confidence cone, $\alpha = 0.05$, $e_\alpha = -\log 0.05 = 2.996$, and (6.12) becomes $q = \arcsin(1.731\hat{\sigma})$. (If $n < 25$, use bootstrap methods - e.g. §3.6.1.)

> **Example 6.12** For the data (set B11) of Example 6.1, the estimate of the polar axis $\mathbf{u}_3$ has direction cosines (0.6592, −0.7516, 0.0251), corresponding to a plunge of 1.4° in the direction 228.7°.
>
> In Example 6.11, it was found that the data could reasonably be regarded as coming from a symmetric bipolar distribution, so we calculate a confidence cone for $\mathbf{u}_3$. The spherical standard error of $\hat{\mathbf{u}}_3$ is $\hat{\sigma} = 0.02384$, and a large-sample 95% confidence cone for $\mathbf{u}_3$, centred on the sample principal axis, has semi-vertical angle 3.4°. (By way of comparison, the elliptical confidence cone, computed using the methods of **6.3.1(ii)** above if the assumption of symmetry is not made, has major and minor semi-axes 3.6° and 3.1°.)

References and footnotes See Watson (1983a, Chapter 5), Prentice (1984).

Figure 6.17 Uniform Q-Q plot of the azimuths of the data in Figure 6.1 measured relative to the sample principal axis. See Example 6.11.

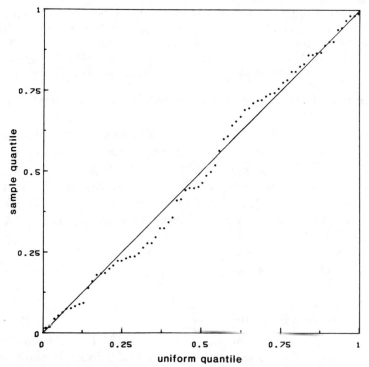

6.3.2. Analysis of a Watson bipolar sample

6.3.1(vi) *Test for a specified value of the principal axis of a symmetric bipolar distribution.* Suppose it is hypothesised that the principal axis $\mathbf{u}_3$ has direction cosines $(\lambda_0, \mu_0, \nu_0)$. To test this hypothesis, calculate the sample principal axis $\hat{\mathbf{u}}'_3 = (\hat{\lambda}, \hat{\mu}, \hat{\nu})$ as in **6.3.1(v)** above, and the quantity Γ defined in (6.11). The test statistic is

$$J_n = 2n(\bar{\tau}_3 - \bar{\tau}_2)^2[1 - (\hat{\lambda}\lambda_0 + \hat{\mu}\mu_0 + \hat{\nu}\nu_0)^2]/(\bar{\tau}_3 - \Gamma) \qquad (6.14)$$

and the hypothesis is rejected if J_n is too large. If the spherical standard error $\hat{\sigma}$ (cf. (6.12)) has already been computed, a more convenient form for J_n is

$$J_n = 2[1 - (\hat{\lambda}\lambda_0 + \hat{\mu}\mu_0 + \hat{\nu}\nu_0)^2]/\hat{\sigma}^2. \qquad (6.15)$$

The test is a large-sample test.

Critical values $n \geq 25$ For a test at the level $100\alpha\%$, the critical value is $-2\log\alpha$. The significance probability of an observed value, $J_n = j_n$ say, is $\exp(-\tfrac{1}{2}j_n)$.

> **Example 6.13** For the data (set B11) of Example 6.1, analysed in Examples 6.11 and 6.12, consider testing the hypothesis that the true principal axis has plunge 5° in the direction 230°. We find that $J_n = 7.62$, corresponding to a significance probability of 0.022, so there is evidence at the 5% level that the hypothesis is not true.
>
> Note that we could have tested the hypothesis equally well at the 5% level by checking whether (5°, 230°) lay inside the confidence cone computed in Example 6.12.

References and footnotes See Watson (1983a, Chapter 5), Watson (1983c).

6.3.2 Parametric models: bipolar distributions with rotational symmetry

The principal model for bipolar axial data possessing rotational symmetry about the principal axis is the Watson bipolar distribution, the properties of which are discussed in §4.4.4. We shall consider the following problems in relation to a sample of data

$$\mathbf{X}'_1 = (x_1, y_1, z_1), \ldots, \mathbf{X}'_n = (x_n, y_n, z_n).$$

(i) Testing the sample for uniformity, against the alternative that it was drawn from a Watson bipolar distribution.

(ii) Testing to see whether the Watson bipolar distribution is a suitable model for the data.

(iii) Testing individual observations which are unusually far from the main data mass to see if they are to be judged discordant (statistically inconsistent with the main data mass), assuming that the underlying distribution is the Watson bipolar model.

168   6. Analysis of a single sample of undirected lines

(iv) Estimating the parameters of the underlying Watson bipolar distribution, including confidence regions for the principal axis and the concentration parameter.

(v) Testing whether the principal axis of the distribution has a specified value.

(vi) Testing whether the concentration parameter of the distribution has a specified value.

6.3.2(i) *Test for uniformity against a Watson* $W\{\lambda, \mu, \nu\}, \kappa\}$ *bipolar alternative*. Since uniformity corresponds to $\kappa = 0$, this is equivalent to the procedures given in §6.3.1(i), the test statistic being the largest eigenvalue $\bar{\tau}_3$, or S_n in (6.2), depending on whether the principal axis (λ, μ, ν) of the alternative model is unknown or known.

6.3.2(ii) *Goodness-of-fit of the Watson* $W\{(\lambda, \mu, \nu), \kappa\}$ *bipolar model.* We present both graphical and formal methods for assessing the adequacy of fit of this model to a set of data. The graphical methods allow a general assessment of the fit, as well as identification of isolated points which may disturb the fit; the formal procedures permit precise statements to be made about the quality of the fit. The methods may not be reliable if $\kappa < 5$.

Graphical procedures Calculate the sample principal axis $\hat{u}_3 = (\hat{\lambda}, \hat{\mu}, \hat{\nu})$, and rotate the data to $\hat{u}_3$ using the method described in §6.2.1(iv); define the rotated direction cosines $(x'_1, y'_1, z'_1), \ldots, (x'_n, y'_n, z'_n)$ so that $z'_1 \geq 0, \ldots, z'_n \geq 0$. Then calculate $(\theta'_1, \phi'_1), \ldots, (\theta'_n, \phi'_n)$ from $(x'_1, y'_1, z'_1), \ldots, (x'_n, y'_n, z'_n)$ using relations of the form (3.4)-(3.6.).

(a) Colatitude plot Let $X_i = 1 - \cos^2 \theta'_i$, $i = 1, \ldots, n$, and re-arrange the X_i's in increasing order, obtaining $X_{(1)} \leq \ldots \leq X_{(n)}$. Define $a_i = F^{-1}[(i - \frac{1}{2})/n]$, where F is the exponential distribution function defined in Table 3.2. Perform a Q-Q plot for the exponential model as described in §3.5.1. The plot should be approximately linear, passing through the origin. The slope of the plot gives an estimate of $1/\kappa$. (Note that, whereas the colatitude plot for the Fisher distribution (§5.3.2(ii)) can in fact be used for κ as low as 2, for the Watson bipolar distribution κ should be at least 5, because the approximation of the distribution of $1 - \cos^2 \theta'_i$ by $E(\kappa)$ is not of the same quality).

(b) Longitude plot Let $X_i = \phi'_i/2\pi$, $i = 1, \ldots, n$, and re-arrange the X_i's in increasing order, obtaining $X_{(1)} \leq \ldots \leq X_{(n)}$. Perform a Q-Q plot using the uniform model, as described in §3.5.1. The plot should be approximately linear, passing through the origin, with slope 45°.

6.3.2(ii). Goodness-of-fit of Watson bipolar model

Outliers (with respect to the Watson bipolar model) in the data set will manifest themselves in the first plot as points deviating markedly from a best-fitting straight line. The second plot is a check on rotational symmetry about the principal axis.

> **Example 6.14** The data listed in Appendix B18 comprise 35 measurements of L_0^1 axes (intersections between cleavage and bedding planes of F_1 folds) in Ordovician turbidites. Consider the subset of these data comprising all but the two data points marked with asterisks (* and **); plots of the data are shown in Figure 6.18, and the corresponding colatitude and longitude Q-Q plots are shown in Figure 6.19. The colatitude plot is reasonably linear, having regard to the degree of accuracy of the method. The slope of a best-fitting line is approximately 0.15/3.0, yielding the graphical estimate $\hat{\kappa}_{gr} = 20$. The longitude plot reveals nothing untoward.

Formal testing procedures Corresponding to (a) and (b) above, we have formal tests as follows.

(a) Colatitude test Calculate the Kolmogorov-Smirnov statistic D_n defined by (3.38), using $X'_1, \ldots, X'_n$ and as $F(x)$ the exponential distribution function in Table 3.2 with κ estimated by (6.22). The test statistic is

$$M_B(D_n) = (D_n - 0.2)(\sqrt{n} + 0.26 + 0.5/\sqrt{n}). \tag{6.16}$$

The hypothesis that the Watson bipolar distribution is a suitable model is rejected if $M_B(D_n)$ is too large. (This is the same statistic as M_E in (5.15)).

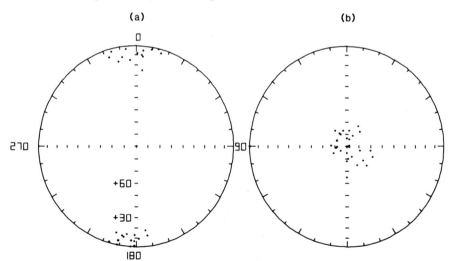

Figure 6.18 Equal-area projection of 33 axes: (a) unrotated, (b) rotated to centralise the sample principal axis. The coordinate system is (Plunge, Plunge azimuth). See Example 6.14.

Critical values Use critical values in Appendix A8.

(b) Longitude test Use the test described in §6.3.1(iv).

> **Example 6.15** For the data in Example 6.14, we obtain $D_n^+ = 0.0800$, $D_n^- = 0.1293$, and the modified value 0.7610 from (6.16) for the colatitude test. From Appendix A8, this is not a significantly large value, at the 5% level. For the longitudes, $P_n = 2.73$, from (6.10), with significance probability 0.065, so the hypothesis of rotational symmetry is acceptable.

References and footnotes See Best & Fisher (1986).

6.3.2(iii) Outlier test for discordancy. The reader is referred to the corresponding section in the chapter on analysing unit vectors, §**5.3.2(iii)**, for a discussion of various aspects of outlier problems; specifically, how they might arise, what their effect might be, and how to deal with them. The remarks made there hold, with appropriate adjustment to formulae, for axial data.

The tests proposed below are appropriate provided that $\kappa > 10$.

> Figure 6.19 Probability plots to check on the goodness-of-fit of the Watson bipolar distribution to the data in Figure 6.18: (a) colatitude plot, (b) longitude plot. See Example 6.14.

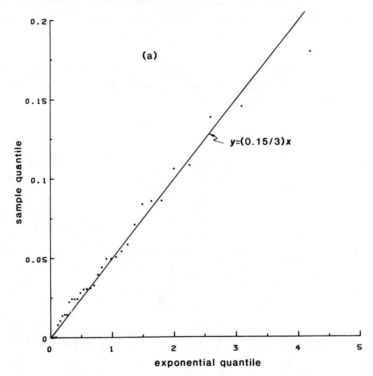

6.3.2(iii). Outlier test for discordancy

Suppose then, that on the basis of a plot of the data, or by using graphical procedure (a) in **6.3.2(ii)** above, that the observation (x_n, y_n, z_n) say is considered to be an outlier, that is, suspiciously far from the main data mass. For definiteness, write $\mathbf{T}_n$ for the orientation matrix (3.15) computed using the complete sample, and $\mathbf{T}_{n-1}$ for the orientation matrix computed with the axis (x_n, y_n, z_n) omitted:

$$\mathbf{T}_{n-1} = \mathbf{T}_n - \begin{pmatrix} x_n^2 & x_n y_n & x_n z_n \\ x_n y_n & y_n^2 & y_n z_n \\ x_n z_n & y_n z_n & z_n^2 \end{pmatrix}. \qquad (6.17)$$

Let $\hat{\tau}_{3,n}$ and $\hat{\tau}_{3,n-1}$ be the largest eigenvalues of $\mathbf{T}_n$ and $\mathbf{T}_{n-1}$ respectively (cf. §**3.2.4**), and calculate

$$H_n = (n-2)(1 + \hat{\tau}_{3,n-1} - \hat{\tau}_{3,n})/(n - 1 - \hat{\tau}_{3,n-1}). \qquad (6.18)$$

The outlier (x_n, y_n, z_n) is judged discordant if H_n is too large.

Critical values For an observed value $H_n = x$, the significance probability can be calculated from (5.20).

Figure 6.19 (*cont.*)

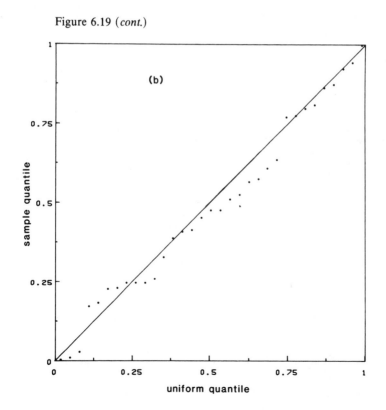

172   6. Analysis of a single sample of undirected lines

***Example* 6.16** Suppose that we add to the data set analysed in Example 6.14 the point marked (*) in Appendix B18. A plot of the data set, rotated to the sample principal axis (plunge 0.1°, azimuth 183.0°) of the reduced data set in Example 6.14, is shown in Figure 6.20, with the

Figure 6.20 Equal-area projection of 34 axes (the data in Figure 6.18 plus one more point), rotated as in Figure 6.18(b). See Example 6.16.

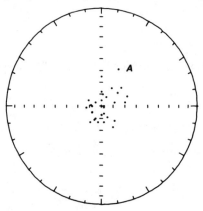

Figure 6.21 Probability plots to check on the goodness-of-fit of the Watson distribution to the data in Figure 6.20: (a) colatitude plot, (b) longitude plot. See Example 6.16.

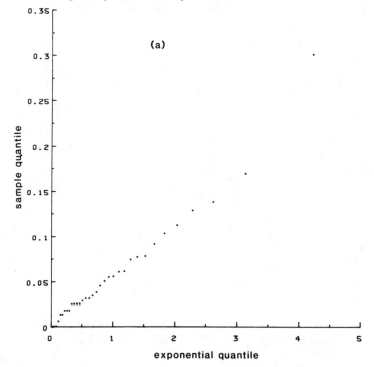

6.3.2(iii). Outlier test for discordancy

new point appearing as an outlier (A) in the top right quadrant. The colatitude plot for these data (Figure 6.21(a)) reveals a somewhat suspect point. To test for discordancy, we have, from (6.18), $H_n = 5.93$. Using (5.20), we obtain a significance probability of 0.15, so the outlier cannot reasonably be adjudged discordant. However, in estimating κ, we may wish to downweight the contribution to this point: see §6.3.2(iv) below.

Sometimes there appear to be two or more suspiciously distant observations. *Again, we refer the reader to the discussion in* §5.3.2(iii), and, as there, confine ourselves to testing the outliers *en bloc* for discordancy. We assume that the outliers do not appear to form a separate group, otherwise the methods of §7.3 would be suitable.

Suppose there are t outlying axes which we assume, for convenience, are $\mathbf{X}_{n-t+1}, \ldots, \mathbf{X}_n$. Corresponding to (6.17), let $\mathbf{T}_{n-t}$ be the orientation matrix computed from $\mathbf{X}_1, \ldots, \mathbf{X}_{n-t}$, with largest sample eigenvalue $\hat{\tau}_{3,n-t}$. Calculate

$$H_n^{(t)} = \{(n-t-1)/t\}(t + \hat{\tau}_{3,n-t} - \hat{\tau}_{3,n})/(n - t - \hat{\tau}_{3,n-t}). \tag{6.19}$$

The outliers are classified as discordant if $H_n^{(t)}$ is too large.

Figure 6.21 (*cont.*)

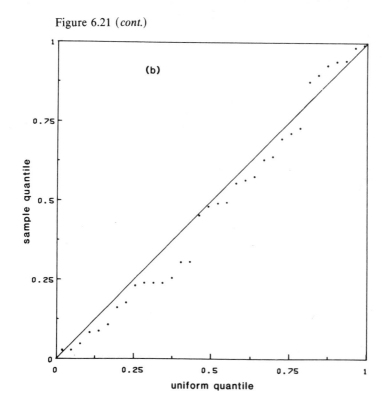

174   6. Analysis of a single sample of undirected lines

Figure 6.22 Equal-area projection of 35 axes (the data in Figure 6.18 plus two more points) rotated as in Figure 6.18(b). See Example 6.17.

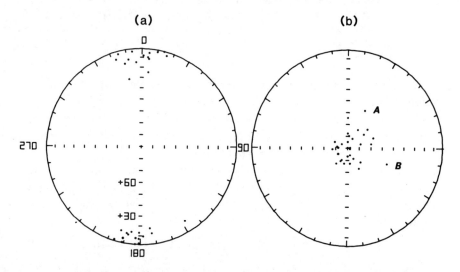

Figure 6.23 Probability plots to check on the goodness-of-fit of the Watson bipolar distribution to the data in Figure 6.22: (a) colatitude plot, (b) longitude plot. See Example 6.17.

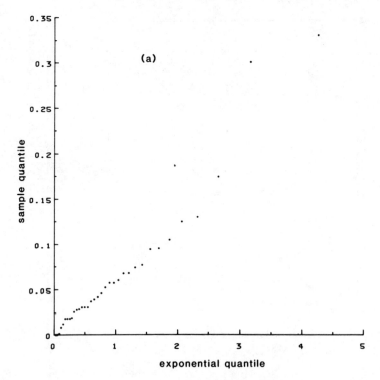

6.3.2(iv). Parameter estimation for Watson bipolar distribution

Critical values Selected critical values for $t=2$ and $t=3$ are given in Appendix A9.

> **Example 6.17** The entire sample of data in Appendix B18 is shown in Figure 6.22(a), and also in Figure 6.22(b) after rotation to the sample principal axis (plunge 0.1°, azimuth 183.0°) of the reduced data set used in Example 14. The corresponding colatitude and longitude Q-Q plots, in Figure 6.23, both show aberrant behaviour. To test the two outliers (marked A and B in Figure 6.22(b)) for discordancy, we obtain, from (6.19), $H_n^{(2)} = 6.22$, corresponding to a significance probability of between 0.1 and 0.05, using Appendix A9. So there is not strong evidence that these points require special treatment, although we might consider using a robust estimate of κ (see Example 6.18 for the effect of this). This example illustrates the desirability of complementing a graphical assessment of a statistical model with a formal test.

References and footnotes See Best & Fisher (1986).

6.3.2(iv) *Parameter estimation for the Watson bipolar distribution.* The usual (maximum likelihood) estimate $(\hat{\lambda}, \hat{\mu}, \hat{\nu})$ of the principal axis (λ, μ, ν) of the Watson bipolar distribution is the sample principal axis $\hat{\mathbf{u}}_3$ of the orientation matrix **T** (3.15). The maximum likelihood estimate

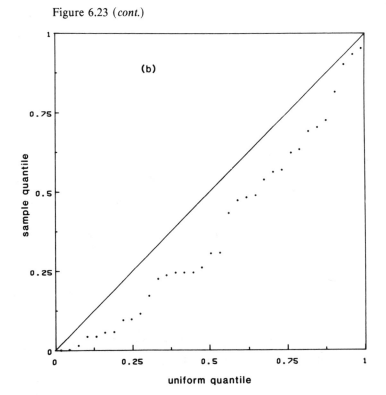

Figure 6.23 (*cont.*)

of κ is the solution of the equation
$$D(-\hat{\kappa}) = \bar{\tau}_3 \tag{6.20}$$
where
$$D(z) = \int_0^1 x^2 \exp(zx^2) \, dx \bigg/ \int_0^1 \exp(zx^2) \, dx \tag{6.21}$$

A reasonable approximation to the solution of (6.20) is:
$$\hat{\kappa} = \begin{cases} 3.75 \times (3\tau_3 - 1) & 0.333 < \tau_3 \leq 0.36 \\ 3.34 \times (3\bar{\tau} - 1) & 0.38 < \bar{\tau}_3 \leq 0.65 \\ 0.7 + 1/(1 - \bar{\tau}_3) & 0.65 < \bar{\tau}_3 \leq 0.99 \\ 1/(1 - \bar{\tau}_3) & \bar{\tau}_3 \geq 0.99 \end{cases} \tag{6.22}$$

The exact solution is tabulated for a range of values, in Appendix A10.

The estimate of the principal axis is unlikely to be substantially affected by the presence of one or two outlying observations, but the estimate of κ is not so robust. For this latter case, suppose there are t points which appear as outliers relative to the main data mass, and suppose for convenience that they are $\mathbf{X}_{n-t+1}, \ldots, \mathbf{X}_n$. (These points may have been identified from a plot of the data, or using the graphical procedures in **6.3.2(ii)** above, or may have been adjudged discordant using the test in **6.3.2(iii)** above). Calculate the values

$$\delta_i = 1 - (x_i \hat{\lambda} + y_i \hat{\mu} + z_i \hat{\nu})^2, \quad i = 1, \ldots, n \tag{6.23}$$

and re-arrange them in increasing order: $\delta_{(1)} \leq \ldots \leq \delta_{(n)}$. The outlying values should correspond to the largest t δ's, $\delta_{(n-t+1)}, \ldots, \delta_{(n)}$. Calculate an adjusted value $\bar{\tau}_3^*$ for the maximum normalised eigenvalue -

$$\bar{\tau}_3^* = \left\{ \sum_{i=1}^{n-t-1} \delta_{(i)} + (t+1)\delta_{(n-t)} \right\} \bigg/ n \tag{6.24}$$

- and solve the equation (6.20), with $\bar{\tau}_3$ replaced by $\bar{\tau}_3^*$, to obtain $\hat{\kappa}_R$.

Example 6.18 For the data of Example 6.14 (a subset of B18), which can reasonably be taken to be from a Watson bipolar distribution, we find that the principal axis has direction cosines (0.9986, 0.0321, 0.0185), corresponding to a plunge of 0.1° in the direction 183.0°. For $\bar{\tau}_3 = 0.9494$, $\hat{\kappa} = 20.5$. By way of comparison, when the outliers referred to in Example 6.17 are included, the estimate of the principal axis has a dip of $-0.3°$ in the direction 181.9° and the new estimate of κ, corresponding to a new $\bar{\tau}_3$ of 0.9337, is 15.8. In this case, the robust estimate of κ is $\hat{\kappa}_R = 17.9$. Thus, the robust method provides a compromise between leaving the outliers out ($\hat{\kappa} = 20.5$) and leaving them in unadjusted ($\hat{\kappa} = 15.8$).

A $100(1-\alpha)\%$ confidence cone for the true principal axis $\mathbf{u}_3$ may be calculated exactly as was done in §**6.3.1(v)**, for a more general symmetric distribution. It is centred on $\hat{\mathbf{u}}_3$, with semi-vertical angle q given by (6.13).

6.3.2(v). Test for specified principal axis

For κ, a confidence interval can be based on the fact that $2\kappa(n-\hat{\tau}_3)$ is distributed approximately as χ^2_{2n-2}, so that a one-sided $100(1-\alpha)\%$ confidence interval of the form $(0, \kappa_\alpha)$ is given by

$$\kappa_\alpha = \chi^2_{2n-2}(\alpha)/[2(n-\hat{\tau}_3)] \tag{6.25}$$

and a two-sided interval of the form $(\kappa_{1-\frac{1}{2}\alpha}, \kappa_{\frac{1}{2}\alpha})$ is given by

$$\begin{aligned}\kappa_{\frac{1}{2}\alpha} &= \chi^2_{2n-2}(\tfrac{1}{2}\alpha)/[2(n-\hat{\tau}_3)], \\ \kappa_{1-\frac{1}{2}\alpha} &= \chi^2_{2n-2}(1-\tfrac{1}{2}\alpha)/[2(n-\hat{\tau}_3)].\end{aligned} \tag{6.26}$$

(Note: $\hat{\tau}_3 = n\bar{\tau}_3$ and Prob $(\chi^2_\nu \geq \chi^2_\nu(\beta)) = \beta$).

Example 6.19 For the data of Example 6.14 (a subset of B18), a 95% confidence cone for $\mathbf{u}_3$ has semi-angle 4.0°, and a 95% two-sided confidence interval for κ is (13.1, 26.4).

References and footnotes Estimation of the parameters was considered in the more general context of the Bingham distribution by Bingham (1964, 1974). Watson (1966) mentions that estimation can be done as for the girdle case. Mardia (1972, Section 9.2.2) discusses parameter estimation, but the confidence cone proposed by Mardia (*op. cit.*, page 281) appears to be based on an incorrect approximation to the distribution of $1-\cos^2\Theta$ for large κ (Mardia's equation (9.7.21): see §4.4.4 and Best & Fisher (1985)). Watson (1983a, Chapter 5) discusses point and interval estimation of the parameters in the general context of Watson distributions on the p-sphere $(p = 3, 4, \ldots)$. The form of estimate (6.22) for κ is due to Best & Fisher (1986).

6.3.2(v) *Test for a specified value of the principal axis of the Watson bipolar distribution.* Suppose that the hypothesised value of $\mathbf{u}_3$ is $\mathbf{u}$. Let $\hat{\tau}_3$ be the largest eigenvalue of the sample orientation matrix $\mathbf{T}$ (cf. §3.2.4) and calculate

$$\begin{aligned}h_n &= (n-1)(\hat{\tau}_3 - \mathbf{u}'\mathbf{T}\mathbf{u})/(n-\hat{\tau}_3) \\ &= (n-1)[\hat{\tau}_3 - \hat{\tau}_1(\mathbf{u}'\hat{\mathbf{u}}_1)^2 - \hat{\tau}_2(\mathbf{u}'\hat{\mathbf{u}}_2)^2 - \hat{\tau}_3(\mathbf{u}'\hat{\mathbf{u}}_3)^2]/(n-\hat{\tau}_3).\end{aligned} \tag{6.27}$$

The hypothesis is rejected if h_n is too large.

Critical values Compare h_n with the appropriate percentile of the $F_{2,2n-2}$ distribution in Appendix A5.

Example 6.20 For the data of Example 6.14 (a subset of B18), consider testing the hypothesis that the true principal axis is (plunge 0°, azimuth 180°), at the 95% level. We obtain, from (6.27), $h_n = 1.62$, which is much smaller than the 95% critical point. (Alternatively, note that (0°, 180°) lies inside the 95% confidence cone calculated in Example 6.19).

6.3.2(vi) *Test for a specified value of the concentration parameter of a Watson bipolar distribution.* Suppose that the hypothesised value of κ is κ_0. The test statistic is

$$K_n = 2\kappa_0(n - \hat{\tau}_3) \tag{6.28}$$

To test whether $\kappa = \kappa_0$ against $\kappa > \kappa_0$, reject the hypothesis if K_n is too small; to test against the alternative $\kappa < \kappa_0$, reject $\kappa = \kappa_0$ if K_n is too large. To test against $\kappa \neq \kappa_0$ reject $\kappa = \kappa_0$ if K_n is too small or too large.

Critical values

To test $\kappa = \kappa_0$ against $\kappa > \kappa_0$ use $\chi^2_{2n-2}(1 - \alpha)$

To test $\kappa = \kappa_0$ against $\kappa < \kappa_0$ use $\chi^2_{2n-2}(\alpha)$

To test $\kappa = \kappa_0$ against $\kappa \neq \kappa_0$ use $\chi^2_{2n-2}(1 - \tfrac{1}{2}\alpha), \chi^2_{2n-2}(\tfrac{1}{2}\alpha)$.

Example 6.21 For the data of Example 6.14 (a subset of B18), consider testing the hypothesis $\kappa = 25.0$ against $\kappa < 25.0$, at the 95% level. From (6.28), $K_n = 83.49$, which is just less than $\chi^2_{64}(.05) = 83.67$. So the hypothesis that $\kappa = 25$ is just acceptable at the 5% level.

6.4 Analysis of a sample of undirected lines from a girdle (equatorial) distribution

6.4.1 Introduction

As with the analysis of axial bipolar data (§**6.3**), the methods of this section are based on the eigenvalues and eigenvectors of the orientation matrix **T** as defined in §**3.2.4**. Whereas with bipolar data we were interested in the principal axis, with girdle data we are interested in the normal or *pole* to the plane of the girdle as a summary quantity. Granted this difference, however, there are close parallels between many of the methods in §**6.3** and those in this section; and, where convenient, we shall simply point out the small modifications required to the appropriate bipolar method to obtain the corresponding girdle method.

There are two sorts of symmetry which can be distinguished for girdle distributions. The first is rotational symmetry about the polar axis. The second, which relates only to equatorial distributions, is symmetry about the girdle or equator along which the data are concentrated: the distribution in the hemisphere "below" the mean equatorial plane of the data is the mirror image of that in the "upper" hemisphere. Existence of the first type of symmetry implies that the second kind must also be present, but not *vice versa*. The Watson (§**4.4**) axial girdle distribution is a model for the first type, whereas the general Bingham distribution (§**4.5**) is a model for the second, for a suitable configuration of its concentration parameters.

There is no clear-cut distinction between bipolar distributions and

6.4.2(i). Uniformity test against girdle alternative

girdle distributions, because of the continuum of transitional forms as exemplified in Figure 3.15. That being the case, for some data sets it may be a matter of taste as to which set of methods is adopted for analysis; for example, for summary purposes, one may wish to use the bipolar model consistently. For some experimental situations, both methods of analysis are of interest, as, for example, with the data typified in Example 6.3.

6.4.2 Simple procedures

We retain the notation of §6.3.1 to describe the data. Simple procedures for the following problems are considered herein.

(i) Testing the sample to see whether it has been drawn from a uniform distribution, against the alternative that it has been drawn from a girdle distribution (with specified or unspecified polar axis).
(ii) Estimating the polar axis, and assigning an error to this estimate.
(iii) Testing whether the polar axis takes a specified value.
(iv) Testing the sample to see if it is drawn from a rotationally symmetric girdle distribution.
(v) Estimating the polar axis of a rotationally symmetric distribution, and assigning an error to this estimate.
(vi) Testing whether the polar axis of a rotationally symmetric girdle distribution takes a specified value.

The remark at the beginning of §6.3.1 about the validity of these methods for small samples applies equally here.

6.4.2(i) *Test for uniformity against a girdle alternative.* As in §6.3.1(i), we distinguish between the case in which the alternative hypothesis is specified completely (by specifying the pole $(\lambda_0, \mu_0, \nu_0)$ to the equatorial plane of the alternative girdle) and the case in which the pole is not specified. The tests are precise analogues of those in §6.3.1(i), so we simply record the differences here.

When $(\lambda_0, \mu_0, \nu_0)$ is *not* known, the *smallest* normalised eigenvalue $\bar{\tau}_1$ of the orientation matrix is used, rather than the largest ($\bar{\tau}_3$), and the hypothesis rejected if $\bar{\tau}_1$ is too *small*. Critical values are again given in Appendix A16.

When $(\lambda_0, \mu_0, \nu_0)$ *is* known, the statistic S_n in (6.2) is computed, and the hypothesis of uniformity rejected if S_n is too *small*. Critical values are as described in §6.3.1(i).

References and footnotes The test based on $\bar{\tau}_1$ is due to Watson (1965); its distribution was studied in detail by Anderson & Stephens (1972).

180   6. Analysis of a single sample of undirected lines

The test based on S_n was proposed by Watson (1965), and its distribution tabulated by Stephens (1966). See also Stephens (1972) and Watson (1983a, Chapter 2).

6.4.2(ii) Estimation of the polar axis of a girdle distribution. The usual estimate of the polar axis is the axis $\hat{\mathbf{u}}_1$ corresponding to the smallest eigenvalue $\hat{\tau}_1$ of the orientation matrix **T**, as described in §3.2.4. Calculation of a large-sample confidence region for the unknown polar axis $\mathbf{u}_1$ parallels exactly the analysis in §6.3.1(ii), the only modifications being to the definitions of **H** (6.3) and **E** (6.4), which here become

$$\mathbf{H} = (\hat{\mathbf{u}}_3, \hat{\mathbf{u}}_2, \hat{\mathbf{u}}_1) \qquad (6.29)$$

and

$$\left. \begin{array}{l} e_{11} = [n(\bar{\tau}_1 - \bar{\tau}_3)^2]^{-1} \sum_{i=1}^{n} (\hat{\mathbf{u}}_1' \mathbf{X}_i)^2 (\hat{\mathbf{u}}_3' \mathbf{X}_i)^2 \\[6pt] e_{22} = [n(\bar{\tau}_1 - \bar{\tau}_2)^2]^{-1} \sum_{i=1}^{n} (\hat{\mathbf{u}}_1' \mathbf{X}_i)^2 (\mathbf{u}_2' \mathbf{X}_i)^2 \\[6pt] e_{12} = e_{21} = [n(\bar{\tau}_1 - \bar{\tau}_2)(\bar{\tau}_1 - \bar{\tau}_3)]^{-1} \sum_{i=1}^{n} (\hat{\mathbf{u}}_1' \mathbf{X}_i)^2 (\hat{\mathbf{u}}_2' \mathbf{X}_i)(\hat{\mathbf{u}}_3' \mathbf{X}_i) \end{array} \right\} \qquad (6.30)$$

Example 6.22 Figures 6.24 and 6.25 are, respectively, a raw data plot and a contour plot of 64 poles to bedding planes of F_1 folds. (The data are listed as orientations of planes in Appendix B19). The estimate of the pole to the best-fitting great circle has direction cosines (0.7743, −0.6298, −0.0612), corresponding to a plunge of 3.5° in the direction

Figure 6.24 Equal-area projection of 64 poles to bedding planes. The coordinate system is (Plunge, Plunge azimuth). See Example 6.22.

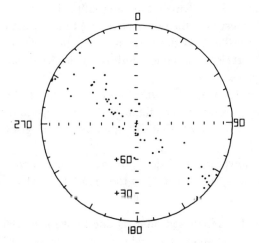

6.4.2(iii). Test for specified polar axis

39.1°. A large-sample 95% elliptical confidence cone for the unknown polar axis, centred on this estimate, has semi-axes 4.1° and 3.1° and a plot of this confidence region is shown in Figure 6.26.

References and footnotes The estimate $\hat{\mathbf{u}}_1$ appears to date at least from Fara & Scheidegger (1963). For the confidence region, see Watson (1983a, Chapter 5), Watson (1984a) and Prentice (1984).

6.4.2(iii) *Test for a specified value of the polar axis.* To test whether the polar axis takes the specific value $\mathbf{u}_0$, with direction cosines $(\lambda_0, \mu_0, \nu_0)$, use the procedure in §**6.3.1(iii)** but with the matrices **H** and **E** defined as in §**6.4.2(ii)** above.

Figure 6.25 Contour plot of the data in Figure 6.24. See Example 6.22.

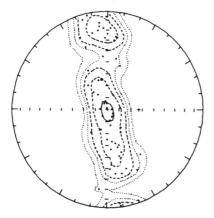

Figure 6.26 Approximate 95% elliptical confidence cone for the polar axis, using the data in Figure 6.24. See Example 6.22.

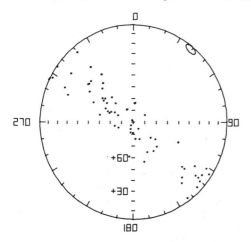

Example **6.23** For the data (Set B19) of Example 6.22, suppose that it is hypothesized that the pole to the plane of the distribution plunges 5° in the direction 35°. The value of the statistic X^2 (cf. (6.9)) is 9.18. corresponding to a significance probability of 0.01. So this hypothesis seems unreasonable. (An alternative way of testing this hypothesis at the 5% level is to check whether the hypothesized axis falls inside the 95% elliptical confidence cone calculated in Example 6.22).

References and footnotes See Watson (1983a, Chapter 5), Watson (1983c), and Prentice (1984).

6.4.2(*iv*) *Test for rotational symmetry about the polar axis.* The graphical and formal procedures for bipolar data in §**6.3.1(iv)** apply equally here, the only alteration required being that the (known or unknown) reference axis is the polar axis. The test statistic P_n in (6.10) becomes

$$P_n^* = 2n(\bar{\tau}_3 - \bar{\tau}_2)^2/(1 - 2\bar{\tau}_1 + \Gamma) \tag{6.31}$$

where Γ is defined by (6.11) and $(\hat{\lambda}, \hat{\mu}, \hat{\nu})$ are the direction cosines of the sample polar axis $\hat{\mathbf{u}}_1$ (cf. §**6.4.1(iv)**).

Example **6.24** For the data (Set B12) of Example 6.2 a Q-Q plot of the quantities $\phi_1', \ldots, \phi_n'$ is shown in Figure 6.27. The hypothesis of rotational symmetry appears reasonable. For a formal test, we obtain, from (6.31) and the statistics in Table 6.1, $P_n^* = 1.81$, corresponding to a significance probability of 0.16, and so confirming the conclusion from the graphical analysis.

References and footnotes See Watson (1983a, Chapter 5), Watson (1983c).

6.4.2(*v*) *Estimation of the polar axis of a rotationally symmetric girdle distribution.* By analogy with §**6.3.1(v)**, we use the estimate $\hat{\mathbf{u}}_1$ for the polar axis described in §**6.4.2(ii)** above, and associate with it, in large samples, a *circular* confidence cone (rather than an *elliptical* confidence cone as derived in the more general setting of §**6.3.1(ii)**). The discussion in §**6.3.1(v)** holds good, subject to the alterations that $(\hat{\lambda}, \hat{\mu}, \hat{\nu})$ are now the direction cosines of $\hat{\mathbf{u}}_1$, with consequent change to Γ in (6.11), and the spherical standard error of $\hat{\mathbf{u}}_1$ is estimated by

$$\hat{\sigma} = [(\bar{\tau}_1 - \Gamma)/n]^{1/2}/(\bar{\tau}_2 - \bar{\tau}_1) \tag{6.32}$$

Example **6.25** For the data (Set B12) in Example 6.2, the estimate $\hat{\mathbf{u}}_1$ of the pole $\mathbf{u}_1$ to the plane of the girdle distribution has direction cosines (0.7729, −0.6318, −0.0592) corresponding to a plunge and plunge azimuth of (3.4°, 39.3°). The analysis in Example 6.24 showed that the assumption of rotational symmetry of the distribution about $\mathbf{u}_1$ is reasonable. Hence, we can compute the estimated spherical standard error $\hat{\sigma}$ for $\hat{\mathbf{u}}_1$ from (6.32), obtaining 0.02698, and a semi-vertical angle for a 95% confidence interval for $\mathbf{u}_1$, centred on $\hat{\mathbf{u}}_1$, as 3.8° (from (6.13)).

References and footnotes As for **6.4.2(ii)** above.

6.4.2(vi) Test for a specified value of the polar axis of a rotationally symmetric girdle distribution.

By analogy with §6.4.2(v), the test is essentially the same as that given in §6.3.1(vi) except with the changes made in §6.4.2(v). Thus, the test statistic is (for specified polar axis $(\lambda_0, \mu_0, \nu_0)$)

$$J_n^* = 2n(\bar{\tau}_2 - \bar{\tau}_1)^2[1 - (\hat{\lambda}\lambda_0 + \hat{\mu}\mu_0 + \hat{\nu}\nu_0)^2]/(\bar{\tau}_1 - \Gamma) \quad (6.33)$$
$$= 2[1 - (\hat{\lambda}\lambda_0 + \hat{\mu}\mu_0 + \hat{\nu}\nu_0)^2]/\hat{\sigma}^2$$

if the spherical standard error $\hat{\sigma}$ of $\hat{\mathbf{u}}_1$ has been calculated.

Example 6.26 For the data (Set B12) in Example 6.2, analysed in Examples 6.24 and 6.25, suppose we are interested in testing whether the true polar axis plunges at 5° in the direction 40°. We find that $J_n^* = 1.31$, corresponding to a significance probability of 0.52, so the hypothesis seems very reasonable. (If we had already computed the confidence cone of Example 6.25, it would have sufficed to see whether (5°, 40°) lay within it; for a test at the 5% level.)

References and footnotes See Watson (1983a, Chapter 5), Watson (1983c).

Figure 6.27 Uniform Q-Q plot of the azimuths of the data in Figure 6.2 measured relative to the sample polar axis. See Example 6.24.

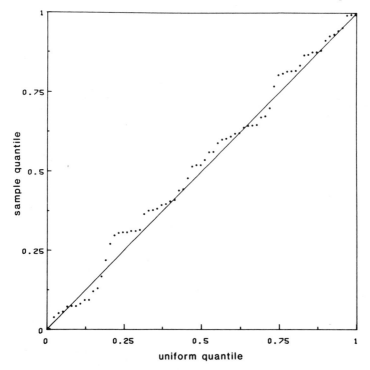

6.4.3 Parametric models: girdle distributions with rotational symmetry

The notation of §6.3.2 will be retained to describe the data. The principal model for girdle axial data possessing rotational symmetry about the polar axis is the Watson girdle distribution, the properties of which are discussed in §4.4.4. The following problems will be considered.

(i) Testing the sample for uniformity, against the alternative that it was drawn from a Watson girdle distribution.
(ii) Testing to see whether the Watson girdle distribution is a suitable model for the data.
(iii) Testing individual observations which are unusually far from the main data mass to see if they are to be judged discordant (statistically inconsistent with the main data mass), assuming that the underlying distribution is the Watson girdle model.
(iv) Estimating the parameters of the underlying Watson girdle distribution, including confidence regions for the polar axis and the concentration parameter.
(v) Testing whether the polar axis of the distribution has a specified value.
(vi) Testing whether the concentration parameter of the distribution has a specified value.

6.4.3(i) *Test for uniformity against a Watson* $W\{(\lambda, \mu, \nu), \kappa\}$ *girdle alternative.* Since uniformity corresponds to $\kappa = 0$, this is equivalent to the procedures given in §6.4.2(i).

6.4.3(ii) *Goodness-of-fit of the Watson* $W\{(\lambda, \mu, \nu), \kappa\}$ *girdle model.* Informal graphical methods, and formal tests, are available for evaluating this model as a fit to data. The informal methods give a general assessment of fit as well as highlighting one or two outlying observations in a sample which otherwise appears to be adequately modelled by the Watson distribution. Both methods are based on rotating the data so that they are measured as deviations from the estimated polar axis, which is the eigenvector $\hat{\mathbf{u}}_1$ corresponding to the smallest eigenvalue $\hat{\tau}_1$ of the orientation matrix **T** (cf. §3.2.4). We denote the direction cosines of $\hat{\mathbf{u}}_1$ by $(\hat{\lambda}, \hat{\mu}, \hat{\nu})$. The methods should be reliable for $\kappa \geq 5$.

Let $(x'_1, y'_1, z'_1), \ldots, (x'_n, y'_n, z'_n)$ be the rotated data points, defined so that $z'_i \geq 0$, $i = 1, \ldots, n$, and let (θ'_i, ϕ'_i) be the colatitude and longitude corresponding to (x'_i, y'_i, z'_i) for each i.

(a) Colatitude plot Let $X_i = \cos^2 \theta'_i = z'^2_i$, $i = 1, \ldots, n$, and re-arrange the X_i's into increasing order to obtain $X_{(1)} \leq \ldots \leq X_{(n)}$. Define $a_i = F^{-1}[(i-\frac{1}{2})/n]$, $i = 1, \ldots, n$, where F is the χ^2_1 distribution function (which

6.4.3(ii). Goodness-of-fit of Watson girdle model

approximates the truncated χ_1^2 distribution) – see Table 3.2. Perform a Q-Q plot as described in §**3.5.1**. The plot should be approximately linear, passing through the origin. The slope of the plot gives an estimate of $|1/(2\kappa)|$: $\kappa < 0$ for the girdle case.

(b) Longitude plot Let $X_i = \phi_i'/2\pi$, $i = 1, \ldots, n$, and re-arrange the ϕ_i''s in increasing order, obtaining $X_{(1)} \leq \ldots \leq X_{(n)}$. Perform a Q-Q plot using the uniform model, as described in §**3.5.1**. The plot should be approximately linear, passing through the origin with slope 45°.

Outliers (with respect to the Watson girdle model) in the data set will manifest themselves in the colatitude plot as points deviating markedly from a best-fitting straight line through the origin. The longitude plot checks rotational symmetry about the polar axis.

> **Example 6.27** Consider the subset of data in Appendix B12 (cf. Example 6.2) comprising the 39 observations marked with an asterisk (*); a plot of these data is shown in Figure 6.28, and the corresponding colatitude and longitude plots are in Figure 6.29. These plots appear reasonable, the colatitude plot having approximate slope $.0625/5 = 0.0125$, leading to the graphical estimate $\hat{\kappa}_{gr} = -40.0$.

Formal testing procedures Corresponding to (a) and (b) above are the following formal tests:

(a) Colatitude test Calculate the Kolmogorov-Smirnov statistic D_n defined by (3.38), using $X_1', \ldots, X_n'$ and as $F(x)$ the distribution function F in Table 3.2 with κ estimated by (6.38). The test statistic is

$$M_G(D_n) = D_n(\sqrt{n} - 0.04 + 0.7/\sqrt{n}) \tag{6.34}$$

Figure 6.28 Equal-area projection of a subset (40 observations) of the data in Figure 6.2. See Example 6.27.

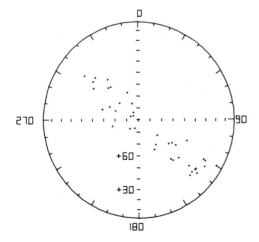

186   6. Analysis of a single sample of undirected lines

The hypothesis that the distribution is a Watson girdle is rejected if M_G is too large.

Critical values Use the critical values in Appendix A18.

(b) Longitude test Use the test described in §**6.4.2(iv)**.

> **Example 6.28** For the data of Example 6.27, $D_n^+ = 0.0462$, $D_n^- = 0.0983$, whence $M_G(D_n) = 0.613$, which is quite acceptable as judged by the percentage points in Appendix A18. For longitudes, the value of P_n^* is 2.14, corresponding to a significance probability of 0.117. Thus the formal tests confirm the informal inferences in Example 6.27. Overall, we conclude that the Watson girdle model is reasonable for these data.

References and footnotes See Best & Fisher (1986).

6.4.3(iii) Outlier test for discordancy. We refer to the general discussion of outliers in §**5.3.2(iii)**, and other comments in §**6.3.2(iii)**. The changes to be made are that the *smallest* eigenvalues $\tau_{1,n}$ and $\tau_{1,n-1}$ are used here, and the statistic corresponding to (6.18) is

$$J_n = (n-3)(\hat{\tau}_{1,n} - \hat{\tau}_{1,n-1})/\hat{\tau}_{1,n-1} \tag{6.35}$$

Figure 6.29 Probability plots to check on the goodness-of-fit of the Watson girdle distribution to the data in Figure 6.28: (a) colatitude plot, (b) longitude plot. See Example 6.27.

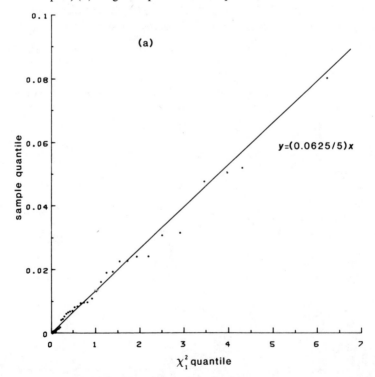

6.4.3(iii). Outlier test for discordancy

and the outlying point (x_n, y_n, z_n) is judged discordant if J_n is too large. This test is adequate for $|\kappa| \geq 10$.

Critical values Compare J_n with the percentage points in Appendix A19.

Example 6.29 Suppose we add to the data set analysed in Example 6.27 the further point marked (**) in Appendix B12. A plot of this data set is shown in Figure 6.30, the new point having a plunge of 58° in the direction 183° (marked A in Figure 6.30). The colatitude plot for these data (Figure 6.31) adds to one's suspicions about whether this new point can reasonably be assumed to be from the same Watson distribution as the rest of the data. To test for discordancy, we obtain, from (6.35), $J_n = 10.94$. From Appendix A19, we conclude that this point is probably not discordant, as J_n barely exceeds the upper 10% point. However, we may choose to use a robust estimate of κ, to lessen the influence of the outlier: see **6.4.3(iv)** below. As we saw in Example 6.17, it is prudent to make a formal test to quantify the evidence of the graphical method about possible discordancy of an outlier.

The remarks which appear in §**5.3.2(iii)** about investigating several outliers should be read in connection with this problem for girdle data. The change required to §**6.3.2(iii)** is that $H_n^{(t)}$ in (6.19) becomes (in an

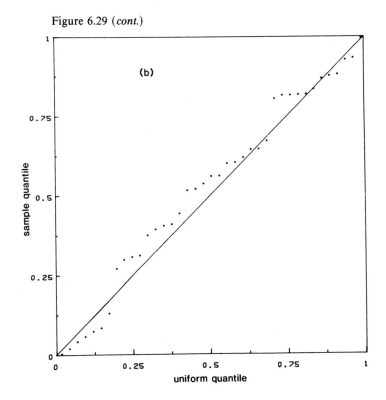

Figure 6.29 (*cont.*)

188   6. Analysis of a single sample of undirected lines

Figure 6.30 Equal-area projection of 41 poles to planes (the data in Figure 6.28 plus one more point). See Example 6.29.

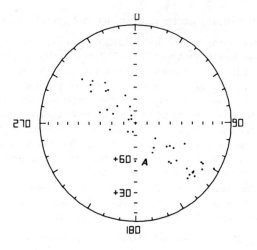

Figure 6.31 Probability plot to check on the goodness-of-fit of the Watson girdle distribution to the data in Figure 6.30. See Example 6.29

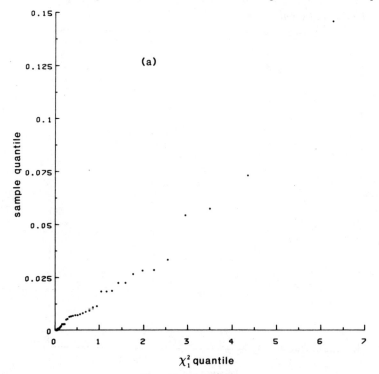

obvious extension of (6.35))

$$J_n^{(t)} = \{(n-t-2)/t\}(\hat{\tau}_{1,n} - \hat{\tau}_{1,n-t})/\hat{\tau}_{1,n-t}. \tag{6.36}$$

The outliers are classified as discordant if $J_n^{(t)}$ is too large.

Critical values Compare $J_n^{(t)}$ with the percentage points in Appendix A19.

References and footnotes See Best & Fisher (1986).

6.4.3(iv) *Parameter estimation for the Watson girdle distribution.* The usual (maximum likelihood) estimate $(\hat{\lambda}, \hat{\mu}, \hat{\nu})$ of the polar axis (λ, μ, ν) of the Watson girdle distribution is the sample axis $\hat{\mathbf{u}}_1$ corresponding to the smallest eigenvalue $\hat{\tau}_1$ of the orientation matrix **T** ((3.15)). The maximum likelihood estimate of κ is the solution of the equation

$$D(\hat{\kappa}) = \bar{\tau}_1 \tag{6.37}$$

where the function D is defined by (6.21). A reasonable approximation to the solution of (6.37) is

$$\hat{\kappa} = \begin{cases} -1/(2\bar{\tau}_1), & \bar{\tau}_1 \leq 0.06 \\ -(0.961 - 7.08\bar{\tau}_1 + 0.466/\bar{\tau}_1), & 0.06 < \bar{\tau}_1 \leq 0.333 \end{cases} \tag{6.38}$$

Figure 6.31 (*cont.*)

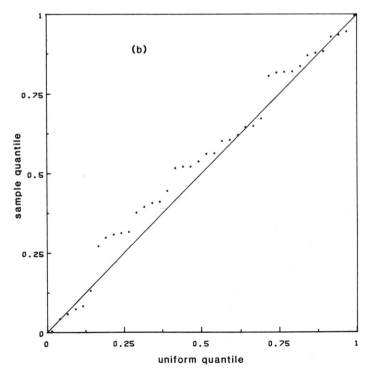

The solution of (6.37) is tabulated for selected values of D in Appendix A10.

Again (as for the bipolar case), the effect of outliers is important when estimating κ, rather than $\mathbf{u}_1$. One way of computing a robust estimate of κ is to calculate the quantities

$$\delta_i = (x_i \hat{\lambda} + y_i \hat{\mu} + z_i \hat{\nu})^2, \quad i = 1, \ldots, n \tag{6.39}$$

and re-arrange them in increasing order, $\delta_{(1)} \leq \ldots \leq \delta_{(n)}$ (say). Suppose there are t outlying values which we would like to down-weight. Calculate the adjusted normalised eigenvalue

$$\bar{\tau}_1^* = \left\{ \sum_{i=1}^{n-t-1} \delta_{(i)} + (t+1)\delta_{(n-t)} \right\} / n \tag{6.40}$$

and then the adjusted estimate $\hat{\kappa}_R$ of κ from (6.38) using $\bar{\tau}_1^*$ instead of $\bar{\tau}_1$.

Example 6.30 For the data (a subset of B12) of Example 6.27, for which the Watson girdle model was found to be reasonable, we obtain $\hat{\mathbf{u}}_1' = (0.7796, -0.6250, -0.0386)$, corresponding to a plunge of 2.2° in the direction 38.7°. For $\bar{\tau}_1 = 0.0132$, $\hat{\kappa} = -38.0$ (cf. the graphical estimate $\hat{\kappa}_{gr} = -40$ from Example 6.27). By way of comparison, when the outlier arising in Example 6.29 is included, the estimated polar axis has a plunge of 3.1° with azimuth 39.1°, a trifling change, whereas the estimate of κ drops to -30.0. The robust estimate of κ is $\hat{\kappa}_R = -35.6$.

A $100(1-\alpha)\%$ confidence cone for the true polar axis $\mathbf{u}_1$, centred on $\hat{\mathbf{u}}_1$, can be obtained using the methods for the more general rotationally symmetric distributions in §6.4.2(v).

For κ, we use the fact that $2|\kappa|\hat{T}_1$ is distributed approximately as χ^2_{n-2}. A one-sided $100(1-\alpha)\%$ confidence interval for κ of the form $(\kappa_\alpha, 0)$ $(\kappa_\alpha < 0)$ is given by

$$|\kappa_\alpha| = \chi^2_{n-2}(\alpha)/(2\hat{T}_1) \tag{6.41}$$

and a two-sided interval of the form $(\kappa_{\frac{1}{2}\alpha}, \kappa_{1-\frac{1}{2}\alpha})$, (where $\kappa_{\frac{1}{2}\alpha}, \kappa_{1-\frac{1}{2}\alpha} < 0$) is given by

$$|\kappa_{\frac{1}{2}\alpha}| = \chi^2_{n-2}(\tfrac{1}{2}\alpha)/(2\hat{T}_1), \quad |\kappa_{1-\frac{1}{2}\alpha}| = \chi^2_{n-2}(1-\tfrac{1}{2}\alpha)/(2\hat{T}_1). \tag{6.42}$$

(Note: $\hat{T}_1 = n\bar{\tau}_1$.)

Example 6.31 For the data (a subset of B12) of Example 6.27, a 95% confidence cone for $\mathbf{u}_1$ has semi-vertical angle 4.5°, and a two-sided 95% confidence interval for κ is $(-57.0, -23.0)$.

References and footnotes The basic (maximum likelihood) estimation theory was done by Watson (1965); see also Mardia (1972, Chapter 9) and Best & Fisher (1986)

6.4.3(v) *Test for a specified value of the polar axis of a Watson girdle distribution.* Suppose the hypothesized value of $\mathbf{u}_1$ is $\mathbf{u}$. Let $\hat{\tau}_1$ be the

6.5. Analysis of small-circle data

smallest eigenvalue of the sample orientation matrix **T** (cf. §3.2.4) and calculate

$$h_n^* = (n-2)(\mathbf{u}'\mathbf{T}\mathbf{u} - \hat{\tau}_1)/(2\hat{\tau}_1) \tag{6.43}$$

The hypothesis is rejected if h_n^* is too large.

Critical values Compare h_n^* with the appropriate percentile of the $F_{2,n-2}$ distribution in Appendix A5.

> **Example 6.32** For the data (a subset of B12) of Example 6.27 suppose we wish to test whether the true pole to the plane of the distribution is (plunge 5°, azimuth 40°). From (6.43), $h_n^* = 2.25$ which is less than the upper 10% point of $F_{2,37}$; so the hypothesis seems reasonable. (Note that an alternative test would be to check whether (5°, 40°) fell inside the 95% confidence cone computed in Example 6.31).

References and footnotes See Watson (1965), Mardia (1972, Chapter 9).

6.4.3(vi) *Test for a specified value of the concentration parameter of a Watson girdle distribution.* Suppose that the hypothesised value of κ is $\kappa_0 (<0)$. The test statistic is:

$$K_n^* = 2|\kappa_0|\hat{\tau}_1. \tag{6.44}$$

To test whether $\kappa = \kappa_0$ against $\kappa < \kappa_0$ (distribution more concentrated), reject the hypothesis if K_n^* is too small; correspondingly, against the alternative $\kappa > \kappa_0$ reject if K_n^* is too large. To test against $\kappa \neq \kappa_0$, reject if K_n^* is too small or too large.

Critical values

> To test $\kappa = \kappa_0$ against $\kappa < \kappa_0$ use $\chi_{n-2}^2(1-\alpha)$
> To test $\kappa = \kappa_0$ against $\kappa > \kappa_0$ use $\chi_{n-2}^2(\alpha)$
> To test $\kappa = \kappa_0$ against $\kappa \neq \kappa_0$ use $\chi_{n-2}^2(1-\tfrac{1}{2}\alpha), \chi_{n-2}^2(\tfrac{1}{2}\alpha)$

> **Example 6.33** For the data of Example 6.27 (a subset of B12), suppose we test $\kappa = -30$ against $\kappa < -30$. From (6.44), $K_n^* = 30.89$, corresponding to a significance probability of 0.75 (since $\chi_{37}^2(0.25) = 30.893$), so K_n^* is not significantly small; hence, we accept the hypothesis.

6.5 Analysis of a sample of measurements from a small-circle distribution

In principle, there is little difference between analysing a sample of axes from a small-circle distribution and analysing a sample of vectors from such a distribution, as far as fitting the small circle is concerned. In §5.4, where the vectorial data problem was considered, the discussion was essentially confined to simple methods for handling the data, since fitting parametric models (cf. §5.4.2) is rather complex. Similarly, for axial data, although a parametric model does exist (the Kelker-Langenberg distribution – see §4.5) which is a generalisation of the Bingham distribution, methods based on it are also rather complicated to implement. On the other hand, the simple methods described in §5.4.2 for vector data can be equally applied to axial data, the sole difference being that the polar axis $\hat{\mathbf{u}}_1$ of the orientation matrix (cf. §6.4) is used instead of the mean direction in *Step 0* of the iterative procedure.

6.6 Analysis of a sample of axes from a multimodal distribution

An illustration of this sort of data set was given in Example 6.4; the raw data plot and contour plots appear in Figures 6.4 and 6.10 respectively. We refer the reader to the discussion of multimodal vector data (§5.5.1) for some general remarks about methods of analysis of multimodal data. Options (ii) and (iii) are not, at present, at a satisfactory stage of development, so the approach described in §5.5.2 is recommended for the time being.

One point not considered in §6.3 or §6.4 is that of obtaining a confidence region for *each* of the three axes $\mathbf{u}_1, \mathbf{u}_2$ and $\mathbf{u}_3$. An elliptical confidence cone for $\mathbf{u}_3$ (the principal axis) is described in §6.3.1(ii) and for $\mathbf{u}_1$ in §6.4.2(ii). To obtain such a cone for $\mathbf{u}_2$, modify the formulae in §6.3.1(ii) as follows:

Equation (6.3) becomes

$$\mathbf{H} = (\hat{\mathbf{u}}_3, \hat{\mathbf{u}}_1, \hat{\mathbf{u}}_2) \tag{6.45}$$

and equation (6.5) becomes

$$\left. \begin{aligned} e_{11} &= [n(\bar{\tau}_2 - \bar{\tau}_3)^2]^{-1} \sum_{i=1}^{n} (\hat{\mathbf{u}}_2'\mathbf{X}_i)^2 (\hat{\mathbf{u}}_3'\mathbf{X}_i)^2 \\ e_{22} &= [n(\bar{\tau}_2 - \bar{\tau}_1)^2]^{-1} \sum_{i=1}^{n} (\hat{\mathbf{u}}_2'\mathbf{X}_i)^2 (\hat{\mathbf{u}}_1'\mathbf{X}_i)^2 \\ e_{12} &= e_{21} = [n(\bar{\tau}_3 - \bar{\tau}_2)(\bar{\tau}_2 - \bar{\tau}_1)]^{-1} \sum_{i=1}^{n} (\hat{\mathbf{u}}_1'\mathbf{X}_i)(\hat{\mathbf{u}}_3'\mathbf{X}_i)(\hat{\mathbf{u}}_2'\mathbf{X}_i)^2 \end{aligned} \right\} \tag{6.46}$$

Then proceed as in §6.3.1(ii). The $100(1-\alpha)\%$ elliptical confidence cone is centred on $\hat{\mathbf{u}}_2$. (If $n < 25$, use bootstrap methods - e.g. §3.6.1.)

References and footnotes See Watson (1983a, Chapter 5), and Prentice (1984).

6.7 A general test of uniformity

We refer the reader to the corresponding discussion in §5.6 for vectorial data. The test statistic for axial data is G_n, as defined by (5.74). The hypothesis of uniformity is rejected if G_n is too large.

6.7. A general test of uniformity

***Example* 6.34** The data (set B15) referred to in Example 6.5 (see Figures 6.5 and 6.11) were collected to investigate whether there were any preferred orientations for the dendritic fields at sites in cats' retinas in response to stimulation by various forms of polarised light. For these data, $G_n = 0.918$, which has significance probability 0.037. So there is some evidence of departure from uniformity.

7

Analysis of two or more samples of vectorial or axial data

7.1 Introduction

Problems of comparing several samples of directional data, and of pooling information from several samples, arise in various ways. Examples of these are:

> **Example 7.1** In a sociological study of the attitudes of 48 individuals to 16 different occupations, judgments were made according to 4 different criteria (Earnings, Social Status, Reward, Social Usefulness), giving rise to 4 samples (each of 48 multivariate measurements). From so-called *external* analysis of the occupational judgments, each multivariate measurement was reduced to a (spherical) unit vector, yielding the 4 samples of unit vectors shown in Figure 7.1. (The data are listed

Figure 7.1 Equal-area projections of occupational judgements according to 4 criteria, the 48 responses in each sample being measured as unit vectors. The coordinate system is (Colatitude, Longitude). (a) earnings (b) social standing (c) rewards (d) social usefulness. See Example 7.1.

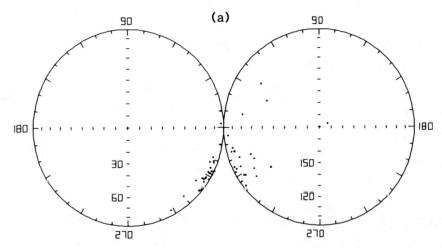

7.1. Introduction

Figure 7.1 (cont.)

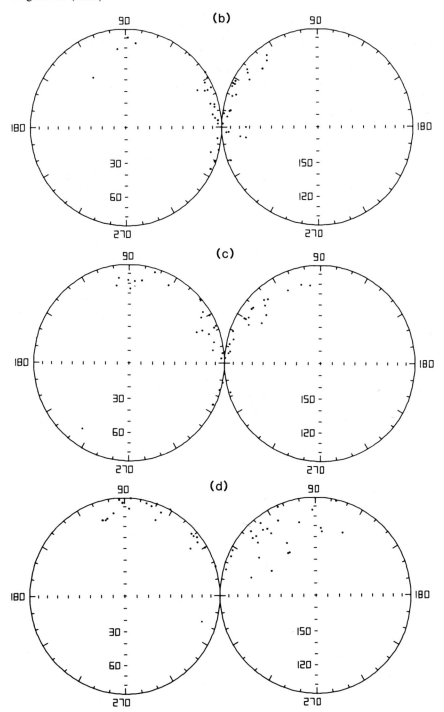

196   7. *Analysis of two or more samples*

in Appendix B20). The question of interest is whether the occupational judgments differ according to the criterion used.

[Because the same individuals were used in each sample (that is, to make judgments for each criterion), one might expect that the responses would be correlated from sample to sample. However, using the method in **§8.2.2** to compare the samples pairwise for correlation of judgments by the same individual, the only correlation of any statistical significance was between responses based on Social Usefulness and responses based on Reward, and this correlation was small. So, for the purpose of analysis, the samples were taken as independent].

Figure 7.2 Equal-area projections of four samples of measurements of magnetic remanence. The coordinate system is (Declination, Inclination). (a) 36 measurements (b) 39 measurements (c) 16 measurements (d) 16 measurements. See Example 7.2.

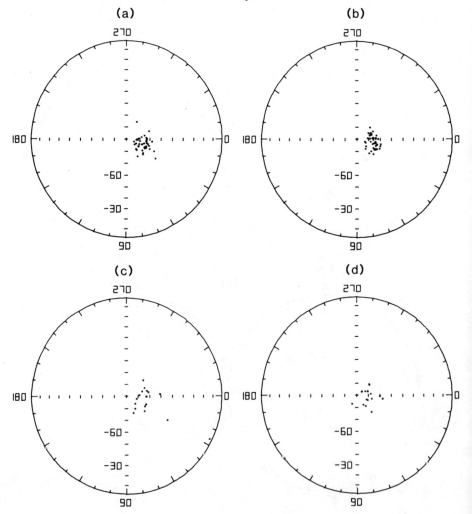

7.1. Introduction

Example **7.2** Measurements of remanent magnetisation made at four locations in Eastern New South Wales. Plots of the data are shown in Figure 7.2. (The data are listed in Appendix B21).

Example **7.3** Two samples of L_0^1 (intersections between cleavage and bedding planes of F_1 folds) in Ordovician turbidites. The data are illustrated in Figure 7.3. (The data are listed in Appendices B18 and B22).

As in Chapters 5 and 6, we have a two-fold classification for the various methods of tackling such problems (according to type of model and type of problem). Figure 7.4 shows schematically the development of this chapter in these terms. §**7.2**, §**7.3** and §**7.4** are concerned respectively

Figure 7.3 Equal-area projections of two samples of measurements of L_0^1 axes. The coordinate system is (Plunge, Plunge azimuth). (a) 35 measurements (b) 42 measurements. See Example 7.3.

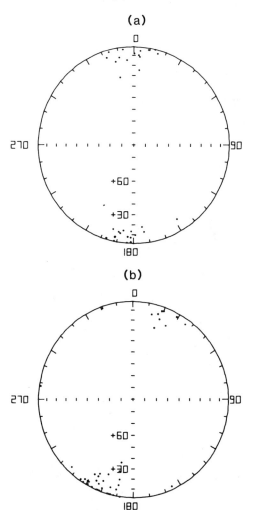

198   7. *Analysis of two or more samples*

with unimodal vector data, axial bipolar data and axial girdle data. Again, for vectorial data, we distinguish between methods based on few assumptions (which we refer to as Simple Procedures) and methods based on a specific probability model; in addition, it will at times be necessary to give special attention to the case in which *two* samples are being compared. For axial data, there has not been such vigorous methodological development, and only a few procedures are available.

7.2   Analysis of two or more samples of unimodal vectorial data

7.2.1   *Introduction*

Suppose we have a sample of unit vectors from each of r unimodal distributions. We denote these samples by

$$(\theta_1, \phi_1), \ldots, (\theta_{n_1}, \phi_{n_1}), (\theta_{n_1+1}, \phi_{n_1+1}), \ldots, (\theta_{n_1+n_2}, \phi_{n_1+n_2}), \ldots$$

n_1 observations n_2 observations
in sample 1   in sample 2   ...

Figure 7.4 Schematic development of Chapter 7.

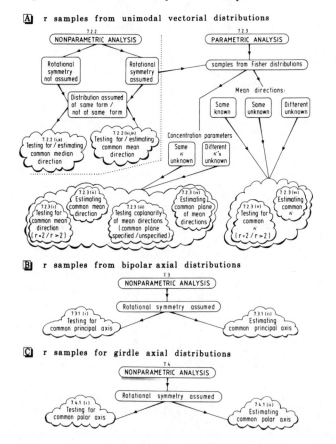

7.2.1. Two or more samples of vectors: Introduction

with corresponding direction cosines

$$(x_1, y_1, z_1), \ldots, (x_{n_1}, y_{n_1}, z_{n_1}), \ldots, (x_N, y_N, z_N)$$

where $N = n_1 + \ldots + n_r$. The summary statistics for the ith data set, $i = 1, \ldots, r$, are shown in Table 7.1.

7.2.2 Simple procedures

Simple procedures are available for the following problems:

(i) Testing whether the r samples are drawn from distributions with a common median direction.
(ii) Estimating a common median direction for the r underlying distributions, and assigning a (large-sample) confidence region to this estimate.
(iii) Testing whether r rotationally symmetric distributions have a common mean direction.
(iv) Estimating the common mean direction of the r distributions, assumed to be symmetric about this mean, and assigning a confidence cone to this estimate.

Table 7.1 *Notation for summary statistics of sample* #i

Quantity	Notation	Reference to earlier chapter
number of observations	n_i	
orientation matrix	$\mathbf{T}_i$	§3.2.4
eigenvalues of orientation matrix	$\hat{\tau}_{1i} \leq \hat{\tau}_{2i} \leq \hat{\tau}_{3i}$ $(\bar{\tau}_{1i} = \hat{\tau}_{1i}/n_i, \bar{\tau}_{2i} = \hat{\tau}_{2i}/n_i, \bar{\tau}_{3i} = \hat{\tau}_{3i}/n_i)$	§3.2.4
eigenvectors of orientation matrix	$\hat{\mathbf{u}}_{1i}, \hat{\mathbf{u}}_{2i}, \hat{\mathbf{u}}_{3i}$	§3.2.4
mean direction (direction cosines)	$(\hat{\alpha}_i, \hat{\beta}_i)$ $(\hat{x}_i, \hat{y}_i, \hat{z}_i)$	§3.2.1
resultant length	R_i ($\bar{R}_i = R_i/n_i$)	§3.2.1
spherical standard error	$\hat{\sigma}_i$	§5.3.1(v) (vectors) §6.3.1(v), §6.4.1(v) (axes)
fourth cosine moment (axial data)	Γ_i	§6.3.1(v)
median direction (direction cosines)	$(\hat{\gamma}_i, \hat{\delta}_i)$ $(\tilde{x}_i, \tilde{y}_i, \tilde{z}_i)$	§5.3.1(ii)
median covariance matrix	$\mathbf{W}_i = \mathbf{C}_i \, \boldsymbol{\Sigma}_i^{-1} \, \mathbf{C}_i$	§5.3.1(ii)

7. Analysis of two or more samples

7.2.2(i) Test for a common median direction of two or more distributions. We distinguish two cases, depending on whether or not the underlying distributions can be assumed to be the same, apart from possibly differing median directions.

(a) Distributions not assumed similar. For the ith sample, $i = 1, \ldots, r$, let $(\theta_{ij}^*, \phi_{ij}^*)$ be the deviation of (θ_{ij}, ϕ_{ij}) from the pooled sample median $(\hat{\gamma}, \hat{\delta})$ (cf. §7.2.2(ii)), obtained by rotating the sample to the pole $(\hat{\gamma}, \hat{\delta})$ using (3.10) with $\mathbf{A}(\hat{\gamma}, \hat{\delta}, 0)$. Similarly, let $(\theta'_{ij}, \phi'_{ij})$ be the deviation of (θ_{ij}, ϕ_{ij}) from $(\hat{\gamma}_i, \hat{\delta}_i)$, the median direction of the ith sample (using (3.10) with $\mathbf{A}(\hat{\gamma}_i, \hat{\delta}_i, 0)$).

Set

$$\mathbf{U}_i = \begin{pmatrix} n_i^{-1/2} \sum_{j=1}^{n_i} \cos \phi_{ij}^* \\ n_i^{-1/2} \sum_{j=1}^{n_i} \sin \phi_{ij}^* \end{pmatrix} \qquad i = 1, \ldots, r \qquad (7.1)$$

and calculate the statistic

$$Y^2 = \sum_{i=1}^{r} \mathbf{U}_i' \mathbf{\Sigma}_i^{-1} \mathbf{U}_i \qquad (7.2)$$

where $\mathbf{\Sigma}_i$ is defined by (5.6) for sample number i. The hypothesis that the distributions have a common median direction is rejected if Y^2 is too large. This is a large-sample test, requiring each sample size $n_i \geq 25$.

Critical values Use appropriate percentile of χ^2_{2r-2} distribution in Appendix A3.

(b) Distributions assumed the same except for median directions Compute $\mathbf{U}_1, \ldots, \mathbf{U}_r$ as in (a) and the pooled estimate

$$\mathbf{\Sigma} = \sum_{i=1}^{r} (n_i - 1) \mathbf{\Sigma}_i / (N - r) \qquad (7.3)$$

The test statistic is

$$Z^2 = \sum_{i=1}^{r} \mathbf{U}_i' \mathbf{\Sigma} \mathbf{U}_i \qquad (7.4)$$

with the same asymptotic distribution as Y^2 (7.2). Then proceed as in (a).

Example 7.4 Figure 7.5 shows the four data sets of Example 7.1 (see Figure 7.1) each rotated to centralise the data mass so that the shapes of the distributions can be appreciated. Summary statistics for each data set are given in Table 7.2. It can be seen that the data sets in Figures 7.5(b) and 7.5(c) are of similar shape, whereas the other two sets are rather different in shape. To test for a common median direction, we first compute a pooled estimate $(\hat{\gamma}, \hat{\delta})$ using the method in **(iib)** below, obtaining (colat. 71.4°, long. 337.5°). The test statistic Y^2 is, from

7.2.2(ii). Estimation of common median direction

(7.2), calculated to be 326.9, unbelievably large as a value of χ_6^2. So, as seems evident from the data displays in Figure 7.1, there are significant differences between the median directions, even allowing for the considerable dispersion in each sample.

References and footnotes See Fisher (1985). A general method of testing whether two samples have been drawn from a common distribution has been published by Wellner (1979).

7.2.2(ii) *Estimation of the common median direction of two or more unimodal distributions.* There are different ways of combining the separate sample estimates of the common median direction (γ, δ) to form a pooled estimate ($\hat{\gamma}$, $\hat{\delta}$) depending on how much we are prepared to assume about the similarity of the underlying distributions. (For sample sizes <25, use bootstrap methods to estimate the confidence region - e.g. §**3.6.1**.)

Figure 7.5 Samples of data in Figure 7.1, each rotated to centralise the data mass. See Example 7.4.

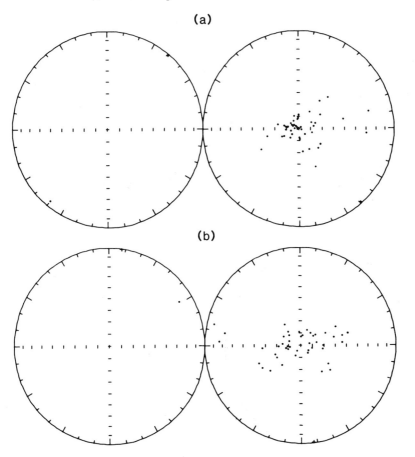

202   7. Analysis of two or more samples

(a) With no assumptions, except that the distributions have a common median direction, the pooled estimate is given by

$$\begin{pmatrix} \sin \hat{\gamma} \cos \hat{\delta} \\ \sin \hat{\gamma} \sin \hat{\delta} \end{pmatrix} = \begin{pmatrix} \tilde{x} \\ \tilde{y} \end{pmatrix} = \mathbf{V}^{-1} \sum_{i=1}^{r} n_i \mathbf{W}_i^{-1} \begin{pmatrix} \tilde{x}_i \\ \tilde{y}_i \end{pmatrix} \qquad (7.5)$$

where

$$\mathbf{V} = \sum_{i=1}^{r} n_i \mathbf{W}_i^{-1}. \qquad (7.6)$$

As in §5.3.1(ii), the asymptotic distribution of $(\tilde{x}, \tilde{y})$ is bivariate normal, with mean value $(\sin \gamma \cos \delta, \sin \gamma \sin \delta)$ and estimated variance-covariance matrix $\mathbf{V}^{-1}$.

Figure 7.5 (cont.)

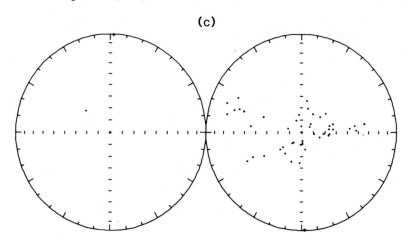

(c)

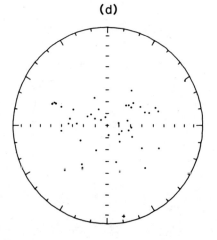

(d)

7.2.2(ii). Estimation of common median direction

Table 7.2 *Summary statistics for data of Example 7.1*

Sample #i	$i=1$ (Earnings)	$i=2$ (Social Standing)	$i=3$ (Rewards)	$i=4$ (Social Usefulness)
n_i	48	48	48	48
$(\hat{\gamma}_i, \hat{\delta}_i)$ (colat., long.)	(89.5°,333.0°)	(89.3°,17.7°)	(89.5°,26.9°)	(96.5°,58.3°)
$\mathbf{W}_i : w_{11}$	0.0237	0.0575	0.1667	0.2813
$\quad : w_{12}$	0.0010	0.0434	0.1318	−0.3220
$\quad : w_{22}$	0.0175	0.4798	0.4478	0.8790
$\mathbf{\Sigma}_i : \sigma_{11}$	0.5048	0.2740	0.3395	0.3668
$\quad : \sigma_{12}$	0.1001	0.0031	0.0748	−0.0865
$\quad : \sigma_{22}$	0.4952	0.7260	0.6605	0.6332

An approximate $100(1-\alpha)\%$ elliptical confidence cone for (γ, δ) is centred on $(\hat{\gamma}, \hat{\delta})$, with points (x, y, z) on the cone's perimeter satisfying

$$V_{11}x^2 + 2V_{12}xy + V_{22}y^2 = -2\log\alpha,$$
$$x^2 + y^2 + z^2 = 1, \ z \geq 0, \quad (7.7)$$

where

$$\mathbf{V} = \begin{pmatrix} V_{11} & V_{12} \\ V_{12} & V_{22} \end{pmatrix}.$$

(b) The weighting of each sample median direction in (7.5) is designed for use when the distributions are thought to have markedly differing dispersions. If, in fact, the distributions are similar in shape and dispersion, a simpler and more efficient estimate of (γ, δ) is given by:

$$\tilde{x} = \sum_{i=1}^{r}(n_i/N)\tilde{x}_i, \quad \tilde{y} = \sum_{i=1}^{r}(n_i/N)\tilde{y}_i, \quad \tilde{z} = (1 - \tilde{x}^2 - \tilde{y}^2)^{1/2} \quad (7.8)$$

and (7.7) becomes

$$\mathbf{V} = \sum_{i=1}^{r} n_i \mathbf{W}_i / N^2. \quad (7.9)$$

Then proceed as in (a).

Example 7.5 For the data of Example 7.1, we found in Example 7.4 that the four samples seem to be drawn from populations with different median directions. However, the data sets in Figures 7.1(b) and 7.1(c) are of similar shape, and have similar median directions. In fact, if we compare them using the test in §**7.2.2(ib)** above, we obtain a significance probability of about 0.01, but for the purposes of demonstration we compute a pooled estimate of a supposed common median direction, here. Using method (b), we obtain $(\hat{\gamma}, \hat{\delta}) = (85.4°, 22.3°)$. An approximate 95% elliptical confidence cone is shown in Figure 7.6: it has major and minor semi-axes 7.3° and 4.7°.

7.2.2(iii) *Test of whether two or more rotationally symmetric distributions have a common mean direction.* We consider graphical and formal methods for comparing the mean directions of r samples. Beginning with graphical methods, let $(\hat{\alpha}_w, \hat{\beta}_w)$ be a pooled estimate of the common mean direction, computed as in §**7.2.2.(iv)**, and, for the ith sample, let $(\theta_{ij}^*, \phi_{ij}^*)$ be the deviation of (θ_{ij}, ϕ_{ij}) from $(\hat{\alpha}_w, \hat{\beta}_w)$, obtained using (3.10) with $A(\hat{\alpha}_w, \hat{\beta}_w, 0)$. Under the hypothesis that all mean directions are the same, the N azimuths

$$\phi_{11}^*, \ldots, \phi_{1n_1}^*, \ldots, \phi_{rN}^* \tag{7.10}$$

should be approximately uniformly distributed on $(0, 2\pi)$. We can test this graphically, using the probability plotting procedure for the uniform distribution in §**3.5.1** (cf. §**5.3.1(iv)**).

> **Example 7.6** For the data of Example 7.2, summary statistics are given in Table 7.3. A pooled estimate of the (hypothesised) common mean direction is calculated below, in Example 7.8, as $(\hat{\alpha}_w, \hat{\beta}_w) = $ (Dec. 12.6°, Inc. −77.2°). A uniform probability plot of the 107 azimuths of the points relative to this direction is shown in Figure 7.7. There is no evidence of departure from uniformity.

For formal testing, let $w_1, \ldots, w_r$ be the weights (defined either by (7.19) or (7.22)) used to compute $(\hat{\alpha}_w, \hat{\beta}_w)$ above, and calculate $\bar{R}_w$ and $\hat{\rho}_w$ using (7.13) and (7.16) respectively.

The test statistic is

$$G_w = 2N(\hat{\rho}_w - \bar{R}_w) \tag{7.11}$$

Figure 7.6 Pooled data from Figures 7.1(b) and 7.1(c), with an approximate 95% elliptical confidence cone for the supposed common median direction. See Example 7.5.

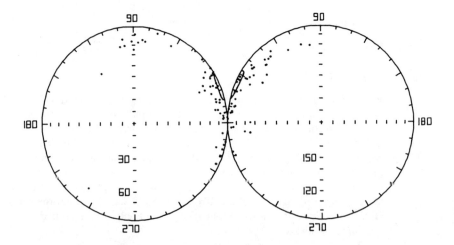

7.2.2(iii). Test for common mean direction

and the hypothesis that the distributions have a common mean direction is rejected if G_w is too large.

Critical values Use appropriate percentile of χ^2_{2r-2} distribution in Appendix A3. (Each sample size should be at least 25).

Table 7.3 *Summary statistics for data of Example 7.2*

Sample #i	$i=1$ (Set A)	$i=2$ (Set B)	$i=3$ (Set C)	$i=4$ (Set D)
n_i	36	39	16	16
R_i	35.69	38.76	15.69	15.83
$\hat{\kappa}_i$	114.6	161.6	48.7	86.0
$(\hat{\alpha}_i, \hat{\beta}_i)$ (Dec., Inc.)	(17.3°, −76.3°)	(10.4°, −77.2)	(11.0°, −75.1°)	(7.6°, −81.0°)
$\hat{\sigma}_i$	0.02178	0.01765	0.04941	0.03713

Figure 7.7 Uniform probability plot for the combined data sets of Figures 7.2, to test for a common mean direction. See Example 7.6.

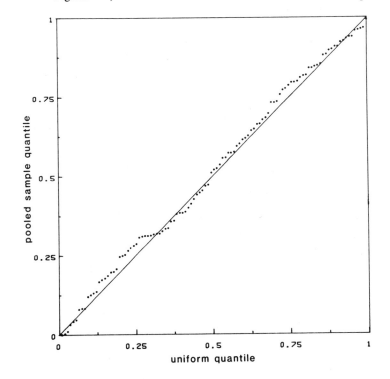

Example 7.7 Suppose we wish to make a formal comparison of the samples compared informally in Example 7.6. Two of the samples are a little small (16) to make the following calculations totally reliable, but we include them here because their dispersions are different, and because they help to illustrate a point in a later example (Example 7.10). For the pooled data, we find that $R_w = 0.99135$, $\rho_w = 0.99177$, and $F_w = 2 \times 108 \times (\rho_w - R_w) = 0.089$, a very small value of χ_6^2. We conclude that the four samples are drawn from distributions with the same mean direction.

References and footnotes See Watson (1983a, Chapter 4) for details of the test based on G_w.

7.2.2(iv) Estimation of a common mean direction of two or more rotationally symmetric distributions.

Suppose we wish to form a pooled estimate of the common mean direction (α, β) of r distributions each rotationally symmetric about this direction. As with estimating the common median direction, how we form the pooled estimate depends on how much we are prepared to assume about the similarities of the distributions. We present first some general formulae using arbitrary weights, and then consider some interesting special cases. The individual sample sizes should each be at least 25 for the confidence cone approximation to be acceptable, otherwise bootstrap methods should be used - e.g. §**3.6.1**.

Let $w_1, \ldots, w_r$ be a set of positive weights summing to unity. We can form a pooled estimate $(\hat{\alpha}_w, \hat{\beta}_w)$ of (α, β) as follows. Let

$$\hat{x}_w = \sum_{i=1}^{r} w_i \bar{R}_i \hat{x}_i, \quad \hat{y}_w = \sum_{i=1}^{r} w_i \bar{R}_i \hat{y}_i, \quad \hat{z}_w = \sum_{i=1}^{r} w_i \bar{R}_i \hat{z}_i \quad (7.12)$$

and

$$\bar{R}_w^2 = \hat{x}_w^2 + \hat{y}_w^2 + \hat{z}_w^2. \quad (7.13)$$

Then

$$\sin \hat{\alpha}_w \cos \hat{\beta}_w = \hat{x}_w / \bar{R}_w,$$
$$\sin \hat{\alpha}_w \sin \hat{\beta}_w = \hat{y}_w / \bar{R}_w, \quad (7.14)$$
$$\cos \hat{\alpha}_w = \hat{z}_w / \bar{R}_w.$$

from which $(\hat{\alpha}_w, \hat{\beta}_w)$ can be obtained using (3.4)–(3.6). Refer to Table 7.1 for definitions of individual sample quantities (e.g. R_i, $\bar{R}_i$).

The spherical variance of $(\hat{\alpha}_w, \hat{\beta}_w)$ is estimated by

$$\hat{V}_w = [1/(2\hat{\rho}_w^2)] \sum_{i=1}^{r} w_i^2 \, d_i / n_i \quad (7.15)$$

where

$$\hat{\rho}_w = \sum_{i=1}^{r} w_i \bar{R}_i \quad (7.16)$$

7.2.2(iv). Estimation of common mean direction

and d_i is defined by (5.11) for the ith sample, $i = 1, \ldots, r$, so that the estimated spherical standard error of $(\hat{\alpha}_w, \hat{\beta}_w)$ is

$$\hat{\sigma}_w = (2\hat{V}_w)^{1/2}. \tag{7.17}$$

The semi-vertical angle, q_w, of the approximate $100(1-\alpha)\%$ confidence cone for (α, β) centred on $(\hat{\alpha}_w, \hat{\beta}_w)$ is then

$$\begin{aligned} q_w &= \arcsin\sqrt{[-\log(\alpha)\hat{\sigma}_w^2 \hat{\rho}_w^2 / \bar{R}_w^2]} \\ &= \arcsin\sqrt{\left[-\log(\alpha)\left\{\sum_{i=1}^{r} w_i^2 d_i / (n_i \bar{R}_w^2)\right\}\right]} \end{aligned} \tag{7.18}$$

There are two special cases of particular interest.

(a) Suppose that the quantities $n_1^{1/2}\hat{\sigma}_1, \ldots, n_r^{1/2}\hat{\sigma}_r$ are not too different, for example, that the largest is not more than twice the smallest. Then we can use weights based on sample size only:

$$w_i = n_i / N, \quad i = 1, \ldots, r. \tag{7.19}$$

This yields the estimates

$$\hat{\rho}_w = \sum_{i=1}^{r} n_i \bar{R}_i / N \tag{7.20}$$

and

$$\hat{V}_w = \left\{\sum_{i=1}^{r} n_i d_i\right\} \bigg/ \left\{2\left(\sum_{i=1}^{r} n_i \bar{R}_i\right)^2\right\} \tag{7.21}$$

from which $\hat{\sigma}_w$ and q_w can be calculated using (7.17) and (7.18).

(b) If $n_1^{1/2}\hat{\sigma}_1, \ldots, n_r^{1/2}\hat{\sigma}_r$ differ markedly, this information can be incorporated into the weights. Set

$$w_i = C_r n_i \bar{R}_i / d_i, \quad i = 1, \ldots, r, \tag{7.22}$$

where

$$C_r = 1 \bigg/ \left\{\sum_{i=1}^{r} n_i \bar{R}_i / d_i\right\}.$$

Then

$$\hat{\rho}_w = \left\{\sum_{i=1}^{r} n_i \bar{R}_i^2 / d_i\right\} \bigg/ \left\{\sum_{i=1}^{r} n_i \bar{R}_i / d_i\right\} \tag{7.12}$$

and

$$\hat{V}_w = 1 \bigg/ \left\{2\sum_{i=1}^{r}\left[n_i \bar{R}_i^2 / d_i\right]\right\}. \tag{7.24}$$

As in (a), $\hat{\sigma}_w$ and q_w are then calculated using (7.17) and (7.18).

208   7. Analysis of two or more samples

> *Example 7.8* Continuing Examples 7.6 and 7.7, the analysis there justifies the calculation of a pooled estimate of a common mean direction (subject to the caveat about two rather small sample sizes). So we compute a pooled estimate of the common mean direction using method (b), obtaining $(\hat{\alpha}_w, \hat{\beta}_w) = $ (Dec. 12.6°, Inc. −77.2°) with standard error 0.01245 and semi-angle of 95% confidence cone $q_w = 1.2°$. (Method (a) gives essentially the same answer: (Dec. 12.7°, Inc. −77.2°) with $q_w = 1.3°$).

References and footnotes See Fisher & Lewis (1983), Watson (1983a, Chapter 4).

7.2.3   Parametric models – rotationally symmetric distributions

The procedures described in this section are based on the assumption that we have random samples from r Fisher distributions, $F\{(\alpha_i, \beta_i), \kappa_i\}$, $i = 1, \ldots, r$. The following problems are discussed:

(i) Testing whether the distributions have a common mean direction.
(ii) Estimating the common mean direction of the r distributions.
(iii) Testing whether the r mean directions lie in a common plane.
(iv) Estimating the common plane of the r mean directions.
(v) Testing whether the r distributions have a common concentration parameter.
(vi) Estimating the common concentration parameter of the r distributions.
(vii) Other problems relating to the Fisher distribution.

7.2.3(*i*) *Test of whether two or more Fisher distributions have a common mean direction.* We can compare samples from r Fisher distributions informally via a graphical method, or by a formal test. The graphical comparison is exactly as described in §7.2.2(iii). Let $(\hat{\alpha}_w, \hat{\beta}_w)$ be a suitable weighted estimate of (α, β) computed as in §7.2.2(iv), rotate all $N = n_1 + \ldots + n_r$ unit vectors so that they are measured relative to $(\hat{\alpha}_w, \hat{\beta}_w)$, and perform a uniform probability plot.

> *Example 7.9* Figure 7.8 shows the combined azimuth plot for data sets (A) and (B) in Example 7.2. There is a strong indication that the samples have different mean directions (see Example 7.10 for further analysis and comment). Such a plot can be enhanced by using different plotting symbols for each sample, if several samples are being compared (as for example, in Figure 7.7, although it was not necessary there).

For formal testing, it is essential to distinguish the case when $\kappa_1, \ldots, \kappa_r$ can be assumed equal from the case when this assumption is not reasonable. The former case presents no difficulties, whereas in the latter, the procedures are only approximate. Also, the two-sample situation requires separate treatment.

7.2.3(i). Test for common mean direction

(a) $\kappa_1, \ldots, \kappa_r$, all equal (to some unknown value κ)
Calculate the statistics
$$Z = (R_1 + \ldots + R_r)/N \qquad (7.25)$$
and
$$\bar{R} = \left[\left(\sum_{i=1}^{r} R_i \hat{x}_i\right)^2 + \left(\sum_{i=1}^{r} R_i \hat{y}_i\right)^2 + \left(\sum_{i=1}^{r} R_i \hat{z}_i\right)^2\right]^{1/2} / N \qquad (7.26)$$
($\bar{R}$ is the mean resultant length of all N unit vectors). The hypothesis that the mean directions are all the same is rejected if Z is too large, $Z > z_0$ say.

Critical values The following description paraphrases that of Stephens (1969a).

Two-sample case $(r = 2)$
$n_1 = n_2$ If $\bar{R} \leq 0.75$, use appropriate critical value z_0 corresponding to $\bar{R}$ and significance level α in Appendix A20.

If $\bar{R} > 0.75$, find the upper $100\alpha\%$ point in the $F_{2,2N-4}$ distribution from Appendix A5, f^* say, and use as critical value
$$z_0 = [\bar{R} + f^*/(N-2)]/[1 + f^*/(N-2)] \qquad (7.27)$$

Figure 7.8 Uniform probability plot for the pooled data of Figures 7.2(a) and 7.2(b). See Example 7.8.

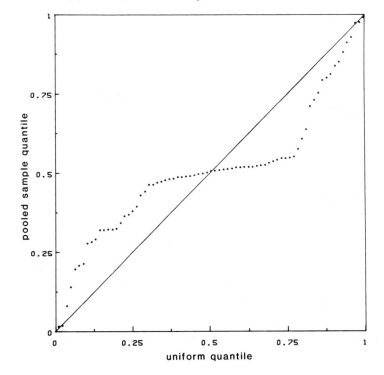

$n_1 \neq n_2$ Set $n_1/n_2 = \gamma$, assumed for convenience to be greater than 1. Appendix A21 comprises two sub-tables, for the values $\gamma = 2$ and $\gamma = 4$. If the calculated γ takes one of these value, and if $\bar{R}$ is small enough, use the sub-table to find the critical value z_0; otherwise, for $\bar{R}$ outside the range of the table, use (7.27), ignoring the value of γ.

If the calculated γ is not equal to 2 or 4, plot the z_0-values from Appendix A20 ($\gamma = 1$) and Appendix A21 ($\gamma = 2$ and $\gamma = 4$) for given $\bar{R}$ and α, and interpolate.

Several-sample case $(r > 2)$

If each $\bar{R}_i \geq 0.55$, find the upper $100\alpha\%$ point of the $F_{2r-2, 2N-2r}$ distribution from Appendix A3, f^* say, calculate

$$f = f^*(r-1)/(N-r) \tag{7.28}$$

and use as critical value

$$z_0 = (\bar{R} + f)/(1 + f) \tag{7.29}$$

For the case where $\bar{R}_i$'s are too small for this, a series of approximations have been given by Mardia (1972, pages 268-269). Alternatively, a randomisation test can be performed, as described in §3.7(ii), using the criterion

$$T = \left(\sum_{i=1}^{r} R_i - R\right) \bigg/ \left(N - \sum_{i=1}^{r} R_i\right) \tag{7.30}$$

(where $R_i = n_i \bar{R}_i$, $R = N\bar{R}$).

Suppose we have performed 1000 randomisations of the N unit vectors, and calculated the statistics, $T_1, \ldots, T_{1000}$ corresponding to each of these. Let T_0 be the value of T for our actual sample. To test at the $100(1-\alpha)\%$ level, calculate

$$m = \text{largest integer not exceeding } 1000(1-\alpha) + 1. \tag{7.31}$$

The hypothesis is rejected if

$$T_0 > T_m \tag{7.32}$$

Example 7.10 For the two samples compared graphically in Figure 7.8, $\bar{R} = 0.99266$, $Z = 0.99279$. Using (7.29), we find that this value has a significance probability of about 0.05, so the hypothesis of a common mean direction is just tenable. Note that when these two samples were two of the four compared jointly, in Example 7.7, there was no evidence of overall difference. Thus, this example highlights the point that individual differences can be swamped when a larger number of comparisons are made jointly.

(b) $\kappa_1, \ldots, \kappa_r$ not all equal

For this case, only an approximate test is available, and then only when all sample sizes are at least 25. Calculate the quantities R and R_w

7.2.3(ii). Estimation of common mean direction

as follows (using $\hat{\kappa}_i = (n_i - 1)/(n_i - R_i)$):

$$R = \hat{\kappa}_1 R_1 + \ldots + \hat{\kappa}_r R_r, \tag{7.33}$$

$$\left. \begin{array}{l} \hat{x} = \hat{\kappa}_1 R_1 \hat{x}_1 + \ldots + \hat{\kappa}_r R_r \hat{x}_r \\ \hat{y} = \hat{\kappa}_1 R_1 \hat{y}_1 + \ldots + \hat{\kappa}_r R_r \hat{y}_r \\ \hat{z} = \hat{\kappa}_1 R_1 \hat{z}_1 + \ldots + \hat{\kappa}_r R_r \hat{z}_r \end{array} \right\} \tag{7.34}$$

and

$$R_w = (\hat{x}^2 + \hat{y}^2 + \hat{z}^2)^{1/2}. \tag{7.35}$$

The test statistic is

$$g_r = 2(R - R_w) \tag{7.36}$$

and the hypothesis is rejected if g_r is too large.

Critical values Compare g_r with the upper $100(1-\alpha)\%$ point of the $\chi^2_{2(r-1)}$ distribution in Appendix A3.

References and footnotes The approximate tests when $\kappa_1, \ldots, \kappa_r$ are equal but unknown, and $R_1, \ldots, R_r$ are large, are due to Watson (1956b); see also Watson (1983b). Exact conditional tests were proposed by Watson & Williams (1956). Stephens implemented the exact tests for comparing two samples and studied in detail the approximate r-sample test. A general summary is given by Mardia (1972, Chapter 9). Watson (1983b) investigated r-sample tests for the cases when the *sample sizes* $n_1, \ldots, n_r$ are large, and proposed tests for cases in which $\kappa_1, \ldots, \kappa_r$ are unknown. Watson (1984b) investigated r-sample tests for the cases when the *concentration* parameters $\kappa_1, \ldots, \kappa_r$ are large, either known (and equal or unequal) or unknown and unequal. In Section 4 of that paper Watson shows that when $\kappa_1, \ldots, \kappa_r$ are unknown *and unequal*, the statistic he has proposed for large samples (Watson, 1983b) does not have a convenient distribution in small samples. So there is at present no satisfactory way of comparing the mean directions of small samples (each having fewer than, say, 25 observations) when the concentration parameters are unequal.

The so-called *fold test* problem in Palaeomagnetism is, in effect, a problem of comparing two mean directions (or more generally, r mean directions if there are r fold limbs) as has been noted, for example, by McFadden & Jones (1981). (McFadden & Jones also note that the original test due to McElhinny (1964) is statistically unsound and can lead to misleading conclusions).

7.2.3(ii) *Estimation of the common mean direction of two or more Fisher distributions.* The method used depends on what we are able to assume about $\kappa_1, \ldots, \kappa_r$.

(a) If $\kappa_1, \ldots, \kappa_r$ can be assumed to be equal, treat the $N = n_1 + \ldots + n_r$ observations as one sample and use the methods in §5.3.2(iv).

(b) If the largest κ_i is no more than 4 times the smallest, we use a method based on §7.2.2(iva), with weights (7.19). We have

$$\hat{x} = \sum_{i=1}^{r} n_i \bar{R}_i \hat{x}_i / N, \qquad \hat{y} = \sum_{i=1}^{r} n_i \bar{R}_i \hat{y}_i / N,$$
$$\hat{z} = \sum_{i=1}^{r} n_i \bar{R}_i \hat{z}_i / N \tag{7.37}$$

and

$$\bar{R}^2 = \hat{x}^2 + \hat{y}^2 + \hat{z}^2 \tag{7.38}$$

so that

$$\left. \begin{array}{l} \sin \hat{\alpha} \cos \hat{\beta} = \hat{x}/\bar{R} \\ \sin \hat{\alpha} \sin \hat{\beta} = \hat{y}/\bar{R} \\ \cos \hat{\alpha} = \hat{z}/\bar{R} \end{array} \right\} \tag{7.39}$$

yielding $(\hat{\alpha}, \hat{\beta})$ from (3.4)–(3.6). Further,

$$\bar{R}_w = \bar{R} \tag{7.40}$$

and

$$d_i = 2\bar{R}_i / \hat{\kappa}_i \tag{7.41}$$

so that the semi-vertical angle q_w of a $100(1-\alpha)\%$ confidence cone for (α, β) centred on $(\hat{\alpha}, \hat{\beta})$ is, from (7.18)

$$q_w = \arcsin\sqrt{[-\log(\alpha) \sum_{i=1}^{r} 2n_i \bar{R}_i / (\hat{\kappa}_i N^2 \bar{R}^2)]} \tag{7.42}$$

(c) If $\kappa_1, \ldots, \kappa_r$ differ more than is allowed in (b), we use §7.2.2(ivb). Using (7.41), (7.22) becomes

$$w_i = n_i \hat{\kappa}_i \Big/ \Big\{ \sum_{i=1}^{r} n_i \hat{\kappa}_i \Big\}, \qquad i = 1, \ldots, r, \tag{7.43}$$

(where $\hat{\kappa}_i = (n_i - 1)/(n_i - R_i)$). Thus

$$\left. \begin{array}{l} \hat{x} = \sum_{i=1}^{r} n_i \hat{\kappa}_i \bar{R}_i \hat{x}_i \Big/ \Big\{ \sum_{i=1}^{r} n_i \hat{\kappa}_i \Big\} \\ \hat{y} = \sum_{i=1}^{r} n_i \hat{\kappa}_i \bar{R}_i \hat{y}_i \Big/ \Big\{ \sum_{i=1}^{r} n_i \hat{\kappa}_i \Big\} \\ \hat{z} = \sum_{i=1}^{r} n_i \hat{\kappa}_i \bar{R}_i \hat{z}_i \Big/ \Big\{ \sum_{i=1}^{r} n_i \hat{\kappa}_i \Big\} \end{array} \right\} \tag{7.44}$$

and

$$\bar{R}^2 = \hat{x}^2 + \hat{y}^2 + \hat{z}^2 \tag{7.45}$$

whence (α, β) is estimated as in (7.45). Using (7.41) and $\bar{R}_w = \bar{R}$ (defined in (7.45)), the semi-angle q_w of a $100(1-\alpha)\%$ confidence cone for (α, β)

7.2.3(iii). Test for coplanarity of mean directions

is, from (7.18),

$$q_w = \arcsin\sqrt{\left[-\log(\alpha)\sum_{i=1}^{r}2n_i\hat{\kappa}_i\bar{R}_i\bigg/\left(\bar{R}^2\left\{\sum_{i=1}^{r}n_i\hat{\kappa}_i\right\}^2\right)\right]}. \quad (7.46)$$

Note that the individual sample sizes should be at least 25 (as mentioned in §7.2.2(iv)) if either (b) or (c) above is used, as far as calculating the confidence cone is concerned.

> **Example 7.11** We calculate the pooled mean direction from data sets (B) and (C) in Example 7.2. Strictly speaking, set (C) is rather too small ($n = 16$) for the associated large-sample confidence cone to be anything but approximate, but the samples serve to illustrate the point that we can get improved precision for our pooled estimate by allowing for differences in dispersion (i.e. in κ). The estimated values of κ are (cf. Table 7.3 and (5.25)) 161.6 for set (B) and 48.65 for (C). Using the weighted method in (c) above, $(\hat{\alpha}_w, \hat{\beta}_w) = $ (Dec. 10.5°, Inc. $-77.0°$) with semi-angle q_w of an approximate 95% confidence cone 1.7°, whereas with method (b), $(\hat{\alpha}_w, \hat{\beta}_w) = $ (Dec. 10.6°, Inc. $-76.6°$) and $q_w = 1.9°$. This exemplifies the fact that method (c) starts to have the edge over (b) when the ratios of κ's is about 4.

References and footnotes See Fisher & Lewis (1983) and Watson (1983a, Chapter 4).

7.2.3(iii) *A test of whether the mean directions of several Fisher distributions are coplanar.* We consider two cases, depending on whether the plane is specified (by its normal $\mathbf{u}_0$ say). In general, one tests first for coplanarity of the r mean directions and then, if required, for whether the normal to the plane takes a specified value.

(a) Test for coplanarity only

If we can assume that $\kappa_1,...,\kappa_r$ are of comparable size (say, the largest is not more than twice the smallest), compute the matrix

$$\mathbf{U} = \begin{pmatrix} \sum_{i=1}^{r}R_i\hat{x}_i^2 & \sum_{i=1}^{r}R_i\hat{x}_i\hat{y}_i & \sum_{i=1}^{r}R_i\hat{x}_i\hat{z}_i \\ \sum_{i=1}^{r}R_i\hat{x}_i\hat{y}_i & \sum_{i=1}^{r}R_i\hat{y}_i^2 & \sum_{i=1}^{r}R_i\hat{y}_i\hat{z}_i \\ \sum_{i=1}^{r}R_i\hat{x}_i\hat{z}_i & \sum_{i=1}^{r}R_i\hat{y}_i\hat{z}_i & \sum_{i=1}^{r}R_i\hat{z}_i^2 \end{pmatrix} \quad (7.47)$$

and find its smallest eigenvalue t_1 say (cf. §3.2.4).

The test statistic is

$$F_r = (N-r)t_1 \bigg/ \left\{(r-2)\sum_{i=1}^{r}(n_i - R_i)\right\} \quad (7.48)$$

and the hypothesis that the r mean directions are coplanar is rejected if F_r is too large.

214   7. Analysis of two or more samples

Critical values   Use the appropriate percentile of the $F_{r-2, 2N-2r}$ distribution in Appendix A5.

If we cannot assume that $\kappa_1, \ldots, \kappa_r$ are of comparable size, calculate the smallest eigenvalue, t_1^* say (cf. §3.2.4), of

$$\mathbf{U} = \begin{vmatrix} \sum_{i=1}^{r} \hat{\kappa}_i R_i \hat{x}_i^2 & \sum_{i=1}^{r} \hat{\kappa}_i R_i \hat{x}_i \hat{y}_i & \sum_{i=1}^{r} \hat{\kappa}_i R_i \hat{x}_i \hat{z}_i \\ \sum_{i=1}^{r} \hat{\kappa}_i R_i \hat{x}_i \hat{y}_i & \sum_{i=1}^{r} \hat{\kappa}_i R_i \hat{y}_i^2 & \sum_{i=1}^{r} \hat{\kappa}_i R_i \hat{y}_i \hat{z}_i \\ \sum_{i=1}^{r} \hat{\kappa}_i R_i \hat{x}_i \hat{z}_i & \sum_{i=1}^{r} \hat{\kappa}_i R_i \hat{y}_i \hat{z}_i & \sum_{i=1}^{r} \hat{\kappa}_i R_i \hat{z}_i^2 \end{vmatrix} \quad (7.49)$$

(where $\hat{\kappa}_i = (n_i - 1)/(n_i - R_i)$). The hypothesis that the r mean directions lie in a common plane is rejected if t_1^* is too large.

Critical values   Use the appropriate percentile of the χ_{r-2}^2 distribution in Appendix A3.

(b) Test for a specified normal to the common plane of several mean directions   Suppose we wish to test the hypothesis that the normal to the common plane of r mean directions of Fisher distributions takes the specified value $\mathbf{u}_0$ (having satisfied ourselves that the mean directions *are* coplanar, using the test in (a), for example).

Under the first assumption in (a) above ($\kappa_1 = \ldots = \kappa_r$), the test statistic is (cf. 7.47))

$$f_r = \left\{ (N-r) \bigg/ 2 \sum_{i=1}^{r} (n_i - R_i) \right\} \{ \mathbf{u}_0' \mathbf{U} \mathbf{u}_0 - t_1 \} \quad (7.50)$$

(where t_1 is the smallest eigenvalue of $\mathbf{U}$). The hypothesis that the common plane has normal $\mathbf{u}_0$ is rejected if f_r is too large.

Critical values   Compare f_r with the appropriate percentile of the $F_{2, 2N-2r}$ distribution in Appendix A5.

If we cannot assume that $\kappa_1, \ldots, \kappa_r$ are of comparable size, the test statistic is (cf. (7.49)).

$$f_r^* = \mathbf{u}_0' \mathbf{W} \mathbf{u}_0 - t_1^* \quad (7.51)$$

(where t_1^* is the smallest eigenvalue of $\mathbf{W}$) and the hypothesis is rejected if f_r^* is too large.

Critical values   The upper $100\alpha\%$ point of the distribution is $-2 \log \alpha$; alternatively, the significance probability of an observed f_r^* value is $\exp(-\frac{1}{2} f_r^*)$.

> **Example 7.12**   The data sets illustrated in Figure 7.9 are samples of natural remanent magnetisation in Old Red Sandstone rocks from two sites in Pembrokeshire. (The data are listed in Appendix B23). Under

7.2.3(iii). Test for coplanarity of mean directions

a certain geophysical model, the two mean directions, and the mean direction of the data set of Example 5.26 (cf. Figure 5.21) but not bedding corrected, should be approximately coplanar. Unfortunately, the same model implies that the individual sets of data will not be quite Fisherian (cf. the analysis in Example 5.27) but we utilise them for the purposes of illustration. Summary statistics for the data sets are given in Table 7.4. Since the estimates of κ are of comparable sizes, we use the unweighted analysis. The eigenvalues and eigenvectors of $\mathbf{U}$ (cf. (7.47)) are

$\hat{\mathbf{u}}_1' = (0.3058, 0.9512, -0.0404)$, $t_1 = 0.0561$,
$\hat{\mathbf{u}}_2' = (0.3711, -0.1581, -0.9150)$, $t_2 = 32.75$
$\hat{\mathbf{u}}_3' = (0.8768, -0.2649, 0.4013)$, $t_3 = 46.28$.

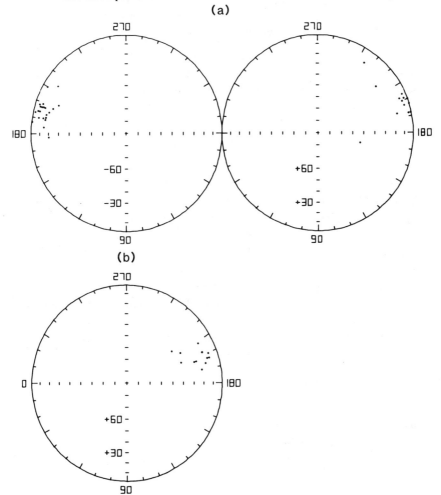

Figure 7.9 Equal-area projection of two samples of directions of natural remanent magnetisation. (a) 35 measurements (b) 13 measurements. See Example 7.12.

Table 7.4 *Summary statistics for data of Example 7.12*

Sample #i	$i=1$ (Modified results from Appendix B9)	$i=2$ (Sample A, Appendix B23)	$i=3$ (Sample B, Appendix B23)
n_i	34	35	13
$(\hat{\alpha}_i, \hat{\beta}_i)$ (Dec., Inc.)	(165.3°, 84.7°)*	(196.9°, −3.8°)	(200.4°, 23.2°)
R_i	33.03	33.33	12.72
$\hat{\kappa}_i$	34.0	20.3	42.9

* Effect of bedding correction removed

To test for goodness-of-fit of a plane to the three mean directions, we calculate F_r (from (7.48)) as 1.52 which is quite an acceptable value of the $F_{1,158}$ distribution. So we accept the hypothesis that the mean directions of the three distributions are coplanar.

References and footnotes The method is due to Watson (1960); it is also described by Mardia (1972, Chapter 9). (There is a small error of sign in the table of direction cosines given by Watson, as noted by Mardia: the third direction cosine of population B should be −0.4460).

7.2.3(iv) *Estimation of the common plane of the mean directions of several Fisher distributions.* Suppose it is reasonable to assume that the mean directions of r ($r \geq 2$) Fisher distributions are coplanar (for example, after having tested this hypothesis using §7.2.3(iii) above). Let the plane be defined by its (unknown) normal $\mathbf{u}_0$, with direction cosines $(\lambda_0, \mu_0, \nu_0)$. As with testing for coplanarity, the method of estimation depends on whether or not the concentration parameters $\kappa_1, \ldots, \kappa_r$ of the Fisher distributions are of comparable size. If they are comparable, calculate the matrix U defined by (7.47), and then its eigenvalues $t_1 \leq t_2 \leq t_3$ say, and corresponding eigenvectors $\hat{\mathbf{u}}_1, \hat{\mathbf{u}}_2, \hat{\mathbf{u}}_3$. Then the estimate of $\mathbf{u}_0$ is $\hat{\mathbf{u}}_1$, (corresponding to the smallest eigenvalue). A $100(1-\alpha)\%$ elliptical confidence cone for $\mathbf{u}_0$, centred on $\hat{\mathbf{u}}_1$, can be based on the fact that f_r (cf. (7.50)) is approximately distributed as an $F_{2,2N-2r}$-variate. The confidence cone is calculated as follows:

Step 0 Form the matrix
$$\mathbf{H} = (\hat{\mathbf{u}}_3, \hat{\mathbf{u}}_2, \hat{\mathbf{u}}_1); \qquad (7.52)$$
also, obtain the upper $100\alpha\%$ point (f_α) of the $F_{2,2N-2r}$ distribution from Appendix A5, and calculate
$$C_\alpha = 2f_\alpha \left[\sum_{i=1}^{r} (n_i - R_i) \right] \bigg/ (N-r). \qquad (7.53)$$

7.2.3(iv). Estimation of common plane of mean directions

Step 1 Set $A = t_3 - t_1$, $B = 0$, $C = t_2 - t_1$, $D = C_\alpha$, and use the method in §**3.2.5**.

If $\kappa_1, \ldots, \kappa_r$ are not of comparable size, calculate the matrix $\mathbf{W}$ (see 7.49)) instead of $\mathbf{U}$, and its eigenvalues $t_1^* \leq t_2^* \leq t_3^*$ and corresponding eigenvectors $\hat{\mathbf{u}}_1^*$, $\hat{\mathbf{u}}_2^*$ and $\hat{\mathbf{u}}_3^*$. The estimate of $\mathbf{u}_0$ is now $\hat{\mathbf{u}}_1^*$, and an approximate $100(1-\alpha)\%$ elliptical confidence cone for $\mathbf{u}_0$, centred on $\hat{\mathbf{u}}_1^*$, can be based on the fact that f_r^* (see (7.51)) has an approximate χ_2^2 distribution. To obtain the confidence cone, proceed as follows:

Step 0 Form the matrix

$$\mathbf{H} = (\hat{\mathbf{u}}_3^*, \hat{\mathbf{u}}_2^*, \hat{\mathbf{u}}_1^*) \tag{7.54}$$

and let $f_\alpha^* = -2 \log \alpha$, and

$$C_\alpha^* = f_\alpha^* \left[\sum_{i=1}^{r} (n_i - R_i) \right] \bigg/ N \tag{7.55}$$

Step 1 Set $A = t_3^* - t_1^*$, $B = 0$, $C = t_2^* - t_1^*$, $D = C_\alpha^*$ and use the method in §**3.2.5**.

> **Example 7.13** In Example 7.12, we found that the mean directions for the three samples of data analysed therein could reasonably be supposed to be drawn from Fisher distributions with coplanar mean directions. Using the unweighted analysis, and the results in Table 7.4, the estimate of $\mathbf{u}_0$ has direction cosines (0.3061, 0.9511, −0.0402), corresponding to a polar axis with one end in the direction (Dec. 287.8°, Inc. 2.3°). The major and minor semi-axes of a 95% elliptical confidence cone are 4.8° and 4.1°; the three mean directions and the confidence cone are shown in Figure 7.10, rotated so that $\hat{\mathbf{u}}_1$ is in the centre, so the perimeter of the figure is in effect the best-fitting plane.

Figure 7.10 95% elliptical confidence cone for the pole to the best-fitting plane for the three sample mean directions in Example 7.13.

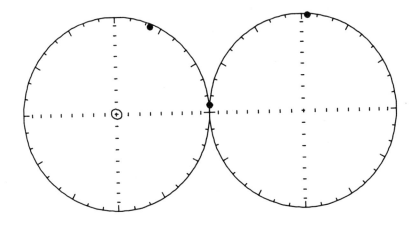

218   7. Analysis of two or more samples

References and footnotes See §7.2.3(iii) above for basic references. (Note that there is a small error in the example given by Watson (1960). The factor $\sum (N_i - R_i) = 2.475$ has been omitted from the right hand side of the first inequality on page 91, which is based on equation (2.17) of that paper. This error also appears in the same example as described by Mardia (1972, page 275)).

7.2.3(v) *A test of whether two or more Fisher distributions have a common concentration parameter.* As with the comparison of mean directions in §7.2.3(i), the comparison of concentration parameters of r distributions can be done both graphically and by a formal test. We consider the cases $r = 2$ and $r > 2$ separately.

(a) $r = 2$ To compare the samples graphically, we apply the procedure in §5.3.2(ii) separately to each sample. Let $(\phi_1', \phi_1'), \ldots, (\theta_{n_1}', \phi_{n_1}')$ be the deviations of the first sample from their sample mean direction $(\hat{\alpha}_1, \hat{\beta}_1)$, and $(\theta_{n_1+1}', \phi_{n_1+1}'), \ldots, (\theta_N', \phi_N')$ (where $N = n_1 + n_2$) the corresponding deviations of the second sample from $(\hat{\alpha}_2, \hat{\beta}_2)$. Then define

$$e_i = 1 - \cos \theta_i', \quad i = 1, \ldots, N,$$

Now perform a two-sample Q-Q plot of $(e_1, \ldots, e_{n_1})$ against $(e_{n_1+1}, \ldots, e_N)$, as described in §3.5.1. The points should plot roughly along a straight line through the origin, whose approximate slope will be κ_1/κ_2. (If they do not plot in this way, at least one of the samples is not from a Fisher distribution, or is contaminated by outliers).

For formal testing, there are several situations to be considered, depending on the sizes of $\bar{R}_1$ and $\bar{R}_2$ and whether or not the mean directions (with direction cosines $(\lambda_1, \mu_1, \nu_1)$ and $(\lambda_2, \mu_2, \nu_2)$) are known. The following discussion largely paraphrases that in Stephens (1969a). Let α be the desired significance level, and $\bar{R}$ the mean resultant length of all $n_1 + n_2$ observations.

(a1) *Mean directions known* $\bar{R}_1 > 0.65$, $\bar{R}_2 > 0.65$
Calculate

$$\begin{aligned} C_1 &= \hat{x}_1 \lambda_1 + \hat{y}_1 \mu_1 + \hat{z}_1 \nu_1, \\ C_2 &= \hat{x}_2 \lambda_2 + \hat{y}_2 \mu_2 + \hat{z}_1 \nu_2 \end{aligned} \quad (7.56)$$

and

$$R_1^* = C_1 R_1, \qquad R_2^* = C_2 R_2 \quad (7.57)$$

Let

$$Z_0^* = (n_2 - 1)(n_1 - R_1^*) / [(n_1 - 1)(n_2 - R_2^*)] \quad (7.58)$$

7.2.3(v). Test for common concentration parameter

The test statistic is

$$Z^* = \begin{cases} Z_0^* & \text{if } Z_0^* \geq 1, \\ 1/Z_0^* & \text{if } Z_0^* < 1. \end{cases} \tag{7.59}$$

The hypothesis is rejected if Z^* is too large.

Critical values Use the upper $100(\tfrac{1}{2}\alpha)$ percentile of the $F_{2n_1, 2n_2}$ distribution in Appendix A5

(a2) Mean directions known; other cases Rotate each sample so that it has its mean direction as reference direction, using the methods in §3.1.2, and consider the problem of testing the hypothesis of equal concentration parameters using the two rotated samples. To do this, use a randomisation procedure as described in §3.7(ii), calculating the quantity Z_0^* in (7.58) for each of, say, 1000 randomisation samples. Let the resulting Z_0^*-values be $Z_{(1)}^* \leq \ldots \leq Z_{(1000)}^*$, in increasing order, and let Z_0^* be the actual sample value. For a test at the $100\alpha\%$ level, let

$$m_U = \text{largest integer not exceeding } 1000(1 - \tfrac{1}{2}\alpha) + 1$$

and

$$m_L = 1000 - m_U + 1.$$

Reject the hypothesis that $\kappa_1 = \kappa_2$ if $Z_0^* < Z_{(m_L)}^*$ or $Z_0^* > Z_{(m_U)}^*$. (Alternatively, if Z_0^* is the mth largest (or the mth smallest, whichever gives a smaller value of m), the significance probability is $2 \times m/1000$ ($= m/500$)). This test can in fact be used in case **(a1)** as well; however the test given there is preferable.

(a3) Mean directions unknown $\bar{R}_1 > 0.65$, $\bar{R}_2 > 0.65$. Let

$$Z_0 = (n_2 - 1)(n_1 - R_1) / [(n_1 - 1)(n_2 - R_2)] \tag{7.60}$$

The test statistic is

$$Z = \begin{cases} Z_0 & \text{if } Z_0 \geq 1, \\ 1/Z_0 & \text{if } Z_0 < 1. \end{cases} \tag{7.61}$$

The hypothesis is rejected if Z is too large.

Critical values Use the upper $100(\tfrac{1}{2}\alpha)$ percentile of the $F_{2n_1-2, 2n_2-2}$ distribution in Appendix A5.

(a4) Mean directions equal but unknown; sample sizes equal Calculate the statistic

$$Z = |R_1 - R_2|/N, \qquad N = 2n_1 = 2n_2 \tag{7.62}$$

For given N, $\bar{R}$ and α, find z_0 from Appendix A22: if $Z > z_0$, reject the hypothesis that $\kappa_1 = \kappa_2$. If $\bar{R}$ is beyond the range of the table, use instead the approximate value

$$z_0 \simeq z_{\frac{1}{2}\alpha}(1-\bar{R})/\sqrt{N}$$

where $z_{\frac{1}{2}\alpha}$ is the upper $100(\frac{1}{2}\alpha)$ percentile of the $N(0,1)$ distribution in Appendix A1.

Mardia (1972, §9.4.2) gives some approximate methods for small values of $\bar{R}_1$ and $\bar{R}_2$.

> **Example 7.14** For data sets (A) and (C) in Example 7.2, a Q-Q colatitude plot is shown in Figure 7.11. From this, we obtain a *graphical* estimate of the ratio of the κ's as $\widehat{\kappa_1/\kappa_2} = 2.2$.
>
> To see if the ratio is significantly different from 1, we obtain from (a3) above and Table 7.3 the sample estimate $Z_0 = 114.6/48.7 = 2.36$, which is an extreme value of the $F_{70,30}$ distribution, whose upper $97\frac{1}{2}\%$ point is approximately 1.93. So the samples seem to come from Fisher distributions with differing dispersions.

(b) $r > 2$ To compare the dispersions of the r samples graphically, for each sample $i, i = 1, \ldots, r$, compute the deviations of the n_i unit vectors from their sample mean direction $(\hat{\alpha}_i, \hat{\beta}_i)$, as described in (a) above for two samples; this yields a total of N rotated points

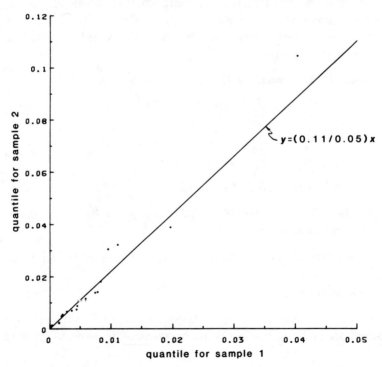

Figure 7.11 Q-Q plot to estimate the ratio of concentration parameters for the data in Figures 7.2(a) and 7.2(c). See Example 7.14.

7.2.3(v). Test for common concentration parameter

$(\theta_1', \phi_1'), \ldots, (\theta_N', \phi_N')$. Next calculate

$$e_j = 1 - \cos\theta_j', \qquad j = 1, \ldots, N$$

and perform a Q-Q plot of $e_1, \ldots, e_N$ against the unit exponential order statistics (cf. §5.3.2(ii) and §3.5.1). This plot should be approximately linear, passing through the origin, with slope estimating the common dispersion κ. If it is not, the plot can be performed separately for each sample, and the differing slopes compared informally.

For formal testing, we again consider separately the cases when the mean directions (with direction cosines $(\lambda_i, \mu_i, \nu_i), i = 1, \ldots, r$) are known, and unknown. Let α be the desired significance level.

(b1) Mean directions known $\bar{R}_i > 0.65$, $i = 1, \ldots, r$, $n_1 = \ldots = n_r$

Let R^*_{min} be the smallest among $R_1^*, \ldots, R_r^*$, and R^*_{max} the largest (where R_i^* is defined analogously to R_1^* and R_2^* in (7.57)), and denote the common sample size by n. The test statistic is

$$Z = (n - R^*_{min})/(n - R^*_{max}) \tag{7.63}$$

and the hypothesis that $\kappa_1 = \ldots = \kappa_r$ is rejected if Z is too large.

Critical values Obtain the appropriate 100α percentile from Appendix A23, using $\nu = 2n$.

(b2) Mean directions known $\bar{R}_i > 0.65$, $i = 1, \ldots, r$, **sample sizes unequal**
Calculate C_i and R_i^* analogously to (7.56) and (7.57), and then

$$Z_0^* = 2N \log\left\{ \sum_{i=1}^{r} (n_i - R_i^*)/2N \right\} \tag{7.64}$$
$$- 2 \sum_{i=1}^{r} n_i \log\{(n_i - R_i^*)/2n_i\}$$

and

$$D = 1 + \left\{ \sum_{i=1}^{r} 1/n_i - 1/N \right\} \Big/ \{3(r-1)\} \tag{7.65}$$

The test statistic is

$$Z^* = Z_0^*/D^* \tag{7.66}$$

and the hypothesis that $\kappa_1 = \ldots = \kappa_r$ is rejected if Z^* is too large.

Critical values Obtain the upper 100α percentile of the χ^2_{r-1} distribution from Appendix A3.

(b3) Mean directions known; other cases Perform a randomisation test analogous to that described in **(a2)** above, using Z^* in (7.66) as the criterion.

(b4) Mean directions unknown $\bar{R}_i > 0.65$, $i = 1, \ldots, r$, $n_1 = \ldots = n_r = n$ say

Use the statistic Z defined in (7.63), except calculated with $R_1, \ldots, R_r$ instead of $R_1^*, \ldots, R_r^*$.

Critical values Obtain the upper 100α percentile from Appendix A23, with $\nu = 2n - 2$.

(b5) Mean directions unknown $\bar{R}_i > 0.65$, $i = 1, \ldots, r$, **unequal sample sizes**

Define $m_i = n_i - 1$, $i = 1, \ldots, r$, $M = \sum_{i=1}^{r} m_i$, and calculate

$$Z_0 = 2M \log \left\{ \sum_{i=1}^{r} (n_i - R_i)/M \right\} \\ - 2 \sum_{i=1}^{r} m_i \log\{(n_i - R_i)/m_i\} \tag{7.67}$$

and

$$D = 1 + \left\{ \sum_{i=1}^{r} 1/m_i - 1/M \right\} / \{3(r-1)\}. \tag{7.68}$$

Figure 7.12 Colatitude probability plots to check whether the data in Figures 7.2(a), 7.2(c) and 7.2(d) have been drawn from distributions with the same dispersion. (a) plot based on combined data (b) plot based on data in (a) with data point corresponding to the outlying point in (a) omitted. See Example 7.15.

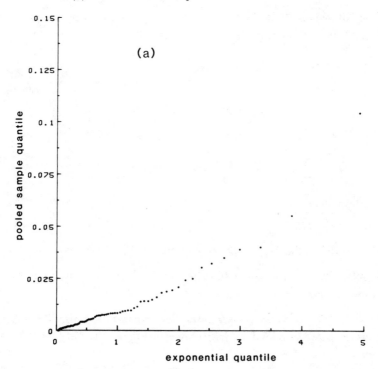

7.2.3(v). Test for common concentration parameter

The test statistic is

$$Z = Z_0/D \tag{7.69}$$

and the hypothesis is rejected if Z is too large.

Critical values Compare Z with the appropriate percentile of the χ^2_{r-1} distribution in Appendix A3.

(b6) *Mean directions unknown*; *other cases* Mardia (1972, §9.5.2) has some results for small values of .

> **Example 7.15** Consider comparing the concentration parameters for samples (A), (C) and (D) in Example 7.2. The graphical procedure described above results in the plot shown in Figure 7.12(a). The outlying point may be affecting our perception of the linearity or otherwise, so we remove it and perform a probability plot on the remaining data, yielding the picture in Figure 7.12(b); here there is some evidence of departure from linearity. For the formal test we have (cf. Table 7.3, and (7.67)-(7.69)) $Z_0 = 8.32$, $D = 1.01$ and $Z = 8.22$. This is to be assessed as a value of χ^2_2. The significance probability is $\exp(-8.22/2) \approx 0.016$, confirming the evidence from the graphical procedure that the concentration parameters of the three populations may well not have a common value.

Figure 7.12 (*cont.*)

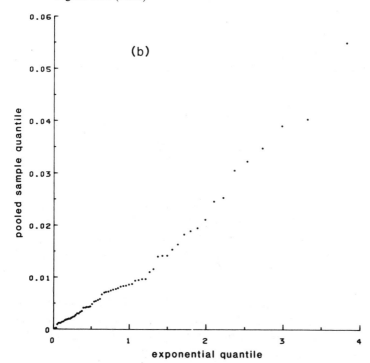

References and footnotes Approximate tests for comparing the dispersions of two or more samples were proposed by Watson & Irving (1957) and Watson & Williams (1956). These were studied in detail by Stephens (1969a), who also proposed the exact two-sample test when the mean vectors are known to be equal. Mardia (1972, Chapter 9) developed some approximate procedures for highly dispersed distributions. [There are some small errors in Stephens' paper and Mardia's book, in this connection. Stephens (1969a, page 178, §5.3) defines $d_i = N_i - 1$, and $Z = CZ_1$; these should be $d_i = 2(N_i - 1)$ and $Z = Z_1/C$ respectively. The term "$+\sum_{i=1}^{r} d_i \ln d_i$" should appear at the end of the right hand side of equation (20). The last sentence of the subsection ("If $Z_1 > z, \ldots$") should be ignored. At the foot of the same page, in subsection 5.5, d_i should become $2N_i$, not N_i. In Mardia (1972, page 270, §9.5.2), in case III, d should be defined as $(\sum \nu_i^{-1} - \nu^{-1})/\{3(q-1)\}$, and the degrees of freedom of χ^2 in (9.5.11) should be $q-1$, not $2q-2$. There are numerical errors in Mardia's Example 9.11 as a result.]

7.2.3(vi) *Estimation of the common concentration parameter of two or more Fisher distributions.* A suitable pooled estimate of the common concentration parameter κ based on the r samples is

$$\hat{\kappa} = (N-r) \Big/ \Big(N - \sum_{i=1}^{r} R_i\Big) \tag{7.70}$$

A $100(1-\alpha)\%$ confidence interval for κ is then obtained using methods analogous to those in §5.3.2(iv): the upper and lower limits for κ are given, respectively, by

and
$$\kappa_U = \tfrac{1}{2}\chi^2_{2N-2r}(\tfrac{1}{2}\alpha) \Big/ \Big(N - \sum_{i=1}^{r} R_i\Big)$$
$$\kappa_L = \tfrac{1}{2}\chi^2_{2N-2r}(1-\tfrac{1}{2}\alpha) \Big/ \Big(N - \sum_{i=1}^{r} R_i\Big) \tag{7.71}$$

Example 7.16 For the data sets (A), (B) and (D) of Example 7.2 (Appendix B12), the homogeneity test in §7.2.3(v) above can be used to show that the sample concentration parameters do not differ significantly. Using the results in Table 7.3, we obtain the pooled estimate $\hat{\kappa} = 123.0$, with 95% confidence interval (98.6, 156.6).

7.2.3(vii) *Other problems of comparing two or more Fisher samples.* Morris and Laycock (1974) have investigated modifications of discriminant analysis methods for two or three Fisher samples. An initial suggestion about discrimination was made by Watson (1956a).

7.3 Analysis of two or more samples of bipolar axial data

We retain the appropriate notation, introduced in §7.2.1 and Table 7.1 to describe the data, and various summary statistics calculated for each sample.

7.3.1 Simple procedures

Simple procedures are available when the r samples may be assumed to be drawn from symmetric bipolar axial distributions (that is, distributions rotationally invariant about their principal axes such as the Watson bipolar distribution) and with comparable dispersions. The problems considered for such data are:

(i) Testing whether the r distributions have the same principal axis.
(ii) Estimating the common principal axis of their distributions.

7.3.1(i) *A test of whether two or more symmetric bipolar distributions with comparable dispersions have the same principal axis.* An informal check on whether the samples have a common principal axis can be made by estimating the supposed common axis as in §7.3.1(ii) below, computing the azimuths of all $n_1 + \ldots + n_r$ data points relative to this axis, and performing a uniform Q-Q plot as described in §7.2.2(iii).

For a formal test, define the pooled orientation matrix $\mathbf{T}$ by

$$\mathbf{T} = \sum_{i=1}^{r} \mathbf{T}_i \tag{7.72}$$

and denote its eigenvalues and eigenvectors by $\hat{\tau}_1 \leq \hat{\tau}_2 \leq \hat{\tau}_3$ and $\hat{\mathbf{u}}_1, \hat{\mathbf{u}}_2, \hat{\mathbf{u}}_3$ respectively. Let Γ be the fourth cosine moment computed for the pooled sample of N data points as in §6.3.1(v). Define

$$g = 2(\bar{\tau}_3 - \Gamma)(3\bar{\tau}_3 - 1), \quad \bar{\tau}_3 = \hat{\tau}_3 / N. \tag{7.73}$$

Also, calculate the largest eigenvalue $\hat{\tau}_{3i}$ of the ith orientation matrix $\mathbf{T}_i$, for each $i = 1, \ldots, r$. The test statistic is

$$N_r = \left\{ \sum_{i=1}^{r} \hat{\tau}_{3i} - \hat{\tau}_3 \right\} \Big/ g \tag{7.74}$$

and the hypothesis of a common principal axis is rejected if N_r is too large.

Critical values Compare N_r with the upper $100\alpha\%$ point of the χ^2_{2r-2} distribution in Appendix A3.

> **Example 7.17** Figure 7.3 shows two samples of L_0^1 axes (intersections between cleavage and bedding planes of F_1 folds) in Ordovician turbidites. The samples may reasonably be taken to be from rotationally symmetric distributions (using §6.3.1(iv)), and to have comparable dispersions. As an initial check on whether the underlying distributions

have the same principal axis, we perform the graphical procedures outlined above. The estimated principal axis for the pooled data has direction cosines (0.9782, −0.2062, 0.0260) (with largest eigenvalue $\hat{\tau}_3 = 68.68$ and fourth cosine moment $\Gamma = 0.8159$). The azimuth plot in Figure 7.13 indicates the possibility that the principal axes are different, because all points lie below the line $x = y$. To test formally, we require in addition the eigenvalues $\hat{\tau}_{3,1} = 32.68$ and $\hat{\tau}_{3,2} = 37.86$. Then $g = 0.0908$ and $N_r = 20.5$, which is a very large value for χ_2^2. So we conclude that the principal axes differ.

References and footnotes The test based on N_r is due to Watson (1983c).

7.3.1(*ii*) *Estimation of the common principal axis of two or more symmetric bipolar distributions with comparable dispersions.* To form a pooled estimate of the common principal axis $\mathbf{u}_3$ of r axial bipolar distributions each rotationally symmetric and approximately equally dispersed about this axis, calculate the pooled orientation matrix **T** and its associated eigenvalues and eigenvectors as defined in **(i)** above, and the values Γ

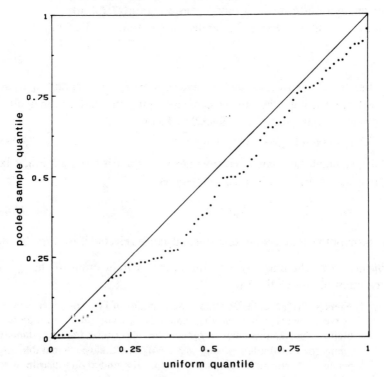

Figure 7.13 Uniform probability plot to check whether the data sets in Figure 7.3 come from rotationally symmetric bipolar distributions with a common principal axis. See Example 7.17.

7.4. Analysis of two or more samples of girdle axis data

and g as defined in §**7.3.1(i)**. The eigenvector $\hat{\mathbf{u}}_3$ is the pooled estimate of $\mathbf{u}_3$. For a large-sample $100(1-\alpha)\%$ confidence cone for $\mathbf{u}_3$, calculate the estimated standard error $\hat{\sigma}$ of $\hat{\mathbf{u}}_3$ by

$$\hat{\sigma} = \sqrt{(g/2N)}. \tag{7.75}$$

The confidence cone is centred on $\hat{\mathbf{u}}_3$, with semi-vertical angle

$$q = \arcsin(\chi_\alpha^{1/2}\hat{\sigma}), \tag{7.76}$$

χ_α being the upper $100\alpha\%$ point of the χ^2_{2r-2} distribution in Appendix A3. (For sample sizes <25, use bootstrap methods to calculate the confidence region - e.g. §**3.6.1**.)

Example 7.18 Figure 7.14 shows two samples of L_0^1 axes collected from the same sub-domain. (The data are listed in Appendix B24). The second sample has some degree of asymmetry (as judged by the test in §**6.3.1(iv)**) but not to a great degree. Summary data are given in Table 7.5. Using the methods in §**7.3.1(i)** above, it seems reasonable to assume that the samples have been drawn from distributions with a common principal axis. The estimate of this axis is (plunge 3.0°, azimuth 220.3°), with semi-angle of an approximate 95% confidence cone 3.6°.

References and footnotes See Watson (1983c).

7.4 Analysis of two or more samples of girdle axial data

As in §**7.3**, the notation of §**7.2.1** and Table 7.1 is retained. Since the methods of this section are simple modifications of those in §**7.3**, we confine ourselves to noting these modifications.

Figure 7.14 Equal-area projection of two samples of L_0^1 axes. The coordinate system is (Plunge, Plunge azimuth). (a) 25 measurements (b) 25 measurements. See Example 7.18.

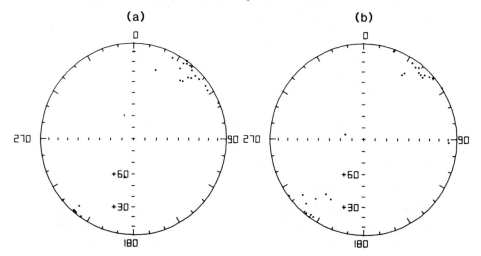

Table 7.5 *Summary statistics for data of Example 7.18*

Sample #i	$i=1$ (Sample A, Appendix B24)	$i=2$ (Sample B, Appendix B24)	Pooled data
n_i	25	25	$N=50$
$\hat{\mathbf{u}}_{3i}$	(5.8°, 219.5°)	(0.3°, 41.1°)	$\hat{\mathbf{u}}_3 = (3.0°, 220.3°)$
$\bar{\tau}_{3i}$	0.9388	0.8730	$\bar{\tau}_3 = 0.9034$
Γ_i	0.8894	0.8095	$\Gamma = 0.8466$

7.4.1 Simple procedures

It is assumed that the samples have been drawn from r girdle (equatorial) distributions rotationally symmetric about their polar axes (for example, Watson girdle distributions), and that the distributions have comparable dispersions.

7.4.1(i) *A test of whether two or more symmetric girdle distributions have a common polar axis.* The graphical test in §7.3.1(i) can be used here, with the single modification that the azimuths are calculated with respect to a pooled estimate $\hat{\mathbf{u}}_1$ (see §7.4.1(ii) below) of $\mathbf{u}_1$. For a formal test, the analogue of that in §7.3.1(i) requires replacing Γ, g and N_r by (respectively)

$$\Gamma^* = (1/N) \sum_{i=1}^{r} (\hat{\lambda}\hat{x}_i + \hat{\mu}\hat{y}_i + \hat{\nu}\hat{z}_i)^4, \qquad (7.77)$$

$(\hat{\lambda}, \hat{\mu}, \hat{\nu})$ being the direction cosines of $\hat{\mathbf{u}}_1$ (the eigenvector corresponding to the smallest eigenvalue $\hat{\tau}_1$ of **T** in (7.72));

$$g^* = 2(\bar{\tau}_1 - \Gamma^*)/(1 - 3\bar{\tau}_1), \quad \bar{\tau}_1 = \hat{\tau}_1/N; \qquad (7.78)$$

and the test statistic is

$$N_r^* = \left\{ \sum_{i=1}^{r} \hat{\tau}_{1i} - \hat{\tau}_1 \right\} \bigg/ g^*. \qquad (7.79)$$

The hypothesis that the distributions have the same polar axis is rejected if N_r^* is too large.

Critical values Compare N_r^* with the upper $100\alpha\%$ point of the χ^2_{2r-2} distribution in Appendix A3.

References and footnotes See Watson (1983c). Watson (1965) proposed a two-sample test under the further assumption that each distribution was a Watson girdle distribution (with the same concentration).

7.4.1(ii) *Estimation of the common polar axis of two or more symmetric girdle distributions.* The changes required to §7.3.1.(ii) are: Γ and g are replaced respectively by Γ^* (see (7.77)) and g^* (see (7.78)), with corresponding change to the estimated standard error $\hat{\sigma}$, and the $100(1-\alpha)\%$ confidence cone with semi-vertical angle given by (7.75) is here centred on $\hat{\mathbf{u}}_1$.

>*Example* **7.19** The data (set B19) shown in Figure 6.24 are measurements of 72 poles to bedding planes of F_1 folds, taken in the same domain as the poles to cleavage surfaces (set B12) depicted in Figure 6.2. We calculate a pooled estimate and confidence cone for the common pole to the plane, since each sample is approximately rotationally symmetric. For the combined sample, $\bar{\tau}_1 = 0.0189$, $\Gamma^* = 0.0010$, and the estimate of the pole is (Plunge 3.5°, azimuth 39.2°), based on all 136 observations. Also $g^* = 0.0399$, and the semi-vertical angle of an approximate 95% confidence cone is 1.7°.

References and footnotes See Watson (1983c).

8
Correlation, regression, and temporal/spatial analysis

8.1 Introduction

In this chapter, we shall be concerned with the relationship between a spherical random variable **X** (a random unit vector or axis in three dimensions) and other variables, which may be random or fixed. Examples of such data are:

Example **8.1** Successive measurements of magnetic remanence in a rock sample, after various stages of partial thermal demagnetisation to investigate the blocking temperature spectrum of components of magnetisation. Figure 8.1 shows such data from measurements of the same two successive demagnetisation steps, in each of 62 specimens. (The

Figure 8.1 Equal-area projection of 62 measurements of magnetic remanence in 62 rock specimens (a) after thermal demagnetisation to 350°C (b) after thermal demagnetisation to 400°C. The coordinate system is (Declination, Inclination). See Example 8.1.

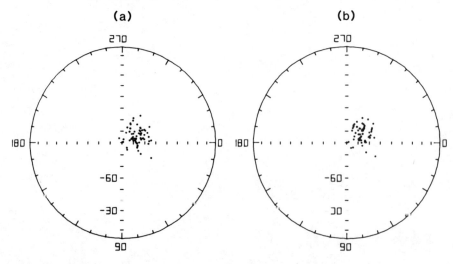

8.2.1. Correlation: Introduction

data are listed in Appendix B8; corresponding measurements in the two data sets may be identified in the data listing, but a convenient correspondence cannot be shown in the figure).

***Example* 8.2** Measurements of wind direction on consecutive days near Singleton, New South Wales, during March, 1981. (They are listed in Appendix B25). The question of interest is whether there is any serial association (i.e. from one day to the next).

***Example* 8.3** A time-ordered sequence of palaeomagnetic pole positions is obtained to gain information about the Apparent Polar Wander Path for a given continent. Questions of interest include: how to fit a smooth curve to the data, together with some estimate of the variability of the fitted curve; and how to compare corresponding data sequences for two different continents.

***Example* 8.4** Samples of fracture plane data are collected at a number of localities throughout a region, yielding measurements of unit axes with spatial variation. A common problem with such data is to obtain a smoothed description of the spatial variation (in particular, by interpolating the axial directions at points on a square grid), with a view to partitioning the region into domains of homogeneous behaviour.

Section **8.2** is concerned with measuring the *association* (specifically, the *correlation*) of **X** with another random variable **Y** which may itself be spherical, or may be circular or linear or something else. There are methods available for dealing with these problems quite adequately.

Sections **8.3** and **8.4** are concerned with relating the behaviour of **X** to other variables such as time or spatial location. This is currently the subject of much research activity, and some relevant papers are referenced in the course of the chapter. However, generally speaking, it is too early to identify a general line of attack on problems such as fitting a curve to a time sequence of unit vectors, smoothing a set of spatially distributed vectors or axes, or comparing two time sequences of unit vectors. Some techniques developed (successfully) for a specific problem are not adaptable to an altered version of the problem. Other techniques offer greater generality, however there is no practical experience yet available to indicate how well they work. Accordingly, we have limited the discussion, especially in §**8.4**, to a short survey of the current status of the subject, and have chosen not to present any analysis of data.

8.2   Correlation

8.2.1   *Introduction*

There are at present only isolated examples in the literature of problems involving the correlation between a spherical random variable (vector or axis) and another variate. Here, we consider separately the problems of correlation between two random unit vectors (§**8.2.2**), or between two random unit axes (§**8.2.3**), and then describe a general measure of correlation between a spherical random variable and another variate (§**8.2.4**). For the correlation between two random unit vectors, it is possible to distinguish positive from negative correlation, in contrast to the correlation between axes. This leads to a different type of statistic for the vector case from that in the axial case.

8.2.2 Correlation of two random unit vectors

Let $(\mathbf{X}_1, \mathbf{X}_1^*), \ldots, (\mathbf{X}_n, \mathbf{X}_n^*)$ be a random sample of bivariate measurements of two random unit vectors $\mathbf{X}$ and $\mathbf{X}^*$, where $\mathbf{X}_i$ has direction cosines (x_i, y_i, z_i) and $\mathbf{X}_i^*$ has direction cosines (x_i^*, y_i^*, z_i^*), $i = 1, \ldots, n$. We can estimate the association between $\mathbf{X}$ and $\mathbf{X}^*$ in terms of how well the $\mathbf{X}_i^*$'s can be matched with the $\mathbf{X}_i$'s by performing an orthogonal transformation of the $\mathbf{X}_i^*$'s. If the match-up is best for an orthogonal transformation $\mathbf{H}\mathbf{X}_i^*$ where $\mathbf{H}$ is a *rotation* ($\det(\mathbf{H}) = 1$) then the association is positive. If the "best" match-up of each $\mathbf{X}_i^*$ with its corresponding $\mathbf{X}_i$ is obtained for $\mathbf{H}$ a *reflection* matrix ($\det(\mathbf{H}) = -1$) then the association is negative. A suitable way of estimating the extent of positive or negative correlation between $\mathbf{X}$ and $\mathbf{X}^*$ is as follows. Calculate the quantities

$$S_{xx^*} = \det\left\{\sum_{i=1}^n \mathbf{X}_i \mathbf{X}_i^{*\prime}\right\} = \det\begin{pmatrix} \sum x_i x_i^* & \sum y_i x_i^* & \sum z_i x_i^* \\ \sum x_i y_i^* & \sum y_i y_i^* & \sum z_i y_i^* \\ \sum x_i z_i^* & \sum y_i z_i^* & \sum z_i z_i^* \end{pmatrix} \quad (8.1)$$

and

$$S_{xx} = \det\left\{\sum_{i=1}^n \mathbf{X}_i \mathbf{X}_i'\right\}, \quad S_{x^*x^*} = \det\left\{\sum_{i=1}^n \mathbf{X}_i^* \mathbf{X}_i^{*\prime}\right\}. \quad (8.2)$$

Then a suitable sample correlation coefficient is given by

$$\hat{\rho}_V = S_{xx^*}/(S_{xx} S_{x^*x^*})^{1/2}, \quad -1 \leq \hat{\rho}_V \leq 1. \quad (8.3)$$

$\hat{\rho}_V$ is an estimate of a population quantity ρ_V which has an intuitive interpretation (Fisher & Lee, 1986) as a measure of correlation. $\rho_V = +1, -1$ correspond respectively to complete positive, and complete negative, correlation.

There are three alternative hypotheses with which we may wish to compare the null hypothesis that the underlying correlation ρ_V is zero. These are (i) $\rho_V \neq 0$, (ii) $\rho_V < 0$, and (iii) $\rho_V > 0$. For any significance level $100\beta\%$, let ρ_β be the upper critical $100\beta\%$ point, and $\rho_{1-\beta}$ the lower critical $100\beta\%$ point. To test at level $100\alpha\%$, we reject the hypothesis $\rho_V = 0$ in favour of $\rho_V \neq 0$ if either $\hat{\rho}_V < \rho_{1-\alpha/2}$ or $\hat{\rho}_V > \rho_{\alpha/2}$; we reject $\rho_V = 0$ in favour of $\rho_V < 0$ if $\hat{\rho}_V < \rho_{1-\alpha}$; and we reject $\rho_V = 0$ in favour of $\rho_V > 0$ if $\hat{\rho}_V > \rho_\alpha$.

8.2.2. Correlation of two random unit vectors

Critical values In general, the significance level p_β has to be obtained using a permutation test of the type described in §**3.7(ii)**. However, there are two important special cases.

If the marginal distributions of $\mathbf{X}$ and $\mathbf{X}^*$ are unimodal and rotationally symmetric about their mean directions and $\mathbf{X}$ is distributed independently of $\mathbf{X}^*$, then the large-sample distribution ($n \geq 25$) of $n\hat{\rho}_V$ is of a form known as double-exponential, with probability density $\frac{1}{2}\exp(-|x|)$. For given β, $0 < \beta < \frac{1}{2}$, $\rho_{1-\beta} = \log(2\beta)/n$, $\rho_\beta = -\log(2\beta)/n$. An example of this would be if both marginal distributions were Fisher distributions (cf. §**4.3**).

If both marginal distributions are uniform, then, in large samples ($n \geq 30$), the quantity $\hat{\rho}^* = n^{\frac{3}{2}}\hat{\rho}_V + 0.691/\sqrt{n} - 9.39/n$ should be computed and compared with the critical values in Appendix A26.

> **Example 8.5** For the 62 pairs of points of Example 8.1, $\hat{\rho}_V = 0.8273$. Consider testing $\rho_V = 0$ against $\rho_V \neq 0$. Using a randomisation test based on 10,000 random permutations yields (for this *specific* set of random permutations) the fact that $\hat{\rho}_V$ is the largest, corresponding to a significance probability of 0.00005 (the test being two-sided). For this example, the marginal samples conform reasonably well to the Fisherian model (using the goodness-of-fit techniques in §**5.3.2(ii)**), so we can also test $n\hat{\rho}_V$ as a double-exponential variate. The significance probability is $\exp(-n|\hat{\rho}_V|) = \exp(-51.3)$, again, a very small quantity. The hypothesis of zero correlation is clearly untenable.

In the event that the sample correlation is significantly different from zero, we can compute an approximate large-sample confidence interval for the unknown ρ_V using the jackknife method to estimate the variance of $\hat{\rho}_V$ (see §**3.7(i)** for a general description). Calculate the n correlations $\hat{\rho}_i^*$, $i = 1, \ldots, n$, where $\hat{\rho}_i^*$ is computed from $(\mathbf{X}_1, \mathbf{X}_1^*), \ldots, (\mathbf{X}_n, \mathbf{X}_n^*)$ with $(\mathbf{X}_i, \mathbf{X}_n^*)$ omitted, and then the pseudo-values $\hat{\rho}_i = n\hat{\rho}_V - (n-1)\hat{\rho}_i^*$, $i = 1, \ldots, n$. The jackknife estimate of ρ_V (which is less biased than the estimate $\hat{\rho}_V$) is

$$\hat{\rho}_{VJ} = (1/n) \sum_{i=1}^{n} \hat{\rho}_i \qquad (8.4)$$

with estimated standard error s_{VJ}, where

$$s_{VJ} = \left\{ [n(n-1)]^{-1} \sum_{i=1}^{n} (\hat{\rho}_i - \hat{\rho}_{VJ})^2 \right\}^{1/2} \qquad (8.5)$$

A large-sample $100(1-\alpha)\%$ confidence interval for ρ_V is then given by

$$(\hat{\rho}_{VJ} - z_{\frac{1}{2}\alpha}s_{VJ}, \hat{\rho}_{VJ} + z_{\frac{1}{2}\alpha}s_{VJ}) \qquad (8.6)$$

where $z_{\frac{1}{2}\alpha}$ is the upper $100(\frac{1}{2}\alpha)\%$ point of the standard normal distribution.

Example 8.6 For the data analysed in Example 8.5, the jackknife estimate of ρ_V is $\hat{\rho}_{VJ} = 0.83$, with estimated standard error $s_{VJ} = 0.052$. An approximate 95% confidence interval for ρ_V is $(0.73, 0.94)$.

To estimate the rotation or reflection matrix **H**, suppose first that $\hat{\rho}_V > 0$. Calculate the quantities

$$\mathbf{A} = (1/n) \sum_{i=1}^{n} \mathbf{X}_i \mathbf{X}_i^{*\prime}, \qquad \mathbf{B} = \mathbf{A}'\mathbf{A} \tag{8.7}$$

and let $\lambda_1 \leq \lambda_2 \leq \lambda_3$ and $\mathbf{u}_1, \mathbf{u}_2, \mathbf{u}_3$ be the eigenvalues and corresponding eigenvectors of **B** computed as in §3.2.4 for the orientation matrix **T**. Define

$$\mathbf{C} = \begin{pmatrix} \mathbf{u}_3' \\ \mathbf{u}_2' \\ \mathbf{u}_1' \end{pmatrix}, \quad \mathbf{Z} = \begin{pmatrix} \lambda_3^{\frac{1}{2}} & 0 & 0 \\ 0 & \lambda_2^{\frac{1}{2}} & 0 \\ 0 & 0 & \lambda_1^{\frac{1}{2}} \end{pmatrix} \tag{8.8}$$

and calculate

$$\mathbf{D} = \mathbf{C}\mathbf{Z}\mathbf{C}' \tag{8.9}$$

Then the estimate of **H** is given in transposed form by

$$\hat{\mathbf{H}}' = \mathbf{D}\mathbf{A}^{-1} \tag{8.10}$$

If $\hat{\rho}_V < 0$, define **Z** as in (8.8) above, except with $-\lambda_1^{\frac{1}{2}}$ instead of $\lambda_1^{\frac{1}{2}}$, and then proceed as for $\hat{\rho}_V > 0$.

The variability of $\hat{\mathbf{H}}$ can be assessed using bootstrap methods to obtain $\hat{\mathbf{H}}_1, \ldots, \hat{\mathbf{H}}_N$ say. Suppose $\mathbf{H}_i$ has rows $\hat{\mathbf{h}}_{1i}', \hat{\mathbf{h}}_{2i}', \hat{\mathbf{h}}_{3i}'$ (three orthogonal vectors). We can plot these three points, for each $i = 1, \ldots, N$, to obtain a visual assessment of the variability.

Example 8.7 Continuing Example 8.5, the estimated rotation matrix is

$$\hat{\mathbf{H}} = \begin{pmatrix} -0.9982 & -0.0590 & 0.0103 \\ -0.0592 & 0.9980 & 0.0203 \\ 0.0091 & 0.0209 & 0.9997 \end{pmatrix}$$

Using 199 bootstrap samples from the original data, 200 estimates of **H** (including the original estimate) were obtained, to gain some idea of the variability of $\hat{\mathbf{H}}$. A plot of the 200 sets of orthogonal axes $\{\hat{\mathbf{h}}_{1i}, \hat{\mathbf{h}}_{2i}, \hat{\mathbf{h}}_{3i}\}$ is shown in Figure 8.2. It reveals that the error associated with the estimated rotation matrix $\hat{\mathbf{H}}$ is very small, with the bootstrap estimates for each axis plotting in small groups about the axes for $\hat{\mathbf{H}}$. (The axes have all been plotted in the lower hemisphere). In fact, since the three sets of estimates overlap the axes $(1, 0, 0)$, $(0, 1, 0)$, $(0, 0, 1)$ comprising the columns of the unit matrix **I**, it is reasonable to conclude that no rotation is required to match up each $\mathbf{X}_i$ with its corresponding $\mathbf{X}_i^*$ as well as possible.

References and footnotes The correlation coefficient $\hat{\rho}_V$ is due to Fisher & Lee (1986), and is one of several in the literature: see e.g. Stephens

8.2.3. Correlation of two random unit axes

(1979); Jupp & Mardia (1980), also §**8.2.3** below; and Saw (1983). Estimation of a rotation is due to Mackenzie (1957): see also Stephens (*op. cit.*). Moran (1976) considered estimation of a rotation in the situation in which $x_1, \ldots, x_n$ are known unit vectors to each of which an unknown rotation is applied, so that the resulting unit vectors $X_1, \ldots, X_n$ are measured with error.

8.2.3 Correlation of two random unit axes

Let $(X_1, X_1^*), \ldots, (X_n, X_n^*)$ be a random sample of bivariate measurements of two random unit axes X and X^*, where X_i has direction cosines (x_i, y_i, z_i) and X_i^* has direction cosines (x_i^*, y_i^*, z_i^*), $i = 1, \ldots, n$. We can estimate association between X and X^* in terms of how well the X_i^*'s can be matched to their corresponding X_i's by performing a rotation of the set $(X_1^*, \ldots, X_n^*)$. Calculate the 3×3 matrices

$$\hat{\Sigma}_{11} = (1/n) \sum_{i=1}^{n} X_i X_i', \quad \hat{\Sigma}_{12} = (1/n) \sum_{i=1}^{n} X_i X_i^{*\prime},$$

$$\hat{\Sigma}_{22} = (1/n) \sum_{i=1}^{n} X_i^* X_i^{*\prime} \tag{8.11}$$

and then $\hat{\Sigma}_{11}^{-1}$ and $\hat{\Sigma}_{22}^{-1}$. Now calculate the matrix product

$$\hat{\Sigma} = \hat{\Sigma}_{11}^{-1} \hat{\Sigma}_{12} \hat{\Sigma}_{22}^{-1} \hat{\Sigma}_{12}' \tag{8.12}$$

$$= \begin{pmatrix} \hat{\sigma}_{11} & \hat{\sigma}_{12} & \hat{\sigma}_{13} \\ \hat{\sigma}_{12} & \hat{\sigma}_{22} & \hat{\sigma}_{23} \\ \hat{\sigma}_{13} & \hat{\sigma}_{23} & \hat{\sigma}_{33} \end{pmatrix} \quad \text{say.} \tag{8.13}$$

Figure 8.2 200 bootstrap estimates of the rotation matrix H for Example 8.7, (represented by plotting the 200 sets of orthogonal axes $(\hat{h}_{1i}, \hat{h}_{2i}, \hat{h}_{3i})$) to assess the variability of $\hat{H}$. (The axes are plotted in the lower hemisphere).

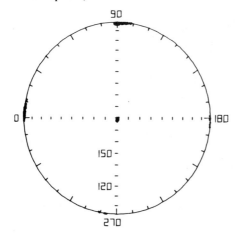

236   8. Correlation, regression, and temporal/spatial analysis

Then the estimate of correlation is

$$\hat{\rho}_A = \tfrac{1}{3}(\hat{\sigma}_{11} + \hat{\sigma}_{22} + \hat{\sigma}_{33}). \tag{8.14}$$

The hypothesis that $\mathbf{X}$ and $\mathbf{X}^*$ are uncorrelated is rejected if $\hat{\rho}_A$ is too large.

Critical values

sample size $n \leq 25$ Use a permutation test, as described in §**3.7(ii)** and §**8.2.2**, (Note that the matrices $\hat{\mathbf{\Sigma}}_{11}$ and $\hat{\mathbf{\Sigma}}_{22}$ do not require re-calculation each time).

sample size $n > 25$ Compare $3n\hat{\rho}_A$ with the upper $100\alpha\%$ point of the χ_9^2 distribution in Appendix A4.

If the hypothesis that $\mathbf{X}$ and $\mathbf{X}^*$ are uncorrelated is rejected, we may wish to estimate the underlying correlation ρ_A. This is done exactly as for vector correlation (§**8.2.2**) using jackknife methods (cf. §**3.7(i)** to obtain a jackknife estimate $\hat{\rho}_{AJ}$ (cf. (8.4)) and estimated standard error s_{AJ} (cf. (8.5)), and a large-sample confidence interval for ρ_A (cf. (8.6)). The maximum value of $\hat{\rho}_A$ is 1, which is attained if the set of $\mathbf{X}_i$'s can be obtained by a simple rotation of the set of $\mathbf{X}_i^*$'s (each $\mathbf{X}_i^*$ rotating exactly onto its corresponding $\mathbf{X}_i$).

Example 8.8 Figure 8.3 shows measurements of the longest axis and shortest axis orientations for each of 101 of tabular stones measured on a slope at Windy Hills, Scotland. (The data are listed in Appendix B26.) We are interested in whether there is any correlation between the

Figure 8.3 Equal-area projection of the longest and shortest axes of each of 101 tabular stones. (a) longest axis (b) shortest axis. The coordinate system is (Plunge, Plunge azimuth). See Example 8.8.

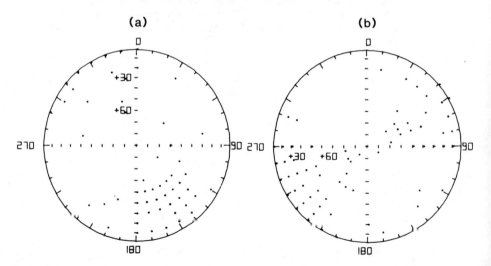

8.2.4. Correlation: other cases

orientations of the longest axis and the shortest axis. Figure 8.4 shows each data set rotated to its principal axis. (Note that rotating either data set will not change the correlation with the other set.) The regular spacings between the data points are due to their being recorded to the nearest 10°.

From (8.14) we obtain the estimated correlation as $\hat{\rho}_A = 0.16$, so that $3n\hat{\rho}_A \approx 50$, which is much too high a value to be credible for a χ_9^2 variate. We conclude that there is some correlation between the axes. The jackknifed estimate of ρ_A is $\hat{\rho}_{AJ} = 0.14$, with estimated standard error $s_{AJ} = 0.037$. An approximate 95% confidence interval for ρ_A is then (0.07, 0.22).

References and footnotes This is a special case of the general correlation coefficient of Jupp & Mardia (1980), described in §**8.2.3**.

8.2.4 *Correlation between a random unit vector or axis and another random variable*

Suppose $(X_1, Y_1), \ldots, (X_n, Y_n)$ is a random sample of bivariate measurements of two random quantities X and Y, where X is a random unit vector or axis, and Y is another random quantity of interest, not necessarily a spherical random variable. In general, we shall suppose that Y has p components. For example, Y may be a single linear variable (i.e. $p = 1$) such as height; or a bivariate variable (i.e. $p = 2$) such as height and age, or a different type of spherical variable so that if X is a unit vector, Y is a unit axis, or *vice versa*. X, of course, has three components, namely its direction cosines.

Figure 8.4 Data sets in Figure 8.3 rotated to centralise their respective principal axes. (a) longest axis (b) shortest axis.

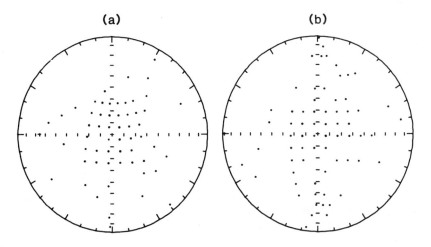

To test whether **X** and **Y** are correlated, calculate the matrices

$$\hat{\Sigma}_{11} = \sum_{i=1}^{n} X_i X_i' \quad \text{(a } 3 \times 3 \text{ matrix)}$$
$$\hat{\Sigma}_{12} = \sum_{i=1}^{n} X_i Y_i' \quad \text{(a } 3 \times p \text{ matrix)} \qquad (8.15)$$
$$\hat{\Sigma}_{22} = \sum_{i=1}^{n} Y_i Y_i' \quad \text{(a } p \times p \text{ matrix)}$$

and

$$\hat{\Sigma} = \hat{\Sigma}_{11}^{-1} \hat{\Sigma}_{12} \hat{\Sigma}_{22}^{-1} \hat{\Sigma}_{12}' \quad \text{(a } 3 \times 3 \text{ matrix)} \qquad (8.16)$$

$$= \begin{pmatrix} \hat{\sigma}_{11} & \hat{\sigma}_{12} & \hat{\sigma}_{13} \\ \hat{\sigma}_{12} & \hat{\sigma}_{22} & \hat{\sigma}_{23} \\ \hat{\sigma}_{13} & \hat{\sigma}_{23} & \hat{\sigma}_{33} \end{pmatrix} \quad \text{say.} \qquad (8.17)$$

The correlation coefficient is

$$\hat{\rho}_G = (\hat{\sigma}_{11} + \hat{\sigma}_{22} + \hat{\sigma}_{33})/q, \qquad q = \text{smaller of 3 and } p. \qquad (8.13)$$

The hypothesis that **X** and **Y** are uncorrelated is rejected if $\hat{\rho}_G$ is too large.

Critical values

sample size $n \leq 25$ Use a permutation test as described in §3.7(ii) and §8.2.2. (Note that the matrices $\hat{\Sigma}_{11}$ and $\hat{\Sigma}_{22}$ do not require re-calculation each time).

sample size $n > 25$ Compare $qn\hat{\rho}_G$ with the upper $100\alpha\%$ point of the χ^2_{3p} distribution in Appendix A4.

If the hypothesis that **X** and **Y** are uncorrelated is rejected we may wish to estimate the underlying correlation ρ_G. This is done exactly analogously to the method in §8.2.2 (cf. §8.2.3 as well). The maximum value of $\hat{\rho}_G$ is 1, corresponding to a form of linear or rotational association between **X** and **Y**, depending on the nature of **Y** (linear, directional, etc.).

References and footnotes See Jupp & Mardia (1980).

8.3 Regression
8.3.1 Introduction

There have been few applications in the statistical literature of regression models involving a spherical random variable **X**. We shall give brief descriptions of the problems of estimating the regression of **X** on a circular variable Θ, and on a set of linear variables. Other aspects of this latter problem are touched on in §8.4.

8.3.2 Simple regression of a random unit vector on a circular variable

The model we seek to fit is of the form

$$\mathbf{x} = \mathbf{A}\mathbf{u} + \mathbf{b} \tag{8.19}$$

$$= \begin{pmatrix} a_{11} & a_{12} \\ a_{21} & a_{22} \\ a_{31} & a_{32} \end{pmatrix} \begin{pmatrix} \cos\theta \\ \sin\theta \end{pmatrix} + \begin{pmatrix} b_1 \\ b_2 \\ b_3 \end{pmatrix}, \quad \text{with } \mathbf{A}'\mathbf{b} = \begin{pmatrix} 0 \\ 0 \end{pmatrix} \tag{8.20}$$

where $\mathbf{x}$ is a vector of three direction cosines. Thus, for a given value θ (i.e. $\mathbf{u}$), the prediction $\mathbf{x}$ lies on a small circle. For convenience in fitting the model, we change the unknown model parameters $\mathbf{A}$ and $\mathbf{b}$ so that this circle becomes the equator of a sphere, i.e. a great circle. Define

$$\mathbf{v} = \begin{pmatrix} \cos\theta \\ \sin\theta \\ 0 \end{pmatrix}, \quad \mathbf{w} = \begin{pmatrix} 0 \\ 0 \\ 1 \end{pmatrix} \tag{8.21}$$

and re-write the model as

$$\mathbf{x} = \mathbf{U}(\mathbf{v}\sin\alpha + \mathbf{w}\cos\alpha) \tag{8.22}$$

where α is the semi-vertical angle of the small circle and $\mathbf{U}$ is an orthogonal matrix, and α and $\mathbf{U}$ are the new parameters to be estimated.

Given a random sample of data $(\mathbf{X}_1, \theta_1), \ldots, (\mathbf{X}_n, \theta_n)$, α and $\mathbf{U}$ have to be estimated iteratively. Let $\mathbf{v}'_i = (\cos\theta_i, \sin\theta_i, 0)$, $i = 1, \ldots, n$, and calculate

$$\bar{\mathbf{X}} = (1/n)\sum_{i=1}^{n} \mathbf{X}_i, \quad \mathbf{S}_{vx} = (1/n)\sum_{i=1}^{n} \mathbf{v}_i \mathbf{X}'_i, \quad \mathbf{S}_{wx} = \mathbf{w}\bar{\mathbf{X}}' \tag{8.23}$$

Let $\hat{\alpha}_0$ be an initial estimate of α. Calculate the matrix

$$\mathbf{B}_1 = \mathbf{S}_{vx}\sin\hat{\alpha}_0 + \mathbf{S}_{wx}\cos\hat{\alpha}_0 \tag{8.24}$$

and then the estimates

$$\hat{\mathbf{U}}_1 = (\hat{\mathbf{B}}'_1\hat{\mathbf{B}}_1)^{-\frac{1}{2}}\hat{\mathbf{B}}'_1, \quad \tan\hat{\alpha}_1 = \bar{\mathbf{X}}'\hat{\mathbf{U}}_1\mathbf{w}/\text{trace}(\mathbf{S}_{vx}\hat{\mathbf{U}}'_1) \tag{8.25}$$

(see §3.3.3 for calculation of a matrix of the form $\mathbf{C}^{\frac{1}{2}}$ and §2.3 for the definition of the trace of a matrix). Substitute $\hat{\alpha}_1$ for $\hat{\alpha}_0$ in (8.24) to obtain an updated value of $\hat{\mathbf{B}}_1$, and then recalculate $\hat{\mathbf{U}}_1$ and $\hat{\alpha}_1$ from (8.25), and so on until sufficient convergence has been obtained. Denote the final estimates by $\hat{\mathbf{U}}$ and $\hat{\alpha}$. The fitted model is

$$\hat{\mathbf{X}} = \hat{\mathbf{U}}(\mathbf{v}\sin\hat{\alpha} + \mathbf{w}\cos\hat{\alpha}) \tag{8.26}$$

yielding, for a given θ (i.e. $\mathbf{v}$) a predicted value $\hat{\mathbf{X}}$. Bootstrap methods (cf. §3.6.1) can be used to estimate the variability of the parameter estimates $\mathbf{U}$ and α, and also the error in $\mathbf{X}$, given θ.

A situation in which this model may be applicable is in studying the diurnal variation of the geomagnetic field.

References and footnotes See Jupp & Mardia (1980).

8.3.3 *Regression of a random unit vector on a set of linear variables*
The model we seek to fit is of the form

$$\mathbf{x} \equiv \begin{pmatrix} \sin\theta\cos\phi \\ \sin\theta\sin\phi \\ \cos\theta \end{pmatrix}$$

$$= \begin{pmatrix} \sin(\alpha_0+\alpha_1 t_1+\ldots+\alpha_p t_p)\cos(\beta_0+\beta_1 w_1+\ldots+\beta_q w_q) \\ \sin(\alpha_0+\alpha_1 t_1+\ldots+\alpha_p t_p)\sin(\beta_0+\beta_1 w_1+\ldots+\beta_q w_q) \\ \cos(\alpha_0+\alpha_1 t_1+\ldots+\alpha_p t_p) \end{pmatrix}$$

(8.27)

where $t_1,\ldots,t_p$ and $u_1,\ldots,u_q$ are linear variables, and $\alpha_0,\ldots,\alpha_p$ and $\beta_0,\ldots,\beta_q$ are parameters to be estimated. For convenience, we write

$$\boldsymbol{\alpha} = \begin{pmatrix} \alpha_1 \\ \alpha_2 \\ \vdots \\ \alpha_p \end{pmatrix}, \quad \mathbf{T} = \begin{pmatrix} t_1 \\ t_2 \\ \vdots \\ t_p \end{pmatrix}, \quad \boldsymbol{\beta} = \begin{pmatrix} \beta_1 \\ \beta_2 \\ \vdots \\ \beta_q \end{pmatrix}, \quad \mathbf{W} = \begin{pmatrix} w_1 \\ w_2 \\ \vdots \\ w_q \end{pmatrix}, \quad (8.28)$$

so that the form becomes

$$\mathbf{x} = \begin{pmatrix} \sin(\alpha_0+\boldsymbol{\alpha}'\mathbf{T})\cos(\beta_0+\boldsymbol{\beta}'\mathbf{W}) \\ \sin(\alpha_0+\boldsymbol{\alpha}'\mathbf{T})\sin(\beta_0+\boldsymbol{\beta}'\mathbf{W}) \\ \cos(\alpha_0+\boldsymbol{\alpha}'\mathbf{T}) \end{pmatrix} \quad (8.29)$$

Given a random sample of data $(\mathbf{X}_1, \mathbf{T}_1, \mathbf{W}_1),\ldots,(\mathbf{X}_n, \mathbf{T}_n, \mathbf{W}_n)$, write

$$\mathbf{Y}_i = \begin{pmatrix} \sin(\alpha_0+\boldsymbol{\alpha}'\mathbf{T}_i)\cos(\beta_0+\boldsymbol{\beta}'\mathbf{W}_i) \\ \sin(\alpha_0+\boldsymbol{\alpha}'\mathbf{T}_i)\sin(\beta_0+\boldsymbol{\beta}'\mathbf{W}_i) \\ \cos(\alpha_0+\boldsymbol{\alpha}'\mathbf{T}_i) \end{pmatrix}, \quad i=1,\ldots,n, \quad (8.30)$$

so that $\mathbf{Y}_1,\ldots,\mathbf{Y}_n$ are functions of the known variables $\mathbf{T}_i, \mathbf{W}_i$, and the unknown parameters α_0, β_0 and $\boldsymbol{\beta}$. The parameter estimates $\hat{\alpha}_0, \hat{\boldsymbol{\alpha}}, \hat{\beta}_0$ and $\hat{\boldsymbol{\beta}}$ are obtained by minimising

$$S = \sum_{i=1}^{n} (\mathbf{X}_i - \mathbf{Y}_i)'(\mathbf{X}_i - \mathbf{Y}_i) \quad (8.31)$$

using a standard optimisation routine (see references in §5.3.1 following (5.2)). The form of the model can often be simplified if an initial rotation of the data $\mathbf{X}_1,\ldots,\mathbf{X}_n$ is made. Calculate the eigenvectors of the orientation matrix $(1/n)\sum \mathbf{X}_i\mathbf{X}_i'$ as in §3.2.4, and then the rotation matrix

$$\mathbf{H} = \begin{pmatrix} \mathbf{u}_1' \\ \mathbf{u}_2' \\ \mathbf{u}_3' \end{pmatrix}. \quad (8.32)$$

8.3.3. Regression of unit vector on linear variables

Calculate $X_i^* = HX_i$, $i = 1, \ldots, n$, and minimise S in (8.31) using X_i^* instead of X_i. The fitted values $\hat{X}_i$ are given by $H'\hat{X}_i^*$ in an obvious notation.

For large samples, the variance-covariance matrix (of the estimates $\hat{\alpha}_0, \hat{\boldsymbol{\alpha}}, \hat{\beta}_0, \hat{\boldsymbol{\beta}}$) output by the optimisation routine can be used, if available. Otherwise, bootstrap methods (cf. §3.6.1) can be utilised to assess the variability of $\hat{\alpha}_0, \hat{\boldsymbol{\alpha}}, \hat{\beta}_0, \hat{\boldsymbol{\beta}}$, and of a prediction $\hat{X}$ given by

$$\hat{X} = \begin{pmatrix} \sin(\hat{\alpha}_0 + \hat{\boldsymbol{\alpha}}'T_0) \cos(\hat{\beta}_0 + \hat{\boldsymbol{\beta}}'W_0) \\ \sin(\hat{\alpha}_0 + \hat{\boldsymbol{\alpha}}'T_0) \sin(\hat{\beta}_0 + \hat{\boldsymbol{\beta}}'W_0) \\ \cos(\hat{\alpha}_0 + \hat{\boldsymbol{\alpha}}_0'T_0) \end{pmatrix} \qquad (8.33)$$

for some specified values $T = T_0$, $W = W_0$.

Under the assumption that $\hat{X}$, given T and W, has a Fisher distribution (cf. §4.3) with mean direction given by the right hand side of (8.29) and (unknown) concentration parameter κ, we can assess the adequacy of the fitted model by calculating the deviation of each datum X_i from its fitted value $\hat{X}_i$ (using (8.33) with $(T_0, W_0) = (T_i, W_i)$), $d_i = \hat{X}_i'X_i$, $i = 1, \ldots, n$, and performing a Q-Q plot of $1 - d_1, \ldots, 1 - d_n$ to test for exponentiality (cf. §5.3.2(ii) and §3.5.1).

A formal test, under the same assumption, that some of the parameters $\alpha_1, \ldots, \alpha_p$ and $\beta_1, \ldots, \beta_q$ take specified values can be done as follows. Suppose we wish to test whether

and
$$\left. \begin{array}{l} \alpha_1 = \alpha_{1,0}, \ldots, \alpha_r = \alpha_{r,0}, \quad r \leq p \\[6pt] \beta_1 = \beta_{1,0}, \ldots, \beta_s = \beta_{s,0}, \quad s \leq q \end{array} \right\} \qquad (8.34)$$

(An obvious example of this would be testing whether all α_i's except α_p were zero, and similarly for the β_j's).

Estimate the parameters under this model, and calculate the fitted values $\hat{X}_{i,0}$, $i = 1, \ldots, n$ analogously to $\hat{X}_i$, $i = 1, \ldots, n$ in the previous paragraph. Calculate

$$R = \sum_{i=1}^{n} \hat{X}_i'X_i, \qquad R_0 = \sum_{i=1}^{n} \hat{X}_i'X_{i,0} \qquad (8.35)$$

and the test statistic

$$G_{r,s} = [(2n - 2 - p - q)/(r + s)](R - R_0)/(n - R) \qquad (8.36)$$

The hypothesis is rejected if $G_{r,s}$ is too large.

Critical values Use appropriate percentile of the F distribution with $r + s$ and $2n - p - q$ degrees of freedom in Appendix A5.

References and footnotes See Gould (1969).

8.4 Temporal and spatial analysis
8.4.1 Introduction

As mentioned in §8.1, there is at present considerable active development of methods for analysing spherical data which have some temporal or spatial variation, and it is too early to recommend techniques as proven by experience. So we have chosen to describe some of the procedures, without in general giving examples of their application to real data sets.

8.4.2 Time series

By a *time series*, we mean here a successive sequence of vectors or axes $X_1, X_2, \ldots$, where $X_1, X_2, \ldots$ are observations at equally-spaced intervals (of time, or length, or of some other linear variable). Example 8.2 is a time series, with the interval being 24 hours. We consider the following problems:

(i) Testing for serial association of a time series of unit vectors.
(ii) Testing for serial association of a time series of unit axes.

8.4.2(i) A test for serial association in a time series of unit vectors. Given a time series $X_1, \ldots, X_n$ of unit vectors from a unimodal distribution, suppose we wish to test the hypothesis that the observations are independent (and hence constitute a random sample from this distribution) against the alternative hypothesis that any two successive measurements are associated.

The test statistic is

$$S = \sum_{i=1}^{n-1} X_i' X_{i+1} \tag{8.37}$$

and the hypothesis of independence is rejected if S is too large.

For samples of size 30 or more, we require some preliminary calculations before defining the critical values. Calculate

$$X = \sum_{i=1}^{n} X_i = \begin{pmatrix} x_1 \\ x_2 \\ x_3 \end{pmatrix} \text{ say}; \quad R = (x_1^2 + x_2^2 + x_3^2)^{1/2}, \tag{8.38}$$

$$T = \sum_{i=1}^{n} X_i X_i' = \begin{pmatrix} t_{11} & t_{12} & t_{13} \\ t_{12} & t_{22} & t_{23} \\ t_{13} & t_{23} & t_{33} \end{pmatrix}, \tag{8.39}$$

$$S_1 = \text{trace}(T^2) = \sum_{i=1}^{3} \sum_{j=1}^{3} t_{ij}^2, \tag{8.40}$$

$$S_2 = X'TX = \sum_{i=1}^{3} \sum_{k=1}^{3} x_i x_j t_{ij}. \tag{8.41}$$

8.4.2(ii). Test for serial association of unit axes

Finally, define quantities

$$S_E = R^2/n - 1, \tag{8.42}$$

$$S_V = S_1/n - 2S_2/[n(n-1)] + R^4/[n^2(n-1)]$$
$$+ 2S_E/(n-1) - 2 + n/(n-1), \tag{8.43}$$

and

$$S^* = (S - S_E)/S_V^{1/2}. \tag{8.44}$$

Critical values

$n < 30$ Use a permutation test, as described in §**3.7(ii)**.

$n \geq 30$ Calculate S^*, and compare with the appropriate percentile of the normal distribution in Appendix A1.

> **Example 8.9** For the data in Example 8.2, we can test for serial association even though there are two missing values. Suppose we test for positive serial correlation. A suitable computational device for handling the two missing values when performing a permutation test is to insert the zero vectors, $(0, 0, 0)$ and $(0, 0, 0)$, at the 13th and 16th places in the time series, and then calculate the quantities in (8.37)-(8.44). However, when n appears as a factor, use $n = 29$, not $n = 31$. We obtain $S = 8.945$. Using a permutation test based on 2500 random permutations of the data, this was found to be number 2407 in the sequence of ordered S-values (number 1 being smallest, number 2500 being largest), and corresponds to a significance probability of 0.037. So there is some evidence for serial association of these daily wind directions.

References and footnotes The test is due to Watson & Beran (1967); see also Epp, Tukey & Watson (1971) for computational aspects.

8.4.2(ii) A test for serial association in a time series of unit axes. Given a time series $X_1, \ldots, X_n$ of unit axes from a bipolar distribution, suppose we wish to test the hypothesis that the axes are independent (and hence constitute a random sample from this distribution) against the alternative hypothesis that any two successive measurements are correlated. The test statistic is

$$N = \sum_{i=1}^{n-1} (\mathbf{X}_i' \mathbf{X}_{i+1})^2 \tag{8.45}$$

and the hypothesis of independence is rejected if N is too large.

Critical values Use a permutation test, as described in §**3.7(ii)**.

(In large samples, an alternative is to calculate the permutation mean $N_E = (S_1/n) - 1$ from (8.39) and (8.40), estimate the permutation variance from a small number of permutations and use a normal approximation as in §**8.4.2(i)** above.)

8.4.3 Time-ordered sequences of unit vectors

It is convenient to discuss this subject in the context of the Apparent Polar Wander problem (cf. Example 8.3). Suppose our data sequence is $X_1, \ldots, X_n$, where X_i is measured at, or corresponds to, age (or distance) T_i, $i = 1, \ldots, n$. The information available about X_i and T_i can vary considerably:

(a) X_i may just be a single mean direction, with no estimate of its error; or it may have an associated confidence cone based on, say, a Fisherian model (cf. §5.3.2(iv)) or it may be the mean direction of a sample $X_{i1}, \ldots, X_{in_i}$ which is also available.

(b) T_i may be known very accurately; or may be an estimate but without an associated estimate of its error; or it may have an associated estimate of error. In some situations, the only information available about T_i is that $T_i > T_{i-1}$ and $T_i < T_{i+1}$, that is, the chronological order of $X_1, \ldots, X_n$ is known, but nothing more about their temporal pattern or structure.

In the worst case, in which only the relative chronology of $X_1, \ldots, X_n$ is known, there appears to be no useful analysis to be performed on $X_1, \ldots, X_n$. It is possible, however, to match up two such sequences $X_1, \ldots, X_n$ and $Y_1, \ldots, Y_m$, if each sequence is estimating the same shape of curve on the surface of the unit sphere over the same time interval. In this case, we are interested in finding a rotation matrix H such that the rotated sequence $HY_1, \ldots, HY_m$ matches up "best" with $X_1, \ldots, X_n$. Let $Y_i^* = H^*Y_i$, $i = 1, \ldots, m$, be some rotation of the Y-path, and suppose

$$X_1, X_2, Y_1^*, X_3, Y_2^*, Y_3^*, \ldots, Y_m^*, X_n$$

is some slotting of the Y^*-sequence into the X-sequence which respects the chronology of both X's and Y's. The *combined path length* of this sequence is defined as the sum of all the arcs between successive points, i.e.

$$\arccos(X_1'X_2) + \arccos(X_2'Y_1^*) + \ldots + \arccos(Y_m^{*'}X_n)$$

The *minimum combined path length* for H^* is the minimum over all possible sequence slottings of the Y^*'s into the X-sequence (which respect each chronology). So, an iterative procedure is required to find the "best" H, based on calculating the minimum combined path length (MCPL) for a sequence of H's. This technique was used by Embleton, Fisher and Schmidt (1983). However, the method they used to find the MCPL for a given H was inefficient and only approximate, whereas the MCPL can be computed simply using dynamic programming. A suitable computer program for this has been developed by Clark (1985).

Usually, we have at least a point estimate T_i of the age of X_i, $i = 1, \ldots, n$, although without an estimate of the error associated with T_i, and wish

8.4.3. Time-ordered sequences of unit vectors

to fit some smooth curve to the sequence $(\mathbf{X}_1, T_1), \ldots, (\mathbf{X}_n, T_n)$. Suppose for the moment that $T_1, \ldots, T_n$ have negligible errors. Clark & Thompson (1978, 1979, 1984) and Thompson & Clark (1981, 1982) considered methods suitable for data $\mathbf{X}_1, \ldots, \mathbf{X}_n$ concentrated on a small area of the surface of the unit sphere. Their methods involve projecting the data onto a plane tangent to the mean direction of $\mathbf{X}_1, \ldots, \mathbf{X}_n$, so that $\mathbf{X}_i = (\theta_i, \phi_i)$ say $= (\theta_i^*, \phi_i^*)$ in rectangular coordinates on the plane. Spline methods are then used to fit the colatitude θ^* and longitude ϕ^* separately, as functions of time T. Provided that the area on the sphere containing all the $\mathbf{X}_i$ values is sufficiently small, the distortion introduced by this projection is negligible. However, the method is clearly not coordinate-free, and cannot readily be adapted to less concentrated data sets.

Parker and Denham (1979) proposed a spline method for fitting a curve $\mathbf{X} = \mathbf{X}(t)$ through the data points (an *interpolating* curve) by treating the three coordinates of $\mathbf{X}$ independently; however in general, the points on the curve are not unit vectors unless t is one of the observation points T_i.

The most promising approach appears to be that of Watson (1983d), who has developed a general spline technique for smoothing and interpolation which respects the geometry of the sphere. No practical application of these methods has yet been reported.

In the absence of a general model-based approach, the simplest way of smoothing the sequence is to use a moving average: the direction $\mathbf{X}$ at time t is simply the mean direction of all unit vectors $\mathbf{X}_i$ such that T_i is within some time interval D of t. This has been used by Irving (1977). It can be enhanced by using a *weighted* mean direction (cf. §**7.2.2(iii)**) with the weights given to different data points reflecting the differing precisions with which the $\mathbf{X}_i$'s may be recorded, the time-distance of T_i from t, and possibly, differing variabilities of the T_i's. This might be done as follows.

Suppose we have a sequence of r palaeomagnetic pole positions $P_1, \ldots, P_r$, with the following information on each P_i:

$(\hat{\theta}_i, \hat{\phi}_i)$	mean direction
$\hat{\kappa}_i$	estimate of concentration parameter of Fisher distribution
n_i	number of unit vectors used to calculate $(\hat{\theta}_i, \hat{\phi}_i)$ and $\hat{\kappa}_i$
$\bar{R}_i$	mean resultant length of the n_i unit vectors
T_i	age of pole
d_i	estimate of error of T_i

We wish to smooth the sequence of pole positions using an *age-window* of half-width D. We shall assume that the true age of P_i is equally likely to be anywhere in the interval $(T_i - d_i, T_i + d_i)$.

Suppose we wish to estimate the smoothed path at time points $t = t_1, \ldots, t_n$. (For example, $t_1, \ldots, t_n$ may be a sequence of equally-spaced ages: 500 my, 490 my, 480 my, ...). The smoothing window around t then contains all poles whose age could possibly be between $t - D$ and $t + D$. However, we would want to weight the effect of any pole according to the probability of its lying in this interval. So, define

$P_i(t)$ = probability that ith pole could lie inside window of half-width D centred on t

$$= \frac{1}{2d_i} \times [\text{amount of overlap of } (T_i - d_i, T_i + d_i) \text{ and}$$

$$(t - D, t + D)] \quad (8.46)$$

Thus $P_i(t)$ may well be zero for a pole whose age T_i is rather different from t, and whose age error d_i is not large.

Apart from the error in age, we would want to weight the pole according to the quality of the pole position. To do this, let

$$M_i = n_i \hat{\kappa}_i \text{ for the } i\text{th pole} \quad (8.47)$$

Finally, define the overall weight for P_i at time t as

$$w_i(t) = P_i(t) M_i / w(t), \quad i = 1, \ldots, r \quad (8.48)$$

where

$$w(t) = P_1(t) M_1 + \ldots + P_r(t) M_r \quad (8.49)$$

The smoothed pole position $(\hat{\theta}(t), \hat{\phi}(t))$ at t is then obtained as follows. Let

$$\left. \begin{array}{l} S_x(t) = w_1(t) \sin \hat{\theta}_1 \cos \hat{\phi}_1 + \ldots + w_r(t) \sin \hat{\theta}_r \cos \hat{\phi}_r \\ S_y(t) = w_1(t) \sin \hat{\theta}_1 \sin \hat{\phi}_1 + \ldots + w_r(t) \sin \hat{\theta}_r \sin \hat{\phi}_r \\ S_z(t) = w_1(t) \cos \hat{\theta}_1 + \ldots + w_r(t) \cos \hat{\theta}_r \end{array} \right\} \quad (8.50)$$

and

$$\bar{R}^2(t) = S_x^2(t) + S_y^2(t) + S_z^2(t). \quad (8.51)$$

Then the direction cosines of $(\theta(t), \phi(t))$ are

$$\left. \begin{array}{l} \sin \hat{\theta}(t) \cos \hat{\phi}(t) = S_x(t) / \bar{R}(t) \\ \sin \hat{\theta}(t) \sin \hat{\phi}(t) = S_y(t) / \bar{R}(t) \\ \cos \hat{\theta}(t) = S_z(t) / \bar{R}(t) \end{array} \right\} \quad (8.52)$$

whence $\hat{\theta}(t)$ and $\hat{\phi}(t)$ are obtained using (3.3)-(3.5).

As presented in this form, smoothed pole positions can be computed at any desired age t, and the procedure is easily programmed. The unresolved matter is, of course, how to choose the half-width D. There may be a geological basis for this selection. In any case, one would wish to look at a range of values, to see the effect of varying the choice.

8.4.3. Time-ordered sequences of unit vectors

It is difficult to form an estimate of the error in $(\hat{\theta}(t), \hat{\phi}(t))$ theoretically. However, some guidance can be obtained by simulation methods. Consider the pole P_1, which was estimated from n_1 unit vectors assumed to be from a Fisher distribution with estimated concentration κ_1; it has age T_1, which is assumed to be distributed uniformly over an age interval of length $2d_1$. We can simulate a new pole and age as follows:

1. Simulate n_1 unit vectors from a Fisher distribution with mean direction $(\hat{\theta}_1, \hat{\phi}_1)$ and concentration parameter $\hat{\kappa}_1$, and compute the mean direction $(\hat{\theta}_1^*, \hat{\phi}_1^*)$ of these n_1 vectors. This gives a pole position P_1^* say.
2. Simulate an age T_1^* from the uniform distribution on $(T_1 - d_1, T_1 + d_1)$.

(See §3.6.2 for details of how to perform these simulations). Similarly, one can obtain P_2^* with age $T_2^*, \ldots, P_r^*$ with age T_r^*. So, to assess the error in the smoothed path, we take the following steps:

A. Simulate $(P_1^*, T_1^*), \ldots, (P_r^*, T_r^*)$.
B. Calculate $(\hat{\theta}^*(t), \hat{\phi}^*(t))$ for chosen $t = t_1, \ldots, t_n$.
C. Repeat (A) and (B) a large number of times (say N times).

The end result is that, at each of the time points $t = t_i$, we have N simulated values of $\hat{\theta}^*(t_i), \hat{\phi}^*(t_i)$. There are now two possibilities:

(i) Plot each smoothed path $(\hat{\theta}^*(t_1), \hat{\phi}^*(t_1)), \ldots, (\hat{\theta}^*(t_n), \hat{\phi}^*(t_n))$ obtained in step (B), joining up the points in the path. This will give a ribbon of paths around the actual fitted path
$(\hat{\theta}(t_1), \hat{\phi}(t_1)), \ldots, (\hat{\theta}(t_n), \hat{\phi}(t_n))$.

(ii) Look at the individual scatter of the N simulated pole positions at a given time point t_i. If the age-window of width $2D$ is not too large, these points may be approximately Fisher-distributed about their mean direction. In this case, calculate an estimate $\hat{\kappa}(t_i)$ of the concentration parameter based on these N points. Then the semi-angle q_α of an approximate $100(1-\alpha)\%$ confidence cone, centred on $(\hat{\theta}(t_i), \hat{\phi}(t_i))$, may be calculated as in §5.3.2(iv) (e.g. from (5.34)).

The Clark-Thompson papers offer, as part of the end product, a pointwise estimate of the error of their fitted function. A corresponding estimate for a path fitted by Watson's (1983d) method will probably become available. Alternatively, simulation methods like the one described above can be used to produce an error ribbon or envelope.

8. Correlation, regression, and temporal/spatial analysis

8.4.4 *Spatial analysis of unit axes*

As in §**8.4.3**, we consider this in the context of an application, in this case Example 8.4. No satisfactory general technique is in current use. Possible approaches include a modification of the method of Fisher *et al.* (1985) for circular orientation data; a simple weighted moving average method using §**7.3.1(ii)**, with weights being simply sample sizes; and the general method described by Watson (1983d), which appears to be the most promising, although no examples of its application are yet available.

Appendix A

Tables and charts

A1 Percentage points of the standardised normal $N(0, 1)$ distribution.

This table gives percentage points $x(\alpha)$ defined by the equation

$$\alpha = (2\pi)^{-\frac{1}{2}} \int_{x(\alpha)}^{\infty} \exp\left(-\tfrac{1}{2}t^2\right) dt.$$

If X is a $N(0, 1)$ random variable, Prob $(X \geq x(\alpha)) = \alpha$; $x(\alpha)$ is the upper P percent point, $P = 100\alpha$. The lower P percent point $x(1-\alpha)$ is given by symmetry as $x(1-\alpha) = -x(\alpha)$, and Prob $(|X| \geq x(\alpha)) = 2\alpha = 2P/100$.

Figure A1

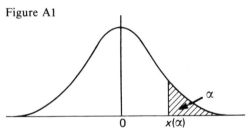

Table A1

P	$x(P/100)$	P	$x(P/100)$	P	$x(P/100)$	P	$x(P/100)$	P	$x(P/100)$	P	$x(P/100)$
50	0·0000	5·0	1·6449	3·0	1·8808	2·0	2·0537	1·0	2·3263	0·10	3·0902
45	0·1257	4·8	1·6646	2·9	1·8957	1·9	2·0749	0·9	2·3656	0·09	3·1214
40	0·2533	4·6	1·6849	2·8	1·9110	1·8	2·0969	0·8	2·4089	0·08	3·1559
35	0·3853	4·4	1·7060	2·7	1·9268	1·7	2·1201	0·7	2·4573	0·07	3·1947
30	0·5244	4·2	1·7279	2·6	1·9431	1·6	2·1444	0·6	2·5121	0·06	3·2389
25	0·6745	4·0	1·7507	2·5	1·9600	1·5	2·1701	0·5	2·5758	0·05	3·2905
20	0·8416	3·8	1·7744	2·4	1·9774	1·4	2·1973	0·4	2·6521	0·01	3·7190
15	1·0364	3·6	1·7991	2·3	1·9954	1·3	2·2262	0·3	2·7478	0·005	3·8906
10	1·2816	3·4	1·8250	2·2	2·0141	1·2	2·2571	0·2	2·8782	0·001	4·2649
5	1·6449	3·2	1·8522	2·1	2·0335	1·1	2·2904	0·1	3·0902	0·0005	4·4172

Source: Lindley & Scott (1984, Table 5)

Appendix A

A2 The standardised normal $N(0, 1)$ distribution.

The function tabulated is

$$\Phi(x) = (2\pi)^{-\frac{1}{2}} \int_{-\infty}^{x} \exp\left(-\tfrac{1}{2}t^2\right) dt.$$

If X is a $N(0, 1)$ random variable, $\Phi(x) = \text{Prob } (X \leq x)$.
When $x < 0$, use $\Phi(x) = 1 - \Phi(-x)$.

Figure A2

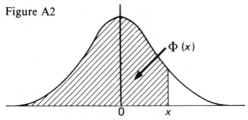

Table A2

x	$\Phi(x)$	x	$\Phi(x)$	x	$\Phi(x)$	x	$\Phi(x)$	x	$\Phi(x)$	x	$\Phi(x)$
0·00	0·5000	0·40	0·6554	0·80	0·7881	1·20	0·8849	1·60	0·9452	2·00	0·97725
·01	·5040	·41	·6591	·81	·7910	·21	·8869	·61	·9463	·01	·97778
·02	·5080	·42	·6628	·82	·7939	·22	·8888	·62	·9474	·02	·97831
·03	·5120	·43	·6664	·83	·7967	·23	·8907	·63	·9484	·03	·97882
·04	·5160	·44	·6700	·84	·7995	·24	·8925	·64	·9495	·04	·97932
0·05	0·5199	0·45	0·6736	0·85	0·8023	1·25	0·8944	1·65	0·9505	2·05	0·97982
·06	·5239	·46	·6772	·86	·8051	·26	·8962	·66	·9515	·06	·98030
·07	·5279	·47	·6808	·87	·8078	·27	·8980	·67	·9525	·07	·98077
·08	·5319	·48	·6844	·88	·8106	·28	·8997	·68	·9535	·08	·98124
·09	·5359	·49	·6879	·89	·8133	·29	·9015	·69	·9545	·09	·98169
0·10	0·5398	0·50	0·6915	0·90	0·8159	1·30	0·9032	1·70	0·9554	2·10	0·98214
·11	·5438	·51	·6950	·91	·8186	·31	·9049	·71	·9564	·11	·98257
·12	·5478	·52	·6985	·92	·8212	·32	·9066	·72	·9573	·12	·98300
·13	·5517	·53	·7019	·93	·8238	·33	·9082	·73	·9582	·13	·98341
·14	·5557	·54	·7054	·94	·8264	·34	·9099	·74	·9591	·14	·98382
0·15	0·5596	0·55	0·7088	0·95	0·8289	1·35	0·9115	1·75	0·9599	2·15	0·98422
·16	·5636	·56	·7123	·96	·8315	·36	·9131	·76	·9608	·16	·98461
·17	·5675	·57	·7157	·97	·8340	·37	·9147	·77	·9616	·17	·98500
·18	·5714	·58	·7190	·98	·8365	·38	·9162	·78	·9625	·18	·98537
·19	·5753	·59	·7224	·99	·8389	·39	·9177	·79	·9633	·19	·98574
0·20	0·5793	0·60	0·7257	1·00	0·8413	1·40	0·9192	1·80	0·9641	2·20	0·98610
·21	·5832	·61	·7291	·01	·8438	·41	·9207	·81	·9649	·21	·98645
·22	·5871	·62	·7324	·02	·8461	·42	·9222	·82	·9656	·22	·98679
·23	·5910	·63	·7357	·03	·8485	·43	·9236	·83	·9664	·23	·98713
·24	·5948	·64	·7389	·04	·8508	·44	·9251	·84	·9671	·24	·98745
0·25	0·5987	0·65	0·7422	1·05	0·8531	1·45	0·9265	1·85	0·9678	2·25	0·98778
·26	·6026	·66	·7454	·06	·8554	·46	·9279	·86	·9686	·26	·98809
·27	·6064	·67	·7486	·07	·8577	·47	·9292	·87	·9693	·27	·98840
·28	·6103	·68	·7517	·08	·8599	·48	·9306	·88	·9699	·28	·98870
·29	·6141	·69	·7549	·09	·8621	·49	·9319	·89	·9706	·29	·98899
0·30	0·6179	0·70	0·7580	1·10	0·8643	1·50	0·9332	1·90	0·9713	2·30	0·98928
·31	·6217	·71	·7611	·11	·8665	·51	·9345	·91	·9719	·31	·98956
·32	·6255	·72	·7642	·12	·8686	·52	·9357	·92	·9726	·32	·98983
·33	·6293	·73	·7673	·13	·8708	·53	·9370	·93	·9732	·33	·99010
·34	·6331	·74	·7704	·14	·8729	·54	·9382	·94	·9738	·34	·99036
0·35	0·6368	0·75	0·7734	1·15	0·8749	1·55	0·9394	1·95	0·9744	2·35	0·99061
·36	·6406	·76	·7764	·16	·8770	·56	·9406	·96	·9750	·36	·99086
·37	·6443	·77	·7794	·17	·8790	·57	·9418	·97	·9756	·37	·99111
·38	·6480	·78	·7823	·18	·8810	·58	·9429	·98	·9761	·38	·99134
·39	·6517	·79	·7852	·19	·8830	·59	·9441	·99	·9767	·39	·99158
0·40	0·6554	0·80	0·7881	1·20	0·8849	1·60	0·9452	2·00	0·9772	2·40	0·99180

Table A2 (cont.)

x	$\Phi(x)$	x	$\Phi(x)$	x	$\Phi(x)$	x	$\Phi(x)$	x	$\Phi(x)$	x	$\Phi(x)$
2·40	0·99180	2·55	0·99461	2·70	0·99653	2·85	0·99781	3·00	0·99865	3·15	0·99918
·41	·99202	·56	·99477	·71	·99664	·86	·99788	·01	·99869	·16	·99921
·42	·99224	·57	·99492	·72	·99674	·87	·99795	·02	·99874	·17	·99924
·43	·99245	·58	·99506	·73	·99683	·88	·99801	·03	·99878	·18	·99926
·44	·99266	·59	·99520	·74	·99693	·89	·99807	·04	·99882	·19	·99929
2·45	0·99286	2·60	0·99534	2·75	0·99702	2·90	0·99813	3·05	0·99886	3·20	0·99931
·46	·99305	·61	·99547	·76	·99711	·91	·99819	·06	·99889	·21	·99934
·47	·99324	·62	·99560	·77	·99720	·92	·99825	·07	·99893	·22	·99936
·48	·99343	·63	·99573	·78	·99728	·93	·99831	·08	·99896	·23	·99938
·49	·99361	·64	·99585	·79	·99736	·94	·99836	·09	·99900	·24	·99940
2·50	0·99379	2·65	0·99598	2·80	0·99744	2·95	0·99841	3·10	0·99903	3·25	0·99942
·51	·99396	·66	·99609	·81	·99752	·96	·99846	·11	·99906	·26	·99944
·52	·99413	·67	·99621	·82	·99760	·97	·99851	·12	·99910	·27	·99946
·53	·99430	·68	·99632	·83	·99767	·98	·99856	·13	·99913	·28	·99948
·54	·99446	·69	·99643	·84	·99774	·99	·99861	·14	·99916	·29	·99950
2·55	0·99461	2·70	0·99653	2·85	0·99781	3·00	0·99865	3·15	0·99918	3·30	0·99952

The critical table below gives on the left the range of values of x for which $\Phi(x)$ takes the value on the right, correct to the last figure given; in critical cases, take the upper of the two values of $\Phi(x)$ indicated.

x	$\Phi(x)$	x	$\Phi(x)$	x	$\Phi(x)$	x	$\Phi(x)$
3·075	0·9990	3·263	0·9994	3·731	0·99990	3·916	0·99995
3·105	0·9991	3·320	0·9995	3·759	0·99991	3·976	0·99996
3·138	0·9992	3·389	0·9996	3·791	0·99992	4·055	0·99997
3·174	0·9993	3·480	0·9997	3·826	0·99993	4·173	0·99998
3·215	0·9994	3·615	0·9998	3·867	0·99994	4·417	0·99999
			0·9999		0·99995		1·00000

When $x > 3.3$ the formula $1 - \Phi(x) \doteq \dfrac{e^{-\frac{1}{2}x^2}}{x\sqrt{2\pi}}\left[1 - \dfrac{1}{x^2} + \dfrac{3}{x^4} - \dfrac{15}{x^6} + \dfrac{105}{x^8}\right]$ is very accurate, with relative error less than $945/x^{10}$.

Source: Lindley & Scott (1984, Table 4)

A3   Percentage points of the χ^2-distribution.

(a) This table gives *lower* percentage points $\chi^2_\nu(\alpha)$ defined by the equation

$$\alpha = [2^{\frac{1}{2}\nu}\Gamma(\tfrac{1}{2}\nu)]^{-1}\int_{\chi^2_\nu(\alpha)}^\infty t^{\frac{1}{2}\nu-1}\exp(-\tfrac{1}{2}t)\,dt,$$

for values of α greater than $\tfrac{1}{2}$.

If Y is a χ^2 random variable with ν degrees of freedom, Prob $(\chi^2_\nu \geq \chi^2_\nu(\alpha)) = \alpha$; $P = 100\alpha$. $\chi^2_\nu(\alpha)$ corresponds to a lower tail probability of $1-\alpha$, i.e. it is the *lower* $(100-P)\%$ point.

(b) This table gives *upper* percentage points $\chi^2_\nu(\alpha)$ defined by the equation

$$\alpha = [2^{\frac{1}{2}\nu}\Gamma(\tfrac{1}{2}\nu)]^{-1}\int_{\chi^2_\nu(\alpha)}^\infty t^{\frac{1}{2}\nu-1}\exp(-\tfrac{1}{2}t)\,dt,$$

for values of α less than $\tfrac{1}{2}$.

Appendix A

If Y is a χ^2 random variable with ν degrees of freedom, Prob $(\chi_\nu^2 \geq \chi_\nu^2(\alpha)) = \alpha$; $\chi_\nu^2(\alpha)$ is the upper P percent point, $P = 100\alpha$.

Figure A3(a) (The diagram is for illustrative purposes. In fact the above shape applies for $\nu \geq 3$ only; when $\nu < 3$ the mode is at the origin as may be seen in Figure 4.5, Chapter 4).

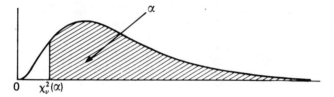

Table A3(a)

P	99.95	99.9	99.5	99	97.5	95	90	80	70	60
$\nu = 1$	0.0⁶3927	0.0⁵1571	0.0⁴3927	0.0³1571	0.0³9821	0.003932	0.01579	0.06418	0.1485	0.2750
2	0.001000	0.002001	0.01003	0.02010	0.05064	0.1026	0.2107	0.4463	0.7133	1.022
3	0.01528	0.02430	0.07172	0.1148	0.2158	0.3518	0.5844	1.005	1.424	1.869
4	0.06392	0.09080	0.2070	0.2971	0.4844	0.7107	1.064	1.649	2.195	2.753
5	0.1581	0.2102	0.4117	0.5543	0.8312	1.145	1.610	2.343	3.000	3.655
6	0.2994	0.3811	0.6757	0.8721	1.237	1.635	2.204	3.070	3.828	4.570
7	0.4849	0.5985	0.9893	1.239	1.690	2.167	2.833	3.822	4.671	5.493
8	0.7104	0.8571	1.344	1.646	2.180	2.733	3.490	4.594	5.527	6.423
9	0.9717	1.152	1.735	2.088	2.700	3.325	4.168	5.380	6.393	7.357
10	1.265	1.479	2.156	2.558	3.247	3.940	4.865	6.179	7.267	8.295
11	1.587	1.834	2.603	3.053	3.816	4.575	5.578	6.989	8.148	9.237
12	1.934	2.214	3.074	3.571	4.404	5.226	6.304	7.807	9.034	10.18
13	2.305	2.617	3.565	4.107	5.009	5.892	7.042	8.634	9.926	11.13
14	2.697	3.041	4.075	4.660	5.629	6.571	7.790	9.467	10.82	12.08
15	3.108	3.483	4.601	5.229	6.262	7.261	8.547	10.31	11.72	13.03
16	3.536	3.942	5.142	5.812	6.908	7.962	9.312	11.15	12.62	13.98
17	3.980	4.416	5.697	6.408	7.564	8.672	10.09	12.00	13.53	14.94
18	4.439	4.905	6.265	7.015	8.231	9.390	10.86	12.86	14.44	15.89
19	4.912	5.407	6.844	7.633	8.907	10.12	11.65	13.72	15.35	16.85
20	5.398	5.921	7.434	8.260	9.591	10.85	12.44	14.58	16.27	17.81
21	5.896	6.447	8.034	8.897	10.28	11.59	13.24	15.44	17.18	18.77
22	6.404	6.983	8.643	9.542	10.98	12.34	14.04	16.31	18.10	19.73
23	6.924	7.529	9.260	10.20	11.69	13.09	14.85	17.19	19.02	20.69
24	7.453	8.085	9.886	10.86	12.40	13.85	15.66	18.06	19.94	21.65
25	7.991	8.649	10.52	11.52	13.12	14.61	16.47	18.94	20.87	22.62
26	8.538	9.222	11.16	12.20	13.84	15.38	17.29	19.82	21.79	23.58
27	9.093	9.803	11.81	12.88	14.57	16.15	18.11	20.70	22.72	24.54
28	9.656	10.39	12.46	13.56	15.31	16.93	18.94	21.59	23.65	25.51
29	10.23	10.99	13.12	14.26	16.05	17.71	19.77	22.48	24.58	26.48
30	10.80	11.59	13.79	14.95	16.79	18.49	20.60	23.36	25.51	27.44
32	11.98	12.81	15.13	16.36	18.29	20.07	22.27	25.15	27.37	29.38
34	13.18	14.06	16.50	17.79	19.81	21.66	23.95	26.94	29.24	31.31
36	14.40	15.32	17.89	19.23	21.34	23.27	25.64	28.73	31.12	33.25
38	15.64	16.61	19.29	20.69	22.88	24.88	27.34	30.54	32.99	35.19
40	16.91	17.92	20.71	22.16	24.43	26.51	29.05	32.34	34.87	37.13
50	23.46	24.67	27.99	29.71	32.36	34.76	37.69	41.45	44.31	46.86
60	31.31	31.74	35.53	37.48	40.48	43.19	46.46	50.64	53.81	56.62
70	37.47	39.04	43.28	45.44	48.76	51.74	55.33	59.90	63.35	66.40
80	44.79	46.52	51.17	53.54	57.15	60.39	64.28	69.21	72.92	76.19
90	52.28	54.16	59.20	61.75	65.65	69.13	73.29	78.56	82.51	85.99
100	59.90	61.92	67.33	70.06	74.22	77.93	82.36	87.95	92.13	95.81

Figure A3(b) (The diagram is for illustrative purposes. In fact the above shape applies for $\nu \geq 3$ only; when $\nu < 3$ the mode is at the origin as may be seen in Figure 4.5, Chapter 4).

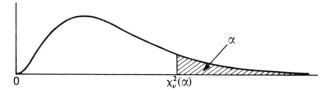

Table A3(b)

P	50	40	30	20	10	5	2·5	1	0·5	0·1	0·05
$\nu = 1$	0·4549	0·7083	1·074	1·642	2·706	3·841	5·024	6·635	7·879	10·83	12·12
2	1·386	1·833	2·408	3·219	4·605	5·991	7·378	9·210	10·60	13·82	15·20
3	2·366	2·946	3·665	4·642	6·251	7·815	9·348	11·34	12·84	16·27	17·73
4	3·357	4·045	4·878	5·989	7·779	9·488	11·14	13·28	14·86	18·47	20·00
5	4·351	5·132	6·064	7·289	9·236	11·07	12·83	15·09	16·75	20·52	22·11
6	5·348	6·211	7·231	8·558	10·64	12·59	14·45	16·81	18·55	22·46	24·10
7	6·346	7·283	8·383	9·803	12·02	14·07	16·01	18·48	20·28	24·32	26·02
8	7·344	8·351	9·524	11·03	13·36	15·51	17·53	20·09	21·95	26·12	27·87
9	8·343	9·414	10·66	12·24	14·68	16·92	19·02	21·67	23·59	27·88	29·67
10	9·342	10·47	11·78	13·44	15·99	18·31	20·48	23·21	25·19	29·59	31·42
11	10·34	11·53	12·90	14·63	17·28	19·68	21·92	24·72	26·76	31·26	33·14
12	11·34	12·58	14·01	15·81	18·55	21·03	23·34	26·22	28·30	32·91	34·82
13	12·34	13·64	15·12	16·98	19·81	22·36	24·74	27·69	29·82	34·53	36·48
14	13·34	14·69	16·22	18·15	21·06	23·68	26·12	29·14	31·32	36·12	38·11
15	14·34	15·73	17·32	19·31	22·31	25·00	27·49	30·58	32·80	37·70	39·72
16	15·34	16·78	18·42	20·47	23·54	26·30	28·85	32·00	34·27	39·25	41·31
17	16·34	17·82	19·51	21·61	24·77	27·59	30·19	33·41	35·72	40·79	42·88
18	17·34	18·87	20·60	22·76	25·99	28·87	31·53	34·81	37·16	42·31	44·43
19	18·34	19·91	21·69	23·90	27·20	30·14	32·85	36·19	38·58	43·82	45·97
20	19·34	20·95	22·77	25·04	28·41	31·41	34·17	37·57	40·00	45·31	47·50
21	20·34	21·99	23·86	26·17	29·62	32·67	35·48	38·93	41·40	46·80	49·01
22	21·34	23·03	24·94	27·30	30·81	33·92	36·78	40·29	42·80	48·27	50·51
23	22·34	24·07	26·02	28·43	32·01	35·17	38·08	41·64	44·18	49·73	52·00
24	23·34	25·11	27·10	29·55	33·20	36·42	39·36	42·98	45·56	51·18	53·48
25	24·34	26·14	28·17	30·68	34·38	37·65	40·65	44·31	46·93	52·62	54·95
26	25·34	27·18	29·25	31·79	35·56	38·89	41·92	45·64	48·29	54·05	56·41
27	26·34	28·21	30·32	32·91	36·74	40·11	43·19	46·96	49·64	55·48	57·86
28	27·34	29·25	31·39	34·03	37·92	41·34	44·46	48·28	50·99	56·89	59·30
29	28·34	30·28	32·46	35·14	39·09	42·56	45·72	49·59	52·34	58·30	60·73
30	29·34	31·32	33·53	36·25	40·26	43·77	46·98	50·89	53·67	59·70	62·16
32	31·34	33·38	35·66	38·47	42·58	46·19	49·48	53·49	56·33	62·49	65·00
34	33·34	35·44	37·80	40·68	44·90	48·60	51·97	56·06	58·96	65·25	67·80
36	35·34	37·50	39·92	42·88	47·21	51·00	54·44	58·62	61·58	67·99	70·59
38	37·34	39·56	42·05	45·08	49·51	53·38	56·90	61·16	64·18	70·70	73·35
40	39·34	41·62	44·16	47·27	51·81	55·76	59·34	63·69	66·77	73·40	76·09
50	49·33	51·89	54·72	58·16	63·17	67·50	71·42	76·15	79·49	86·66	89·56
60	59·33	62·13	65·23	68·97	74·40	79·08	83·30	88·38	91·95	99·61	102·7
70	69·33	72·36	75·69	79·71	85·53	90·53	95·02	100·4	104·2	112·3	115·6
80	79·33	82·57	86·12	90·41	96·58	101·9	106·6	112·3	116·3	124·8	128·3
90	89·33	92·76	96·52	101·1	107·6	113·1	118·1	124·1	128·3	137·2	140·8
100	99·33	102·9	106·9	111·7	118·5	124·3	129·6	135·8	140·2	149·4	153·2

Source: Lindley & Scott (1984, Table 8)

Appendix A

Figure A4 (The diagram is for illustrative purposes. In fact the above shape applies for $\nu \geq 3$ only; when $\nu < 3$ the mode is at the origin as may be seen in Figure 4.5, Chapter 4).

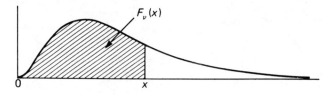

Table A4

$\nu =$	1	$\nu =$	1	$\nu =$	2	$\nu =$	2	$\nu =$	3	$\nu =$	3
x = 0·0	0·0000	x = 4·0	0·9545	x = 0·0	0·0000	x = 4·0	0·8647	x = 0·0	0·0000	x = 4·0	0·7385
·1	·2482	·1	·9571	·1	·0488	·1	·8713	·1	·0082	·2	·7593
·2	·3453	·2	·9596	·2	·0952	·2	·8775	·2	·0224	·4	·7786
·3	·4161	·3	·9619	·3	·1393	·3	·8835	·3	·0400	·6	·7965
·4	·4729	·4	·9641	·4	·1813	·4	·8892	·4	·0598	·8	·8130
0·5	0·5205	4·5	0·9661	0·5	0·2212	4·5	0·8946	0·5	0·0811	5·0	0·8282
·6	·5614	·6	·9680	·6	·2592	·6	·8997	·6	·1036	·2	·8423
·7	·5972	·7	·9698	·7	·2953	·7	·9046	·7	·1268	·4	·8553
·8	·6289	·8	·9715	·8	·3297	·8	·9093	·8	·1505	·6	·8672
·9	·6572	·9	·9731	·9	·3624	·9	·9137	·9	·1746	·8	·8782
1·0	0·6827	5·0	0·9747	1·0	0·3935	5·0	0·9179	1·0	0·1987	6·0	0·8884
·1	·7057	·1	·9761	·1	·4231	·1	·9219	·1	·2229	·2	·8977
·2	·7267	·2	·9774	·2	·4512	·2	·9257	·2	·2470	·4	·9063
·3	·7458	·3	·9787	·3	·4780	·3	·9293	·3	·2709	·6	·9142
·4	·7633	·4	·9799	·4	·5034	·4	·9328	·4	·2945	·8	·9214
1·5	0·7793	5·5	0·9810	1·5	0·5276	5·5	0·9361	1·5	0·3177	7·0	0·9281
·6	·7941	·6	·9820	·6	·5507	·6	·9392	·6	·3406	·2	·9342
·7	·8077	·7	·9830	·7	·5726	·7	·9422	·7	·3631	·4	·9398
·8	·8203	·8	·9840	·8	·5934	·8	·9450	·8	·3851	·6	·9450
·9	·8319	·9	·9849	·9	·6133	·9	·9477	·9	·4066	·8	·9497
2·0	0·8427	6·0	0·9857	2·0	0·6321	6·0	0·9502	2·0	0·4276	8·0	0·9540
·1	·8527	·1	·9865	·1	·6501	·2	·9550	·1	·4481	·2	·9579
·2	·8620	·2	·9872	·2	·6671	·4	·9592	·2	·4681	·4	·9616
·3	·8706	·3	·9879	·3	·6834	·6	·9631	·3	·4875	·6	·9649
·4	·8787	·4	·9886	·4	·6988	·8	·9666	·4	·5064	·8	·9679
2·5	0·8862	6·5	0·9892	2·5	0·7135	7·0	0·9698	2·5	0·5247	9·0	0·9707
·6	·8931	·6	·9898	·6	·7275	·2	·9727	·6	·5425	·2	·9733
·7	·8997	·7	·9904	·7	·7408	·4	·9753	·7	·5598	·4	·9756
·8	·9057	·8	·9909	·8	·7534	·6	·9776	·8	·5765	·6	·9777
·9	·9114	·9	·9914	·9	·7654	·8	·9798	·9	·5927	·8	·9797
3·0	0·9167	7·0	0·9918	3·0	0·7769	8·0	0·9817	3·0	0·6084	10·0	0·9814
·1	·9217	·1	·9923	·1	·7878	·2	·9834	·1	·6235	·2	·9831
·2	·9264	·2	·9927	·2	·7981	·4	·9850	·2	·6382	·4	·9845
·3	·9307	·3	·9931	·3	·8080	·6	·9864	·3	·6524	·6	·9859
·4	·9348	·4	·9935	·4	·8173	·8	·9877	·4	·6660	·8	·9871
3·5	0·9386	7·5	0·9938	3·5	0·8262	9·0	0·9889	3·5	0·6792	11·0	0·9883
·6	·9422	·6	·9942	·6	·8347	·2	·9899	·6	·6920	·2	·9893
·7	·9456	·7	·9945	·7	·8428	·4	·9909	·7	·7043	·4	·9903
·8	·9487	·8	·9948	·8	·8504	·6	·9918	·8	·7161	·6	·9911
·9	·9517	·9	·9951	·9	·8577	·8	·9926	·9	·7275	·8	·9919
1·0	0·9545	8·0	0·9953	4·0	0·8647	10·0	0·9933	4·0	0·7385	12·0	0·9926

Source: Lindley & Scott (1984, Table 7)

Tables and charts

A4 The distribution function of χ^2 for small degrees of freedom.
The function tabulated is

$$F_\nu(x) = [2^{\frac{1}{2}\nu}\Gamma(\tfrac{1}{2}\nu)]^{-1} \int_0^x t^{\frac{1}{2}\nu - 1} \exp(-\tfrac{1}{2}t) \, dt.$$

If Y is a χ^2_ν random variable, $F_\nu(x) = \text{Prob}\,(Y \leq x)$. Note that $F_1(x) = 2\Phi(x^{\frac{1}{2}}) - 1$ (cf. Appendix A2).

For $\nu > 3$, calculate $F_\nu(x)$ for given x as follows (cf. (4.12) in Chapter 4). Compute $u = 1/\{1 - \tfrac{1}{6}\log(x/\nu)\}$, $\mu = 1 - (9\nu)^{-1}$, $\sigma = (18\nu)^{-\frac{1}{2}} + (18\nu)^{-\frac{3}{2}}$, and then $y = (u - \mu)/\sigma$. Then $F_\nu(x) \approx \Phi(y)$ (cf. Appendix A2).

A5 Percentage points of the F-distribution.
The function tabulated is $F_{\nu_1,\nu_2}(\alpha)$, the value of F_{ν_1,ν_2} which has probability α of being exceeded. $F_{\nu_1,\nu_2}(\alpha)$ is the *upper P percent point*, $P = 100\alpha$. The function is tabulated for $P = 10, 5, 2.5, 1, 0.5$ and 0.1. The *lower* percentage point $F_{\nu_1,\nu_2}(1-\alpha)$, i.e. the value of F_{ν_1,ν_2} which has probability $1 - \alpha$ of being exceeded (or which corresponds to a *lower* tail area α) may be found by the formula

$$F_{\nu_1,\nu_2}(1-\alpha) = 1/F_{\nu_2,\nu_1}(\alpha).$$

(a) 10% points of the F-distribution
(b) 5% points of the F-distribution
(c) 2.5% points of the F-distribution
(d) 1% points of the F-distribution
(e) 0.5% points of the F-distribution
(f) 0.1% points of the F-distribution

Figure A5 (The diagram is for illustrative purposes. In fact the above shape applies only when $\nu_1 \geq 3$. When $\nu_1 < 3$ the mode is at the origin).

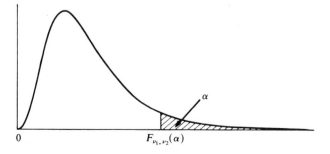

Table A5 (a) 10% points

$v_1 =$	1	2	3	4	5	6	7	8	10	12	24	∞
$v_2 = 1$	39·86	49·50	53·59	55·83	57·24	58·20	58·91	59·44	60·19	60·71	62·00	63·33
2	8·526	9·000	9·162	9·243	9·293	9·326	9·349	9·367	9·392	9·408	9·450	9·491
3	5·538	5·462	5·391	5·343	5·309	5·285	5·266	5·252	5·230	5·216	5·176	5·134
4	4·545	4·325	4·191	4·107	4·051	4·010	3·979	3·955	3·920	3·896	3·831	3·761
5	4·060	3·780	3·619	3·520	3·453	3·405	3·368	3·339	3·297	3·268	3·191	3·105
6	3·776	3·463	3·289	3·181	3·108	3·055	3·014	2·983	2·937	2·905	2·818	2·722
7	3·589	3·257	3·074	2·961	2·883	2·827	2·785	2·752	2·703	2·668	2·575	2·471
8	3·458	3·113	2·924	2·806	2·726	2·668	2·624	2·589	2·538	2·502	2·404	2·293
9	3·360	3·006	2·813	2·693	2·611	2·551	2·505	2·469	2·416	2·379	2·277	2·159
10	3·285	2·924	2·728	2·605	2·522	2·461	2·414	2·377	2·323	2·284	2·178	2·055
11	3·225	2·860	2·660	2·536	2·451	2·389	2·342	2·304	2·248	2·209	2·100	1·972
12	3·177	2·807	2·606	2·480	2·394	2·331	2·283	2·245	2·188	2·147	2·036	1·904
13	3·136	2·763	2·560	2·434	2·347	2·283	2·234	2·195	2·138	2·097	1·983	1·846
14	3·102	2·726	2·522	2·395	2·307	2·243	2·193	2·154	2·095	2·054	1·938	1·797
15	3·073	2·695	2·490	2·361	2·273	2·208	2·158	2·119	2·059	2·017	1·899	1·755
16	3·048	2·668	2·462	2·333	2·244	2·178	2·128	2·088	2·028	1·985	1·866	1·718
17	3·026	2·645	2·437	2·308	2·218	2·152	2·102	2·061	2·001	1·958	1·836	1·686
18	3·007	2·624	2·416	2·286	2·196	2·130	2·079	2·038	1·977	1·933	1·810	1·657
19	2·990	2·606	2·397	2·266	2·176	2·109	2·058	2·017	1·956	1·912	1·787	1·631
20	2·975	2·589	2·380	2·249	2·158	2·091	2·040	1·999	1·937	1·892	1·767	1·607
21	2·961	2·575	2·365	2·233	2·142	2·075	2·023	1·982	1·920	1·875	1·748	1·586
22	2·949	2·561	2·351	2·219	2·128	2·060	2·008	1·967	1·904	1·859	1·731	1·567
23	2·937	2·549	2·339	2·207	2·115	2·047	1·995	1·953	1·890	1·845	1·716	1·549
24	2·927	2·538	2·327	2·195	2·103	2·035	1·983	1·941	1·877	1·832	1·702	1·533
25	2·918	2·528	2·317	2·184	2·092	2·024	1·971	1·929	1·866	1·820	1·689	1·518
26	2·909	2·519	2·307	2·174	2·082	2·014	1·961	1·919	1·855	1·809	1·677	1·504
27	2·901	2·511	2·299	2·165	2·073	2·005	1·952	1·909	1·845	1·799	1·666	1·491
28	2·894	2·503	2·291	2·157	2·064	1·996	1·943	1·900	1·836	1·790	1·656	1·478
29	2·887	2·495	2·283	2·149	2·057	1·988	1·935	1·892	1·827	1·781	1·647	1·467
30	2·881	2·489	2·276	2·142	2·049	1·980	1·927	1·884	1·819	1·773	1·638	1·456
32	2·869	2·477	2·263	2·129	2·036	1·967	1·913	1·870	1·805	1·758	1·622	1·437
34	2·859	2·466	2·252	2·118	2·024	1·955	1·901	1·858	1·793	1·745	1·608	1·419
36	2·850	2·456	2·243	2·108	2·014	1·945	1·891	1·847	1·781	1·734	1·595	1·404
38	2·842	2·448	2·234	2·099	2·005	1·935	1·881	1·838	1·772	1·724	1·584	1·390
40	2·835	2·440	2·226	2·091	1·997	1·927	1·873	1·829	1·763	1·715	1·574	1·377
60	2·791	2·393	2·177	2·041	1·946	1·875	1·819	1·775	1·707	1·657	1·511	1·291
120	2·748	2·347	2·130	1·992	1·896	1·824	1·767	1·722	1·652	1·601	1·447	1·193
∞	2·706	2·303	2·084	1·945	1·847	1·774	1·717	1·670	1·599	1·546	1·383	1·000

Source: Lindley & Scott (1984, Table 12)

Table A5 (cont.) (b) 5% points

$\nu_1 =$	1	2	3	4	5	6	7	8	10	12	24	∞
$\nu_2 = 1$	161.4	199.5	215.7	224.6	230.2	234.0	236.8	238.9	241.9	243.9	249.1	254.3
2	18.51	19.00	19.16	19.25	19.30	19.33	19.35	19.37	19.40	19.41	19.45	19.50
3	10.13	9.552	9.277	9.117	9.013	8.941	8.887	8.845	8.786	8.745	8.639	8.526
4	7.709	6.944	6.591	6.388	6.256	6.163	6.094	6.041	5.964	5.912	5.774	5.628
5	6.608	5.786	5.409	5.192	5.050	4.950	4.876	4.818	4.735	4.678	4.527	4.365
6	5.987	5.143	4.757	4.534	4.387	4.284	4.207	4.147	4.060	4.000	3.841	3.669
7	5.591	4.737	4.347	4.120	3.972	3.866	3.787	3.726	3.637	3.575	3.410	3.230
8	5.318	4.459	4.066	3.838	3.687	3.581	3.500	3.438	3.347	3.284	3.115	2.928
9	5.117	4.256	3.863	3.633	3.482	3.374	3.293	3.230	3.137	3.073	2.900	2.707
10	4.965	4.103	3.708	3.478	3.326	3.217	3.135	3.072	2.978	2.913	2.737	2.538
11	4.844	3.982	3.587	3.357	3.204	3.095	3.012	2.948	2.854	2.788	2.609	2.404
12	4.747	3.885	3.490	3.259	3.106	2.996	2.913	2.849	2.753	2.687	2.505	2.296
13	4.667	3.806	3.411	3.179	3.025	2.915	2.832	2.767	2.671	2.604	2.420	2.206
14	4.600	3.739	3.344	3.112	2.958	2.848	2.764	2.699	2.602	2.534	2.349	2.131
15	4.543	3.682	3.287	3.056	2.901	2.790	2.707	2.641	2.544	2.475	2.288	2.066
16	4.494	3.634	3.239	3.007	2.852	2.741	2.657	2.591	2.494	2.425	2.235	2.010
17	4.451	3.592	3.197	2.965	2.810	2.699	2.614	2.548	2.450	2.381	2.190	1.960
18	4.414	3.555	3.160	2.928	2.773	2.661	2.577	2.510	2.412	2.342	2.150	1.917
19	4.381	3.522	3.127	2.895	2.740	2.628	2.544	2.477	2.378	2.308	2.114	1.878
20	4.351	3.493	3.098	2.866	2.711	2.599	2.514	2.447	2.348	2.278	2.082	1.843
21	4.325	3.467	3.072	2.840	2.685	2.573	2.488	2.420	2.321	2.250	2.054	1.812
22	4.301	3.443	3.049	2.817	2.661	2.549	2.464	2.397	2.297	2.226	2.028	1.783
23	4.279	3.422	3.028	2.796	2.640	2.528	2.442	2.375	2.275	2.204	2.005	1.757
24	4.260	3.403	3.009	2.776	2.621	2.508	2.423	2.355	2.255	2.183	1.984	1.733
25	4.242	3.385	2.991	2.759	2.603	2.490	2.405	2.337	2.236	2.165	1.964	1.711
26	4.225	3.369	2.975	2.743	2.587	2.474	2.388	2.321	2.220	2.148	1.946	1.691
27	4.210	3.354	2.960	2.728	2.572	2.459	2.373	2.305	2.204	2.132	1.930	1.672
28	4.196	3.340	2.947	2.714	2.558	2.445	2.359	2.291	2.190	2.118	1.915	1.654
29	4.183	3.328	2.934	2.701	2.545	2.432	2.346	2.278	2.177	2.104	1.901	1.638
30	4.171	3.316	2.922	2.690	2.534	2.421	2.334	2.266	2.165	2.092	1.887	1.622
32	4.149	3.295	2.901	2.668	2.512	2.399	2.313	2.244	2.142	2.070	1.864	1.594
34	4.130	3.276	2.883	2.650	2.494	2.380	2.294	2.225	2.123	2.050	1.843	1.569
36	4.113	3.259	2.866	2.634	2.477	2.364	2.277	2.209	2.106	2.033	1.824	1.547
38	4.098	3.245	2.852	2.619	2.463	2.349	2.262	2.194	2.091	2.017	1.808	1.527
40	4.085	3.232	2.839	2.606	2.449	2.336	2.249	2.180	2.077	2.003	1.793	1.509
60	4.001	3.150	2.758	2.525	2.368	2.254	2.167	2.097	1.993	1.917	1.700	1.389
120	3.920	3.072	2.680	2.447	2.290	2.175	2.087	2.016	1.910	1.834	1.608	1.254
∞	3.841	2.996	2.605	2.372	2.214	2.099	2.010	1.938	1.831	1.752	1.517	1.000

Table A5 (*cont.*) (c) 2.5% points

$\nu_1 =$	1	2	3	4	5	6	7	8	10	12	24	∞
$\nu_2 = 1$	647·8	799·5	864·2	899·6	921·8	937·1	948·2	956·7	968·6	976·7	997·2	1018
2	38·51	39·00	39·17	39·25	39·30	39·33	39·36	39·37	39·40	39·41	39·46	39·50
3	17·44	16·04	15·44	15·10	14·88	14·73	14·62	14·54	14·42	14·34	14·12	13·90
4	12·22	10·65	9·979	9·605	9·364	9·197	9·074	8·980	8·844	8·751	8·511	8·257
5	10·01	8·434	7·764	7·388	7·146	6·978	6·853	6·757	6·619	6·525	6·278	6·015
6	8·813	7·260	6·599	6·227	5·988	5·820	5·695	5·600	5·461	5·366	5·117	4·849
7	8·073	6·542	5·890	5·523	5·285	5·119	4·995	4·899	4·761	4·666	4·415	4·142
8	7·571	6·059	5·416	5·053	4·817	4·652	4·529	4·433	4·295	4·200	3·947	3·670
9	7·209	5·715	5·078	4·718	4·484	4·320	4·197	4·102	3·964	3·868	3·614	3·333
10	6·937	5·456	4·826	4·468	4·236	4·072	3·950	3·855	3·717	3·621	3·365	3·080
11	6·724	5·256	4·630	4·275	4·044	3·881	3·759	3·664	3·526	3·430	3·173	2·883
12	6·554	5·096	4·474	4·121	3·891	3·728	3·607	3·512	3·374	3·277	3·019	2·725
13	6·414	4·965	4·347	3·996	3·767	3·604	3·483	3·388	3·250	3·153	2·893	2·595
14	6·298	4·857	4·242	3·892	3·663	3·501	3·380	3·285	3·147	3·050	2·789	2·487
15	6·200	4·765	4·153	3·804	3·576	3·415	3·293	3·199	3·060	2·963	2·701	2·395
16	6·115	4·687	4·077	3·729	3·502	3·341	3·219	3·125	2·986	2·889	2·625	2·316
17	6·042	4·619	4·011	3·665	3·438	3·277	3·156	3·061	2·922	2·825	2·560	2·247
18	5·978	4·560	3·954	3·608	3·382	3·221	3·100	3·005	2·866	2·769	2·503	2·187
19	5·922	4·508	3·903	3·559	3·333	3·172	3·051	2·956	2·817	2·720	2·452	2·133
20	5·871	4·461	3·859	3·515	3·289	3·128	3·007	2·913	2·774	2·676	2·408	2·085
21	5·827	4·420	3·819	3·475	3·250	3·090	2·969	2·874	2·735	2·637	2·368	2·042
22	5·786	4·383	3·783	3·440	3·215	3·055	2·934	2·839	2·700	2·602	2·331	2·003
23	5·750	4·349	3·750	3·408	3·183	3·023	2·902	2·808	2·668	2·570	2·299	1·968
24	5·717	4·319	3·721	3·379	3·155	2·995	2·874	2·779	2·640	2·541	2·269	1·935
25	5·686	4·291	3·694	3·353	3·129	2·969	2·848	2·753	2·613	2·515	2·242	1·906
26	5·659	4·265	3·670	3·329	3·105	2·945	2·824	2·729	2·590	2·491	2·217	1·878
27	5·633	4·242	3·647	3·307	3·083	2·923	2·802	2·707	2·568	2·469	2·195	1·853
28	5·610	4·221	3·626	3·286	3·063	2·903	2·782	2·687	2·547	2·448	2·174	1·829
29	5·588	4·201	3·607	3·267	3·044	2·884	2·763	2·669	2·529	2·430	2·154	1·807
30	5·568	4·182	3·589	3·250	3·026	2·867	2·746	2·651	2·511	2·412	2·136	1·787
32	5·531	4·149	3·557	3·218	2·995	2·836	2·715	2·620	2·480	2·381	2·103	1·750
34	5·499	4·120	3·529	3·191	2·968	2·808	2·688	2·593	2·453	2·353	2·075	1·717
36	5·471	4·094	3·505	3·167	2·944	2·785	2·664	2·569	2·429	2·329	2·049	1·687
38	5·446	4·071	3·483	3·145	2·923	2·763	2·643	2·548	2·407	2·307	2·027	1·661
40	5·424	4·051	3·463	3·126	2·904	2·744	2·624	2·529	2·388	2·288	2·007	1·637
60	5·286	3·925	3·343	3·008	2·786	2·627	2·507	2·412	2·270	2·169	1·882	1·482
120	5·152	3·805	3·227	2·894	2·674	2·515	2·395	2·299	2·157	2·055	1·760	1·310
∞	5·024	3·689	3·116	2·786	2·567	2·408	2·288	2·192	2·048	1·945	1·640	1·000

Table A5 (*cont.*) (d) 1% points

$\nu_1 =$	1	2	3	4	5	6	7	8	10	12	24	∞
$\nu_2 = 1$	4052	4999	5403	5625	5764	5859	5928	5981	6056	6106	6235	6366
2	98·50	99·00	99·17	99·25	99·30	99·33	99·36	99·37	99·40	99·42	99·46	99·50
3	34·12	30·82	29·46	28·71	28·24	27·91	27·67	27·49	27·23	27·05	26·60	26·13
4	21·20	18·00	16·69	15·98	15·52	15·21	14·98	14·80	14·55	14·37	13·93	13·46
5	16·26	13·27	12·06	11·39	10·97	10·67	10·46	10·29	10·05	9·888	9·466	9·020
6	13·75	10·92	9·780	9·148	8·746	8·466	8·260	8·102	7·874	7·718	7·313	6·880
7	12·25	9·547	8·451	7·847	7·460	7·191	6·993	6·840	6·620	6·469	6·074	5·650
8	11·26	8·649	7·591	7·006	6·632	6·371	6·178	6·029	5·814	5·667	5·279	4·859
9	10·56	8·022	6·992	6·422	6·057	5·802	5·613	5·467	5·257	5·111	4·729	4·311
10	10·04	7·559	6·552	5·994	5·636	5·386	5·200	5·057	4·849	4·706	4·327	3·909
11	9·646	7·206	6·217	5·668	5·316	5·069	4·886	4·744	4·539	4·397	4·021	3·602
12	9·330	6·927	5·953	5·412	5·064	4·821	4·640	4·499	4·296	4·155	3·780	3·361
13	9·074	6·701	5·739	5·205	4·862	4·620	4·441	4·302	4·100	3·960	3·587	3·165
14	8·862	6·515	5·564	5·035	4·695	4·456	4·278	4·140	3·939	3·800	3·427	3·004
15	8·683	6·359	5·417	4·893	4·556	4·318	4·142	4·004	3·805	3·666	3·294	2·868
16	8·531	6·226	5·292	4·773	4·437	4·202	4·026	3·890	3·691	3·553	3·181	2·753
17	8·400	6·112	5·185	4·669	4·336	4·102	3·927	3·791	3·593	3·455	3·084	2·653
18	8·285	6·013	5·092	4·579	4·248	4·015	3·841	3·705	3·508	3·371	2·999	2·566
19	8·185	5·926	5·010	4·500	4·171	3·939	3·765	3·631	3·434	3·297	2·925	2·489
20	8·096	5·849	4·938	4·431	4·103	3·871	3·699	3·564	3·368	3·231	2·859	2·421
21	8·017	5·780	4·874	4·369	4·042	3·812	3·640	3·506	3·310	3·173	2·801	2·360
22	7·945	5·719	4·817	4·313	3·988	3·758	3·587	3·453	3·258	3·121	2·749	2·305
23	7·881	5·664	4·765	4·264	3·939	3·710	3·539	3·406	3·211	3·074	2·702	2·256
24	7·823	5·614	4·718	4·218	3·895	3·667	3·496	3·363	3·168	3·032	2·659	2·211
25	7·770	5·568	4·675	4·177	3·855	3·627	3·457	3·324	3·129	2·993	2·620	2·169
26	7·721	5·526	4·637	4·140	3·818	3·591	3·421	3·288	3·094	2·958	2·585	2·131
27	7·677	5·488	4·601	4·106	3·785	3·558	3·388	3·256	3·062	2·926	2·552	2·097
28	7·636	5·453	4·568	4·074	3·754	3·528	3·358	3·226	3·032	2·896	2·522	2·064
29	7·598	5·420	4·538	4·045	3·725	3·499	3·330	3·198	3·005	2·868	2·495	2·034
30	7·562	5·390	4·510	4·018	3·699	3·473	3·304	3·173	2·979	2·843	2·469	2·006
32	7·499	5·336	4·459	3·969	3·652	3·427	3·258	3·127	2·934	2·798	2·423	1·956
34	7·444	5·289	4·416	3·927	3·611	3·386	3·218	3·087	2·894	2·758	2·383	1·911
36	7·396	5·248	4·377	3·890	3·574	3·351	3·183	3·052	2·859	2·723	2·347	1·872
38	7·353	5·211	4·343	3·858	3·542	3·319	3·152	3·021	2·828	2·692	2·316	1·837
40	7·314	5·179	4·313	3·828	3·514	3·291	3·124	2·993	2·801	2·665	2·288	1·805
60	7·077	4·977	4·126	3·649	3·339	3·119	2·953	2·823	2·632	2·496	2·115	1·601
120	6·851	4·787	3·949	3·480	3·174	2·956	2·792	2·663	2·472	2·336	1·950	1·381
∞	6·635	4·605	3·782	3·319	3·017	2·802	2·639	2·511	2·321	2·185	1·791	1·000

Table A5 (*cont.*) (e) 0.5% points

$v_1 =$	1	2	3	4	5	6	7	8	10	12	24	∞
$v_2 = 1$	16211	20000	21615	22500	23056	23437	23715	23925	24224	24426	24940	25464
2	198·5	199·0	199·2	199·2	199·3	199·3	199·4	199·4	199·4	199·4	199·5	199·5
3	55·55	49·80	47·47	46·19	45·39	44·84	44·43	44·13	43·69	43·39	42·62	41·83
4	31·33	26·28	24·26	23·15	22·46	21·97	21·62	21·35	20·97	20·70	20·03	19·32
5	22·78	18·31	16·53	15·56	14·94	14·51	14·20	13·96	13·62	13·38	12·78	12·14
6	18·63	14·54	12·92	12·03	11·46	11·07	10·79	10·57	10·25	10·03	9·474	8·879
7	16·24	12·40	10·88	10·05	9·522	9·155	8·885	8·678	8·380	8·176	7·645	7·076
8	14·69	11·04	9·596	8·805	8·302	7·952	7·694	7·496	7·211	7·015	6·503	5·951
9	13·61	10·11	8·717	7·956	7·471	7·134	6·885	6·693	6·417	6·227	5·729	5·188
10	12·83	9·427	8·081	7·343	6·872	6·545	6·302	6·116	5·847	5·661	5·173	4·639
11	12·23	8·912	7·600	6·881	6·422	6·102	5·865	5·682	5·418	5·236	4·756	4·226
12	11·75	8·510	7·226	6·521	6·071	5·757	5·525	5·345	5·085	4·906	4·431	3·904
13	11·37	8·186	6·926	6·233	5·791	5·482	5·253	5·076	4·820	4·643	4·173	3·647
14	11·06	7·922	6·680	5·998	5·562	5·257	5·031	4·857	4·603	4·428	3·961	3·436
15	10·80	7·701	6·476	5·803	5·372	5·071	4·847	4·674	4·424	4·250	3·786	3·260
16	10·58	7·514	6·303	5·638	5·212	4·913	4·692	4·521	4·272	4·099	3·638	3·112
17	10·38	7·354	6·156	5·497	5·075	4·779	4·559	4·389	4·142	3·971	3·511	2·984
18	10·22	7·215	6·028	5·375	4·956	4·663	4·445	4·276	4·030	3·860	3·402	2·873
19	10·07	7·093	5·916	5·268	4·853	4·561	4·345	4·177	3·933	3·763	3·306	2·776
20	9·944	6·986	5·818	5·174	4·762	4·472	4·257	4·090	3·847	3·678	3·222	2·690
21	9·830	6·891	5·730	5·091	4·681	4·393	4·179	4·013	3·771	3·602	3·147	2·614
22	9·727	6·806	5·652	5·017	4·609	4·322	4·109	3·944	3·703	3·535	3·081	2·545
23	9·635	6·730	5·582	4·950	4·544	4·259	4·047	3·882	3·642	3·475	3·021	2·484
24	9·551	6·661	5·519	4·890	4·486	4·202	3·991	3·826	3·587	3·420	2·967	2·428
25	9·475	6·598	5·462	4·835	4·433	4·150	3·939	3·776	3·537	3·370	2·918	2·377
26	9·406	6·541	5·409	4·785	4·384	4·103	3·893	3·730	3·492	3·325	2·873	2·330
27	9·342	6·489	5·361	4·740	4·340	4·059	3·850	3·687	3·450	3·284	2·832	2·287
28	9·284	6·440	5·317	4·698	4·300	4·020	3·811	3·649	3·412	3·246	2·794	2·247
29	9·230	6·396	5·276	4·659	4·262	3·983	3·775	3·613	3·377	3·211	2·759	2·210
30	9·180	6·355	5·239	4·623	4·228	3·949	3·742	3·580	3·344	3·179	2·727	2·176
32	9·090	6·281	5·171	4·559	4·166	3·889	3·682	3·521	3·286	3·121	2·670	2·114
34	9·012	6·217	5·113	4·504	4·112	3·836	3·630	3·470	3·235	3·071	2·620	2·060
36	8·943	6·161	5·062	4·455	4·065	3·790	3·585	3·425	3·191	3·027	2·576	2·013
38	8·882	6·111	5·016	4·412	4·023	3·749	3·545	3·385	3·152	2·988	2·537	1·970
40	8·828	6·066	4·976	4·374	3·986	3·713	3·509	3·350	3·117	2·953	2·502	1·932
60	8·495	5·795	4·729	4·140	3·760	3·492	3·291	3·134	2·904	2·742	2·290	1·689
120	8·179	5·539	4·497	3·921	3·548	3·285	3·087	2·933	2·705	2·544	2·089	1·431
∞	7·879	5·298	4·279	3·715	3·350	3·091	2·897	2·744	2·519	2·358	1·898	1·000

Table A5 (cont.) (f) 0.1% points

$\nu_1 =$	1	2	3	4	5	6	7	8	10	12	24	∞
$\nu_2 = 1$*	4053	5000	5404	5625	5764	5859	5929	5981	6056	6107	6235	6366
2	998.5	999.0	999.2	999.2	999.3	999.3	999.4	999.4	999.4	999.4	999.5	999.5
3	167.0	148.5	141.1	137.1	134.6	132.8	131.6	130.6	129.2	128.3	125.9	123.5
4	74.14	61.25	56.18	53.44	51.71	50.53	49.66	49.00	48.05	47.41	45.77	44.05
5	47.18	37.12	33.20	31.09	29.75	28.83	28.16	27.65	26.92	26.42	25.13	23.79
6	35.51	27.00	23.70	21.92	20.80	20.03	19.46	19.03	18.41	17.99	16.90	15.75
7	29.25	21.69	18.77	17.20	16.21	15.52	15.02	14.63	14.08	13.71	12.73	11.70
8	25.41	18.49	15.83	14.39	13.48	12.86	12.40	12.05	11.54	11.19	10.30	9.334
9	22.86	16.39	13.90	12.56	11.71	11.13	10.70	10.37	9.894	9.570	8.724	7.813
10	21.04	14.91	12.55	11.28	10.48	9.926	9.517	9.204	8.754	8.445	7.638	6.762
11	19.69	13.81	11.56	10.35	9.578	9.047	8.655	8.355	7.922	7.626	6.847	5.998
12	18.64	12.97	10.80	9.633	8.892	8.379	8.001	7.710	7.292	7.005	6.249	5.420
13	17.82	12.31	10.21	9.073	8.354	7.856	7.489	7.206	6.799	6.519	5.781	4.967
14	17.14	11.78	9.729	8.622	7.922	7.436	7.077	6.802	6.404	6.130	5.407	4.604
15	16.59	11.34	9.335	8.253	7.567	7.092	6.741	6.471	6.081	5.812	5.101	4.307
16	16.12	10.97	9.006	7.944	7.272	6.805	6.460	6.195	5.812	5.547	4.846	4.059
17	15.72	10.66	8.727	7.683	7.022	6.562	6.223	5.962	5.584	5.324	4.631	3.850
18	15.38	10.39	8.487	7.459	6.808	6.355	6.021	5.763	5.390	5.132	4.447	3.670
19	15.08	10.16	8.280	7.265	6.622	6.175	5.845	5.590	5.222	4.967	4.288	3.514
20	14.82	9.953	8.098	7.096	6.461	6.019	5.692	5.440	5.075	4.823	4.149	3.378
21	14.59	9.772	7.938	6.947	6.318	5.881	5.557	5.308	4.946	4.696	4.027	3.257
22	14.38	9.612	7.796	6.814	6.191	5.758	5.438	5.190	4.832	4.583	3.919	3.151
23	14.20	9.469	7.669	6.696	6.078	5.649	5.331	5.085	4.730	4.483	3.822	3.055
24	14.03	9.339	7.554	6.589	5.977	5.550	5.235	4.991	4.638	4.393	3.735	2.969
25	13.88	9.223	7.451	6.493	5.885	5.462	5.148	4.906	4.555	4.312	3.657	2.890
26	13.74	9.116	7.357	6.406	5.802	5.381	5.070	4.829	4.480	4.238	3.586	2.819
27	13.61	9.019	7.272	6.326	5.726	5.308	4.998	4.759	4.412	4.171	3.521	2.754
28	13.50	8.931	7.193	6.253	5.656	5.241	4.933	4.695	4.349	4.109	3.462	2.695
29	13.39	8.849	7.121	6.186	5.593	5.179	4.873	4.636	4.292	4.053	3.407	2.640
30	13.29	8.773	7.054	6.125	5.534	5.122	4.817	4.581	4.239	4.001	3.357	2.589
32	13.12	8.639	6.936	6.014	5.429	5.021	4.719	4.485	4.145	3.908	3.268	2.498
34	12.97	8.522	6.833	5.919	5.339	4.934	4.633	4.401	4.063	3.828	3.191	2.419
36	12.83	8.420	6.744	5.836	5.260	4.857	4.559	4.328	3.992	3.758	3.123	2.349
38	12.71	8.331	6.665	5.763	5.190	4.790	4.494	4.264	3.930	3.697	3.064	2.288
40	12.61	8.251	6.595	5.698	5.128	4.731	4.436	4.207	3.874	3.642	3.011	2.233
60	11.97	7.768	6.171	5.307	4.757	4.372	4.086	3.865	3.541	3.315	2.694	1.890
120	11.38	7.321	5.781	4.947	4.416	4.044	3.767	3.552	3.237	3.016	2.402	1.543
∞	10.83	6.908	5.422	4.617	4.103	3.743	3.475	3.266	2.959	2.742	2.132	1.000

* Entries in the row $\nu_2 = 1$ must be multiplied by 100.

A6   Percentage points of D_n^* and V_n^*.

The tabulated values are percentage points of the modified goodness-of-fit statistics D_n^* and V_n^*.

Table A6

Statistic	Percentage point ($100\alpha\%$)			
	15%	10%	5%	1%
D_n^*	1.138	1.224	1.358	1.628
V_n^*	1.537	1.620	1.747	2.001

Source: Stephens (1974), Table 1A.

A7   Percentage points of R for the uniform distribution.

The tabulated values are percentiles ($100\alpha\%$) of the distribution of the resultant length R in samples of size n from the uniform distribution on the sphere.

Table A7

Sample size n	Percentage point ($100\alpha\%$)			
	10%	5%	2%	1%
4	2.85	3.10	3.35	3.49
5	3.19	3.50	3.83	4.02
6	3.50	3.85	4.24	4.48
7	3.78	4.18	4.61	4.89
8	4.05	4.48	4.96	5.26
9	4.30	4.76	5.28	5.61
10	4.54	5.03	5.58	5.94
11	4.76	5.28	5.87	6.25
12	4.97	5.52	6.14	6.55
13	5.18	5.75	6.40	6.83
14	5.38	5.98	6.65	7.10
15	5.57	6.19	6.90	7.37
16	5.75	6.40	7.13	7.62
17	5.93	6.60	7.36	7.86
18	6.10	6.79	7.58	8.10
19	6.27	6.98	7.79	8.33
20	6.44	7.17	8.00	8.55
21	6.60	7.35	8.20	8.77
22	6.75	7.52	8.40	8.99
23	6.90	7.69	8.59	9.19
24	7.05	7.86	8.78	9.40
25	7.20	8.02	8.96	9.60

Source: Adapted from Stephens (1964, Table 1).

Tables and charts

A8 Percentage points of $M_E(D_n)$, $M_U(V_n)$ and $M_N(D_n)$.
The tabulated values are percentage points of the modified statistics for testing goodness-of-fit of a Fisher distribution.

Table A8

Statistic	Percentage point (100α%)		
	10%	5%	1%
$M_E(D_n)$	0.990	1.094	1.308
$M_U(V_n)$	1.138	1.207	1.347
$M_N(D_n)$	0.819	0.895	1.035

Source: Fisher & Best (1984), Stephens (1974, Table 1A).

A9 Percentage points of $E_n^{(t)}$.
The tabulated values are percentage points of the statistic used for testing discordancy of t outliers in a sample of size n from a Fisher distribution; the same percentiles apply to the statistic $H_n^{(t)}$ for testing discordancy of t outliers in a sample of size n from a Watson bipolar distribution.

Table A9

Sample size n	Percentage point (100α%) (t = 2)				
	10%	5%	2.5%	1%	0.1%
10	6.59	7.86	9.16	11.18	15.85
15	5.92	6.79	7.73	8.95	12.90
20	5.84	6.54	7.29	8.42	10.84
30	5.80	6.46	7.06	7.95	10.74
50	5.75	6.41	6.86	7.53	8.39
	Percentage point (t = 3)				
15	6.10	7.03	7.88	9.07	12.58
20	5.79	6.42	7.14	8.16	10.33
30	5.59	6.17	6.70	7.47	9.35
50	5.55	6.01	6.43	7.06	7.47

Source: Best & Fisher (1986, Table 3).

A10 Solution of the equation $A(\kappa_V) = x$ and of the equation $D(\kappa_A) = x$, $0 \le x \le 1$, where $A(\kappa_V) = \coth(\kappa_V) - 1/\kappa_V$ and

$$D(\kappa_A) = \int_0^1 y^2 \exp(\kappa_A y^2) \, dy \bigg/ \int_0^1 \exp(\kappa_A y^2) \, dy.$$

Denote these solutions by $A^{-1}(x)$ and $D^{-1}(x)$ respectively. If x is the mean resultant length $\bar{R}$ of a random sample of vectors from a Fisher distribution, then κ_V is the maximum likelihood estimate of the concentration parameter. If x is the maximum [minimum] normalised eigenvalue $\bar{\tau}_3$ [$\bar{\tau}_1$] of the orientation matrix calculated from a Watson bipolar [girdle] distribution, so that $\frac{1}{3} < x \leq 1$ [$0 \leq x < \frac{1}{3}$], then κ_A is the maximum likelihood estimate of the concentration parameter.

Table A10

x	$A^{-1}(x)$	$D^{-1}(x)$	x	$A^{-1}(x)$	$D^{-1}(x)$
0.0005	0.002	−1000.0	0.145	0.441	−3.149
0.001	0.003	−500.0	0.150	0.456	−3.007
0.002	0.006	−250.0	0.155	0.472	−2.871
0.003	0.009	−166.7	0.160	0.488	−2.742
0.004	0.012	−125.0	0.165	0.503	−2.618
0.005	0.015	−100.0	0.170	0.519	−2.499
0.006	0.018	−83.33	0.175	0.535	−2.385
0.007	0.021	−71.43	0.180	0.551	−2.276
0.008	0.024	−62.56	0.185	0.567	−2.170
0.009	0.027	−55.60	0.190	0.583	−2.068
0.010	0.030	−50.04	0.195	0.599	−1.970
			0.200	0.615	−1.874
0.015	0.045	−33.36	0.205	0.631	−1.782
0.020	0.060	−25.01	0.210	0.647	−1.692
0.025	0.075	−20.01	0.215	0.664	−1.605
0.030	0.090	−16.67	0.220	0.680	−1.520
0.035	0.105	−14.29	0.225	0.697	−1.438
0.040	0.120	−12.50	0.230	0.713	−1.357
0.045	0.135	−11.11	0.235	0.730	−1.279
0.050	0.150	−10.00	0.240	0.746	−1.202
0.055	0.165	−9.091	0.245	0.763	−1.127
0.060	0.180	−8.329	0.250	0.780	−1.053
0.065	0.195	−7.684	0.255	0.797	−0.981
0.070	0.211	−7.128	0.260	0.814	−0.911
0.075	0.226	−6.643	0.265	0.831	−0.842
0.080	0.241	−6.217	0.270	0.848	−0.774
0.085	0.256	−5.837	0.275	0.865	−0.707
0.090	0.271	−5.497	0.280	0.883	−0.642
0.095	0.287	−5.189	0.285	0.900	−0.577
0.100	0.302	−4.909	0.290	0.918	−0.514
0.105	0.317	−4.652	0.295	0.935	−0.451
0.110	0.332	−4.416	0.300	0.953	−0.390
0.115	0.348	−4.197	0.305	0.971	−0.329
0.120	0.363	−3.993	0.310	0.989	−0.269
0.125	0.379	−3.803	0.315	1.007	−0.210
0.130	0.394	−3.625	0.320	1.025	−0.152
0.135	0.410	−3.457	0.325	1.044	−0.094
0.140	0.425	−3.299	0.330	1.062	−0.037

Table A10 (*cont.*)

x	$A^{-1}(x)$	$D^{-1}(x)$	x	$A^{-1}(x)$	$D^{-1}(x)$
0.331	1.066	−0.026	0.720	3.551	4.233
0.332	1.070	−0.015	0.730	3.687	4.392
0.333	1.073	−0.003	0.740	3.832	4.559
0.334	1.077	0.008	0.750	3.989	4.735
0.335	1.081	0.019	0.760	4.158	4.921
0.336	1.085	0.030	0.770	4.341	5.119
0.337	1.088	0.041	0.780	4.541	5.330
0.338	1.092	0.053	0.790	4.759	5.557
0.339	1.096	0.064	0.800	4.998	5.802
0.340	1.100	0.075	0.810	5.262	6.069
			0.820	5.555	6.361
0.350	1.137	0.185	0.830	5.882	6.683
0.360	1.176	0.293	0.840	6.250	7.042
0.370	1.215	0.399	0.850	6.667	7.447
0.380	1.255	0.503	0.860	7.143	7.907
0.390	1.295	0.606	0.870	7.692	8.437
0.400	1.336	0.708	0.880	8.333	9.057
0.410	1.378	0.809	0.890	9.091	9.792
0.420	1.421	0.909	0.900	10.00	10.68
0.430	1.464	1.008			
0.440	1.508	1.107	0.905	10.53	11.20
0.450	1.554	1.205	0.910	11.11	11.77
0.460	1.600	1.303	0.915	11.76	12.42
0.470	1.647	1.400	0.920	12.50	13.14
0.480	1.696	1.498	0.925	13.33	13.97
0.490	1.746	1.595	0.930	14.29	14.92
0.500	1.797	1.693	0.935	15.38	16.02
0.510	1.849	1.791	0.940	16.67	17.30
0.520	1.903	1.890	0.945	18.18	18.82
0.530	1.958	1.989	0.950	20.00	20.64
0.540	2.015	2.088	0.955	22.22	22.87
0.550	2.074	2.189	0.960	25.00	25.67
0.560	2.135	2.291	0.965	28.57	29.27
0.570	2.198	2.394	0.970	33.33	34.09
0.580	2.263	2.498	0.975	40.00	40.85
0.590	2.330	2.603	0.980	50.00	51.03
0.600	2.401	2.711	0.985	66.67	67.18
0.610	2.473	2.820	0.990	100.0	100.5
0.620	2.549	2.932			
0.630	2.628	3.046	0.991	111.1	111.6
0.640	2.711	3.162	0.992	125.0	125.5
0.650	2.798	3.282	0.993	142.9	143.4
0.660	2.888	3.404	0.994	166.7	167.2
0.670	2.984	3.531	0.995	200.0	200.5
0.680	3.085	3.661	0.996	250.0	250.5
0.690	3.191	3.796	0.997	333.3	333.8
0.700	3.304	3.936	0.998	500.0	500.5
0.710	3.423	4.082	0.999	1000.0	1000.5

Source: Specially calculated for this book.

A11 Percentage points of $\bar{R}^*$ for the Fisher distribution.

For given n, α and κ, the tabulated values $\bar{R}^*_\alpha$ are the 100α percent points of $\bar{R}^*$, where $\bar{R}^* = (1/n)\sum_{i=1}^{n} \cos\theta_i$, and $(\theta_1, \phi_1), \ldots, (\theta_n, \phi_n)$ are independent observations from a Fisher $F\{0, \kappa\}$ distribution; that is, Prob $(\bar{R}^* \geq \bar{R}^*_\alpha) = \alpha$. In descending order, the points given are those for which $\alpha = 0.01, 0.05, 0.95$ and 0.99.

Table A11

n \ κ	0·0	0·5	1·0	1·5	2·0	2·5	3·0	3·5	4·0	4·5	5·0
4	0·6500	0·7560	0·8264	0·8708	0·8993	0·9182	0·9315	0·9412	0·9486	0·9543	0·9588
	·4767	·6148	·7177	·7873	·8334	·8645	·8865	·9025	·9146	·9241	·9317
	−·4767	−·3156	−·1462	·0146	·1568	·2765	·3741	·4533	·5178	·5700	·6126
	−·6500	−·5071	−·3465	−·1838	−·0328	·0993	·2115	·3047	·3816	·4455	·4990
5	0·5851	0·7036	0·7859	0·8400	0·8750	0·8985	0·9149	0·9269	0·9360	0·9431	0·9488
	·4261	·5679	·6782	·7558	·8082	·8439	·8691	·8875	·9015	·9124	·9212
	−·4261	−·2630	−·0948	·0629	·2007	·3156	·4090	·4842	·5448	·5940	·6342
	−·5851	−·4369	−·2730	−·1120	·0348	·1618	·2682	·3560	·4282	·4877	·5371
6	0·5366	0·6618	0·7533	0·8143	0·8548	0·8820	0·9011	0·9150	0·9256	0·9338	0·9405
	·3887	·5334	·6482	·7312	·7883	·8276	·8554	·8757	·8911	·9032	·9129
	−·3887	−·2245	−·0573	·0979	·2325	·3440	·4342	·5065	·5645	·6115	·6499
	−·5366	−·3844	−·2203	−·0606	·0838	·2067	·3089	·3927	·4613	·5177	·5643
7	0·4979	0·6286	0·7261	0·7931	0·8377	0·8681	0·8894	0·9050	0·9167	0·9260	0·9333
	·3597	·5066	·6246	·7115	·7722	·8143	·8442	·8661	·8827	·8957	·9061
	−·3597	−·1950	−·0287	·1248	·2569	·3657	·4534	·5235	·5796	·6250	·6620
	−·4979	−·3434	−·1789	−·0207	·1213	·2409	·3398	·4206	·4865	·5404	·5851
8	0·4671	0·6010	0·7030	0·7748	0·8229	0·8561	0·8793	0·8963	0·9092	0·9192	0·9273
	·3364	·4848	·6054	·6953	·7588	·8032	·8348	·8580	·8757	·8894	·9005
	−·3364	−·1713	−·0056	·1463	·2763	·3830	·4687	·5370	·5917	·6357	·6716
	−·4671	−·3104	−·1457	·0112	·1502	·2681	·3643	·4426	·5063	·5583	·6012
9	0·4413	0·5774	0·6838	0·7591	0·8105	0·8456	0·8706	0·8888	0·9026	0·9134	0·9220
	·3171	·4669	·5894	·6817	·7475	·7938	·8269	·8512	·8697	·8841	·8957
	−·3171	−·1516	·0133	·1638	·2923	·3973	·4813	·5482	·6015	·6445	·6796
	−·4413	−·2830	−·1183	·0375	·1746	·2903	·3843	·4606	·5224	·5728	·6144
10	0·4193	0·5583	0·6670	0·7452	0·7993	0·8363	0·8628	0·8822	0·8968	0·9082	0·9174
	·3008	·4517	·5758	·6700	·7378	·7857	·8200	·8452	·8644	·8795	·8915
	−·3008	−·1351	·0293	·1786	·3057	·4092	·4918	·5575	·6098	·6519	·6862
	−·4193	−·2598	−·0952	·0595	·1950	·3083	·4011	·4756	·5359	·5850	·6254
12	0·3837	0·5268	0·6393	0·7224	0·7806	0·8211	0·8497	0·8709	0·8870	0·8995	0·9095
	·2745	·4267	·5538	·6510	·7218	·7723	·8086	·8354	·8558	·8718	·8846
	−·2745	−·1085	·0549	·2023	·3269	·4282	·5086	·5723	·6230	·6637	·6969
	−·3837	−·2223	−·0581	·0947	·2274	·3377	·4271	·4993	·5572	·6042	·6429
16	0·333	0·480	0·599	0·689	0·753	0·798	0·830	0·854	0·872	0·886	0·898
	·238	·392	·522	·624	·699	·753	·792	·821	·843	·861	·875
	−·238	−·071	·090	·235	·356	·454	·532	·593	·641	·680	·712
	−·333	−·170	−·007	·143	·272	·378	·463	·531	·586	·630	·666
20	0·299	0·448	0·571	0·665	0·733	0·781	0·816	0·842	0·862	0·877	0·889
	·212	·368	·501	·605	·683	·739	·780	·811	·834	·853	·867
	−·212	−·046	·115	·257	·376	·472	·547	·607	·653	·691	·722
	−·299	−·134	·028	·176	·302	·405	·487	·553	·605	·648	·682
30	0·246	0·397	0·526	0·627	0·701	0·754	0·793	0·822	0·844	0·861	0·875
	·174	·331	·467	·576	·657	·717	·762	·795	·820	·840	·856
	−·174	−·007	·152	·292	·407	·499	·571	·628	·672	·708	·737
	−·246	−·079	·082	·226	·348	·446	·524	·586	·634	·674	·706
40	0·204	0·367	0·499	0·603	0·680	0·737	0·778	0·809	0·833	0·851	0·866
	·150	·309	·447	·558	·642	·704	·750	·785	·811	·832	·849
	−·150	·016	·174	·312	·425	·515	·585	·640	·683	·718	·746
	−·204	−·046	·114	·256	·375	·470	·545	·604	·651	·689	·720
60	0·173	0·330	0·466	0·574	0·656	0·716	0·760	0·794	0·819	0·839	0·855
	·122	·283	·423	·536	·623	·688	·737	·773	·801	·823	·841
	−·122	·043	·200	·335	·446	·534	·602	·654	·696	·729	·756
	−·173	−·007	·151	·291	·406	·498	·570	·626	·671	·706	·735
100	0·134	0·293	0·432	0·545	0·630	0·694	0·742	0·777	0·805	0·826	0·844
	·175	·256	·398	·515	·604	·672	·722	·761	·790	·813	·832
	−·175	·071	−·226	·359	·467	·552	·618	·669	·709	·740	·766
	−·134	·032	·188	·325	·437	·525	·594	·648	·690	·723	·751
∞	0·000	0·164	0·313	0·438	0·537	0·613	0·672	0·716	0·751	0·778	0·800

Source: Stephens (1967, Table 1); reproduced with kind permission of the author and the Biometrika Trust.

Table A12

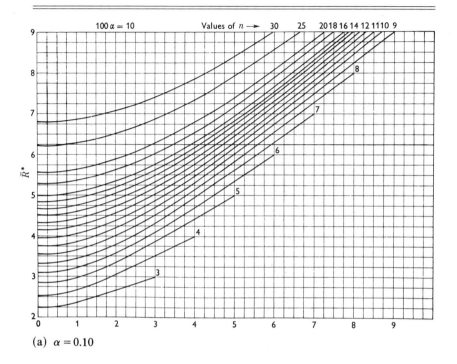

(a) $\alpha = 0.10$

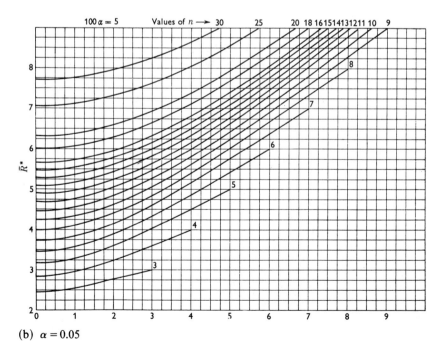

(b) $\alpha = 0.05$

Table A12 (*cont.*)

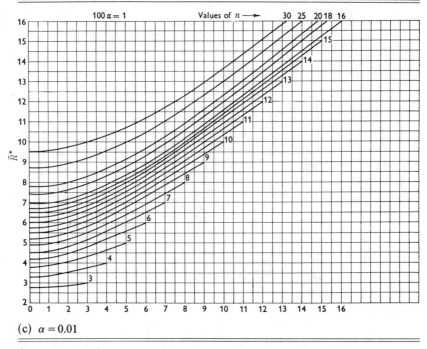

(c) $\alpha = 0.01$

Source: Stephens (1962, Figures 1, 2 and 3); reproduced with kind permission of the author and the Biometrika Trust.

A12 Values of $R|R^*$ for the Fisher distribution.
For given n, α and R^*, the nomograms give values R_α such that $\text{Prob}(R > R_\alpha | R^*) = \alpha$, where R is the resultant length of a random sample of size n from a Fisher distribution and $R^* = R \times$ (cosine of angle between the sample mean direction and the true mean direction).
(a) $\alpha = 0.10$ (b) $\alpha = 0.05$ (c) $\alpha = 0.01$.

A13 Percentage points of $\bar{R}$ for the Fisher distribution.
For given n, α and κ, the tabulated values $\bar{R}_\alpha$ are the 100α percent points of $\bar{R}$, where $\bar{R}$ is the mean resultant length of a random sample of size n from the Fisher distribution with concentration parameter κ; that is, $\text{Prob}(\bar{R} \geq \bar{R}_\alpha) = \alpha$. In descending order, the points given are those for which $\alpha = 0.01$, 0.05, 0.95 and 0.99.

Table A13

n \ κ	0·0	0·5	1·0	1·5	2·0	2·5	3·0	3·5	4·0	4·5	5·0
4	0·8725	0·8863	0·9117	0·9325	0·9469	0·9568	0·9638	0·9689	0·9728	0·9758	0·9782
	·7758	·7969	·8383	·8747	·9008	·9191	·9321	·9417	·9489	·9546	·9591
	·1805	·1908	·2228	·2779	·3498	·4253	·4951	·5569	·6087	·6509	·6854
	·1020	·1079	·1270	·1620	·2165	·2859	·3584	·4254	·4845	·5364	·5811
5	0·8046	0·8275	0·8668	0·8982	0·9197	0·9347	0·9453	0·9530	0·9588	0·9634	0·9670
	·7002	·7288	·7843	·8327	·8675	·8919	·9093	·9220	·9317	·9393	·9454
	·1609	·1721	·2074	·2693	·3496	·4310	·5031	·5640	·6147	·6562	·6901
	·0924	·0990	·1202	·1610	·2240	·3016	·3790	·4480	·5071	·5571	·5995
6	0·7467	0·7767	0·8278	0·8683	0·8963	0·9156	0·9291	0·9391	0·9467	0·9526	0·9574
	·6422	·6764	·7241	·7994	·8410	·8702	·8910	·9063	·9179	·9270	·9343
	·1454	·1577	·1974	·2667	·3531	·4375	·5106	·5713	·6211	·6618	·6952
	·0832	·0905	·1142	·1616	·2341	·3178	·3973	·4665	·5248	·5736	·6145
7	0·6986	0·7346	0·7949	0·8426	0·8760	0·8990	0·9153	0·9271	0·9362	0·9433	0·9490
	·5969	·6355	·7084	·7725	·8194	·8525	·8761	·8935	·9067	·9170	·9253
	·1339	·1472	·1907	·2667	·3579	·4440	·5176	·5780	·6271	·6671	·6999
	·0766	·0843	·1108	·1644	·2447	·3326	·4131	·4819	·5393	·5870	·6267
8	0·6579	0·6990	0·7667	0·8207	0·8585	0·8847	0·9033	0·9169	0·9271	0·9352	0·9417
	·5600	·6024	·6813	·7504	·8015	·8378	·8637	·8829	·8974	·9087	·9179
	·1247	·1389	·1863	·2682	·3630	·4503	·5239	·5839	·6325	·6720	·7043
	·0713	·0796	·1087	·1685	·2553	·3459	·4267	·4950	·5513	·5982	·6369
9	0·6236	0·6691	0·7425	0·8016	0·8431	0·8722	0·8928	0·9079	0·9193	0·9282	0·9353
	·5291	·5751	·6587	·7320	·7865	·8253	·8533	·8736	·8895	·9017	·9115
	·1172	·1323	·1834	·2705	·3680	·4561	·5296	·5892	·6373	·6763	·7081
	·0670	·0758	·1075	·1734	·2654	·3577	·4385	·5061	·5618	·6076	·6456
10	0·5940	0·6432	0·7217	0·7849	0·8299	0·8612	0·8835	0·8999	0·9124	0·9221	0·9298
	·5028	·5518	·6396	·7163	·7735	·8147	·8443	·8661	·8827	·8957	·9061
	·1109	·1269	·1815	·2732	·3729	·4614	·5347	·5940	·6416	·6802	·7116
	·0636	·0727	·1070	·1781	·2742	·3682	·4489	·5158	·5708	·6158	·6530
12	0·5457	0·6012	0·6872	0·7572	0·8075	0·8428	0·8680	0·8866	0·9007	0·9117	0·9205
	·4603	·5147	·6090	·6909	·7526	·7972	·8295	·8534	·8716	·8858	·8972
	·1008	·1185	·1799	·2791	·3817	·4708	·5437	·6021	·6490	·6868	·7177
	·0577	·0680	·1071	·1888	·2907	·3856	·4662	·5319	·5855	·6292	·6652
16	0·476	0·541	0·638	0·717	0·774	0·815	0·845	0·867	0·883	0·896	0·906
	·400	·463	·567	·655	·723	·773	·809	·835	·856	·872	·884
	·087	·107	·182	·291	·397	·486	·557	·615	·660	·697	·727
	·050	·055	·109	·210	·317	·412	·491	·555	·607	·648	·683
20	0·428	0·500	0·604	0·688	0·751	0·796	0·828	0·852	0·871	0·885	0·897
	·358	·428	·538	·631	·703	·755	·794	·823	·845	·862	·876
	·078	·100	·186	·301	·408	·496	·567	·624	·668	·704	·733
	·044	·049	·111	·225	·337	·432	·509	·571	·621	·662	·695
30	0·355	0·436	0·550	0·643	0·713	0·764	0·802	0·829	0·850	0·867	0·880
	·295	·374	·493	·594	·671	·729	·771	·803	·827	·846	·862
	·063	·093	·197	·319	·427	·515	·584	·639	·682	·716	·744
	·036	·043	·133	·257	·369	·463	·538	·597	·645	·683	·714
40	0·307	0·398	0·517	0·616	0·690	0·745	0·785	0·815	0·838	0·856	0·870
	·255	·343	·467	·572	·653	·713	·758	·791	·817	·837	·853
	·054	·090	·206	·332	·440	·527	·595	·648	·690	·724	·751
	·031	·043	·150	·278	·390	·482	·555	·613	·659	·696	·726
60	0·251	0·353	0·479	0·583	0·663	0·722	0·765	0·798	0·823	0·842	0·858
	·208	·307	·437	·546	·631	·694	·742	·777	·805	·826	·844
	·044	·092	·221	·348	·456	·541	·608	·660	·701	·733	·760
	·025	·048	·174	·304	·416	·506	·576	·632	·676	·711	·739
100	0·194	0·308	0·441	0·550	0·635	0·698	0·745	0·780	0·807	0·828	0·845
	·161	·272	·407	·521	·609	·676	·725	·763	·792	·815	·834
	·034	·098	·237	·366	·472	·556	·621	·672	·711	·743	·768
	·020	·062	·201	·333	·442	·529	·598	·651	·693	·726	·753
∞	0·000	0·164	0·313	0·438	0·537	0·613	0·672	0·716	0·751	0·778	0·800

Source: Pearson & Hartley (1972, Table 59).

A14 Selected values of the function $G(\kappa) = I_{\frac{1}{2}}(\kappa)/I_{\frac{3}{2}}(\kappa)$ where $I_{\frac{1}{2}}$ and $I_{\frac{3}{2}}$ are modified Bessel functions.

Table A14

κ	$G(\kappa)$	κ	$G(\kappa)$	κ	$G(\kappa)$
1	16.4	13	1.27	45	1.07
2	5.15	14	1.25	50	1.06
3	3.05	15	1.23	60	1.05
4	2.29	16	1.21	70	1.04
5	1.92	17	1.20	80	1.04
6	1.71	18	1.19	90	1.03
7	1.58	19	1.18	100	1.03
8	1.49	20	1.17	110	1.03
9	1.42	25	1.13	120	1.03
10	1.37	30	1.11	130	1.02
11	1.33	35	1.09	140	1.02
12	1.30	40	1.08	150	1.02

Source: Specially calculated for this book.

A15 Percentage points ($100\alpha\%$) of A_n, G_n and F_n.

The tabulated values are percentage points of the statistics A_n, G_n and F_n for testing the hypothesis of uniformity for spherical data.

Table A15

	Percentage point ($100\alpha\%$)				
Statistic	20%	10%	5%	1%	0.1%
A_n	1.413	1.816	2.207	3.090	4.320
G_n	0.646	0.768	0.883	1.135	1.475
F_n	1.948	2.355	2.748	3.633	4.865

Source: Diggle, Fisher and Lee (1985).

A16 Percentage points of $\bar{\tau}_1$ and $\bar{\tau}_3$ for the Uniform distribution on the sphere.

The tabulated values are lower tail percentage points of $\bar{\tau}_1$, and upper tail percentage points of $\bar{\tau}_3$, where $\bar{\tau}_1$ and $\bar{\tau}_3$ are the smallest and largest normalised eigenvalues calculated from the orientation matrix for a random sample of size n from the uniform U_S distribution.

For $n > 100$, use the approximation $\frac{1}{3} + b/n^{\frac{1}{2}}$, where b is given in the last row.

Table A16

n	lower tail (per cent)				upper tail (per cent)			
	1.0	2.5	5.0	10.0	10.0	5.0	2.5	1.0
5	0.007	0.011	0.019	0.031	0.714	0.751	0.784	0.821
6	.016	.024	.034	.050	.678	.712	.743	.779
7	.026	.038	.050	.067	.651	.685	.712	.746
8	.037	.051	.064	.081	.630	.662	.687	.718
9	.048	.062	.075	.093	.610	.641	.667	.694
10	.058	.073	.087	.105	.596	.625	.650	.677
12	0.076	0.091	0.106	0.123	0.574	0.598	0.621	0.648
14	.091	.107	.120	.137	.554	.578	.599	.623
16	.103	.120	.133	.150	.538	.559	.581	.604
18	.114	.131	.144	.158	.526	.544	.566	.587
20	.124	.140	.152	.167	.515	.535	.553	.575
25	.144	.158	.170	.184	.496	.512	.530	.550
30	0.160	0.172	0.184	0.196	0.479	0.495	0.510	0.528
40	.183	.192	.203	.214	.459	.473	.487	.501
50	.198	.207	.216	.227	.447	.460	.471	.484
60	.208	.217	.226	.235	.438	.449	.458	.470
70	.216	.226	.234	.243	.429	.439	.448	.461
80	.223	.231	.239	.248	.423	.432	.441	.452
100	.233	.242	.248	.257	.413	.422	.430	.440
b	−1.038	−0.948	−0.874	−0.788	0.788	0.874	0.948	1.038

Source: Pearson & Hartley (1972, Table 65).

A17 Percentage points of $(1/n)\sum_{i=1}^{n}\cos^{2}\theta_{i}$ for the Uniform distribution on the sphere.

The tabulated values are lower and upper tail percentage points of the distribution of $(1/n)\sum_{i=1}^{n}\cos^{2}\theta_{i}$, where $(\theta_1, \phi_1), \ldots, (\theta_n, \phi_n)$ are a random sample of size n from the uniform U_S distribution.

Table A17

Sample size n	Lower tail (per cent)					Upper tail (per cent)				
	0.5	1.0	2.5	5.0	10.0	10.0	5.0	2.5	1.0	0.5
3	0.015	0.024	0.044	0.070	0.111	0.569	0.639	0.697	0.761	0.801
4	.032	.045	.071	.101	.142	.534	.594	.641	.703	.740
5	.049	.065	.094	.124	.164	.512	.565	.611	.663	.697
6	0.066	0.083	0.113	0.142	0.180	0.495	0.544	0.585	0.633	0.665
7	.081	.098	.128	.156	.190	.483	.527	.566	.610	.640
8	.094	.111	.140	.168	.200	.472	.514	.550	.592	.620
9	.105	.123	.150	.176	.207	.464	.503	.537	.577	.603
10	.115	.133	.159	.184	.214	.457	.494	.526	.564	.589
12	0.132	0.148	0.173	0.196	0.224	0.446	0.480	0.509	0.543	0.566
14	.145	.160	.185	.206	.232	.438	.468	.495	.527	.548
16	.155	.171	.194	.214	.239	.431	.459	.485	.514	.534
18	.165	.179	.201	.221	.244	.425	.452	.476	.503	.522
20	.173	.187	.208	.226	.249	.420	.446	.468	.494	.512

Table A17 (cont.)

Sample size n	Lower tail (per cent)					Upper tail (per cent)				
	0·5	1·0	2·5	5·0	10·0	10·0	5·0	2·5	1·0	0·5
25	0·188	0·201	0·220	0·237	0·258	0·411	0·434	0·454	0·477	0·493
30	·200	·212	·230	·246	·264	·404	·425	·443	·464	·479
35	·210	·221	·237	·252	·269	·399	·418	·435	·454	·467
40	·217	·228	·243	·257	·273	·394	·412	·428	·446	·459
45	·223	·232	·248	·261	·277	·391	·408	·422	·440	·451
50	0·229	0·238	0·253	0·265	0·280	0·388	0·404	0·418	0·434	0·445
60	·238	·246	·259	·271	·284	·383	·398	·410	·425	·435
70	·244	·253	·265	·275	·288	·379	·393	·404	·418	·427
80	·250	·258	·269	·279	·291	·376	·389	·400	·412	·421
90	·255	·262	·273	·282	·293	·374	·386	·396	·408	·416
100	·258	·265	·276	·285	·295	·372	·383	·393	·404	·412

Source: Pearson & Hartley (1972, Table 64).

A18 Percentage points of $M_G(D_n)$.

The tabulated values are percentage points of the modified statistic $M_G(D_n)$ for testing goodness-of-fit of a Watson girdle distribution.

Table A18

	Percentage point ($100\alpha\%$)		
Statistic	10%	5%	1%
$M_G(D_n)$	1.04	1.15	1.36

Source: Best & Fisher (1986).

A19 Percentage points of J_n and $J_n^{(t)}$.

The tabulated values are percentage points of the statistics J_n, $J_n^{(t)}$ used for testing a single outlier, or t ($=2, 3$) outliers, respectively for discordancy in a sample of size n from a Watson girdle distribution.

Table A19

	Percentage point ($100\alpha\%$)				
Sample size n	10%	5%	2.5%	1%	0.1%
J_n					
10	12.1	16.2	21.2	29.4	61.7
15	10.7	13.3	15.5	20.9	33.8
20	10.4	12.6	14.8	18.3	30.4
30	10.4	12.3	14.1	16.8	25.0
50	10.4	12.3	13.5	15.2	21.7

Table A19 (cont.)

Sample size n	Percentage point ($100\alpha\%$)				
	10%	5%	2.5%	1%	0.1%
$J_n^{(2)}$					
10	16.5	22.0	29.5	43.0	105.8
15	11.6	14.2	16.8	21.1	33.0
20	10.6	12.6	14.5	18.1	26.7
30	10.0	11.4	13.0	14.7	20.0
50	9.9	11.3	12.4	13.3	17.2
$J_n^{(3)}$					
15	13.6	16.5	20.8	25.8	48.3
20	11.3	13.5	15.6	19.3	28.9
30	9.9	11.4	12.9	14.7	20.7
50	9.2	10.1	12.4	11.3	14.4

Source: Best & Fisher (1986, Table 4).

A20 Percentage points of Z for testing equality of the mean directions of two Fisher distributions: equal sample sizes $n_1 = n_2$, $= n$ say. Let $N = n_1 + n_2 = 2n$, $\bar{R} = R/N$, $Z = (R_1 + R_2)/N$. The table entries, preceded by a decimal point, give values, z_0 say, such that $\text{Prob}(Z > z_0) = \alpha$, given the values of $\bar{R}$ and N, for values of $\alpha = 0.01, 0.025, 0.05, 0.10$ in descending order.

Table A20

N (n_1, n_2)	$\bar{R}$ 0·05	0·10	0·15	0·20	0·25	0·30	0·35	0·40	0·45	0·50	0·55	0·60	0·65	0·70
12 (6, 6)	571	573	577	583	592	605	622	642	666	692	720	749	780	811
	523	526	531	539	550	565	584	607	633	662	692	724	758	793
	482	486	492	501	514	531	552	577	606	637	670	704	741	778
	435	439	446	457	473	492	516	544	576	610	646	684	722	762
16 (8, 8)	493	496	502	511	524	542	563	589	617	648	680	714	749	784
	451	455	462	473	488	507	531	559	590	623	658	694	731	769
	415	419	427	440	457	478	505	535	568	603	640	679	718	758
	373	379	328	403	422	447	476	508	544	582	622	662	704	746
20 (10, 10)	440	444	452	464	480	500	525	555	586	620	655	692	730	768
	402	407	416	429	447	470	498	529	563	599	637	676	715	756
	369	375	385	400	420	446	475	509	545	583	622	663	704	746
	332	339	350	368	390	418	451	487	525	565	607	649	693	736
24 (12, 12)	401	407	416	429	448	471	499	531	565	601	639	677	716	757
	366	372	383	398	419	445	475	509	545	583	623	663	704	746
	337	343	355	372	395	423	455	491	529	569	610	652	695	738
	303	310	323	343	368	399	434	472	513	554	597	641	685	730
30 (15, 15)	359	365	376	392	414	441	474	507	544	581	621	662	703	746
	327	334	346	364	389	417	453	488	527	567	608	651	693	737
	300	308	322	341	368	400	435	473	514	555	598	642	686	731
	271	279	295	317	346	380	417	458	500	543	587	632	678	723

Table A20 (cont.)

N (n_1, n_2)	0·05	0·10	0·15	0·20	0·25	0·30	0·35	0·40	0·45	0·50	0·55	0·60	0·65	0·70
40 (20, 20)	311	319	332	352	378	409	444	481	521	562	604	647	691	734
	284	292	307	329	357	390	427	467	508	551	594	638	683	728
	260	270	286	310	340	376	414	455	492	542	586	631	677	723
	234	245	263	290	322	360	401	443	487	532	578	624	671	718
60 (30, 30)	255	265	282	307	339	375	414	455	498	542	586	632	677	723
	232	243	263	290	324	362	402	445	489	534	579	625	672	718
	213	225	247	276	312	351	393	437	482	528	574	621	668	715
	192	206	229	261	299	340	384	429	475	522	569	616	664	712
120 (60, 60)	182	197	223	258	297	339	383	428	474	521	568	616	664	711
	167	183	211	247	288	332	377	423	470	517	565	613	661	709
	153	171	201	239	281	326	372	419	466	514	562	610	659	708
	139	159	191	231	275	320	367	414	462	511	559	608	657	706
240 (120, 120)	133	155	189	230	274	320	367	415	463	511	559	608	657	706
	122	144	182	224	269	316	364	412	460	509	557	606	655	705
	113	139	177	220	267	314	363	410	458	507	556	605	654	704
	103	131	171	216	262	310	360	408	456	505	554	604	653	703
∞	0·05	0·10	0·15	0·20	0·25	0·30	0·35	0·40	0·45	0·50	0·55	0·60	0·65	0·70

Source: Pearson & Hartley (1972, Table 61A).

Table A21

$\bar{R}$ / N	$\gamma = n_1/n_2 = 2$						$\gamma = n_1/n_2 = 4$		
	0·10	0·15	0·20	0·25	0·30	0·35	0·10	0·20	0·30
20	420	436	454	474	497	525	367	423	478
	382	400	421	438	468	498	334	396	457
	355	371	392	414	444	475	313	370	433
	320	341	361	385	417	451	285	338	406
24	385	403	422	445	469	499	340	400	457
	350	370	392	416	443	475	312	372	436
	325	345	366	392	422	455	291	350	416
	295	315	338	366	398	434	265	323	394
30	350	366	387	412	441	472	310	368	435
	321	338	361	387	416	451	285	345	414
	296	314	338	367	399	435	265	328	396
	268	288	314	345	379	417	240	305	378
40	307	325	350	378	409	—	277	337	408
	283	302	327	359	391	—	255	317	387
	261	281	309	340	376	—	237	302	375
	237	258	288	322	360	—	216	283	360
60	257	278	307	339	375	—	238	302	374
	237	260	289	324	361	—	220	285	361
	220	244	275	312	351	—	204	271	350
	200	226	260	299	340	—	187	258	340
120	195	223	258	—	—	—	186	257	339
	181	211	247	—	—	—	173	247	331
	169	201	200	—	—	—	163	240	326
	157	191	231	—	—	—	152	230	320
∞	0·10	0·15	0·20	0·25	0·30	0·35	0·10	0·20	0·30

Source: Pearson & Hartley (1972, Table 61B).

Tables and charts

A21 Percentage points of Z for testing equality of the mean directions of two Fisher distributions: unequal sample sizes n_1, n_2.
Let $N = n_1 + n_2$, $\gamma = n_1/n_2$, $\bar{R} = R/N$, $Z = (R_1 + R_2)/N$. The table entries, preceded by a decimal point, give values, z_0 say, such that Prob $(Z > z_0) = \alpha$, given the values of $\bar{R}$, γ and N, for values of $\alpha = 0.01, 0.025, 0.05, 0.10$ in descending order.

A22 Percentage points of Z for testing equality of the concentration parameters of two Fisher distributions with a common (unknown) mean direction: equal sample sizes $n_1 = n_2, = n$ say.

Table A22

N (n_1, n_2)	\multicolumn{14}{c}{$\bar{R}$}
	0·10	0·15	0·20	0·25	0·30	0·35	0·40	0·45	0·50	0·55	0·60	0·65	0·70	0·75
12 (6, 6)	099	147	195	239	276	301	311	308	295	276	253	226	196	164
	097	144	187	226	254	271	277	273	262	246	226	202	176	148
	094	138	177	209	231	243	246	242	233	219	202	181	158	132
	088	127	159	184	200	208	209	206	199	187	173	155	135	114
16 (8, 8)	099	147	193	233	263	278	281	275	262	245	225	201	174	147
	097	142	184	217	238	248	248	243	232	217	199	179	156	131
	093	136	171	198	213	220	219	214	205	193	177	159	139	116
	087	123	152	172	182	187	186	182	174	164	151	136	118	099
20 (10, 10)	098	146	191	227	250	258	258	250	238	223	204	182	159	134
	096	141	180	208	223	228	227	220	210	197	180	162	141	119
	093	133	166	187	198	202	200	194	185	174	160	143	125	105
	086	121	146	161	168	171	169	164	157	147	135	122	106	090
24 (12, 12)	098	146	188	220	237	242	239	231	220	205	188	168	147	123
	096	140	175	199	210	213	210	203	193	181	166	149	130	109
	092	131	160	178	186	187	185	179	170	159	146	131	115	097
	085	118	140	152	158	159	156	151	144	135	124	111	097	082
30 (15, 15)	098	145	184	208	219	221	217	209	199	185	170	152	133	111
	095	138	169	186	193	194	190	183	174	163	149	134	117	098
	092	127	152	166	170	171	167	161	153	143	132	118	103	087
	084	113	131	141	144	144	141	136	129	121	111	100	087	073
40 (20, 20)	098	142	175	192	197	196	191	187	174	162	148	133	116	098
	095	133	158	169	173	171	167	162	152	142	130	117	102	086
	089	122	141	149	152	150	147	142	134	125	114	103	090	076
	081	107	121	127	128	127	123	119	113	105	097	087	076	064
60 (30, 30)	097	136	158	165	166	164	159	151	143	134	122	110	096	081
	093	124	140	145	145	143	138	135	125	117	107	096	084	071
	087	112	123	127	127	125	121	116	110	103	094	085	074	062
	077	096	104	107	107	105	103	095	092	086	079	071	062	052
120 (60, 60)	092	115	122	123	123	120	115	109	103	097	087	079	068	057
	085	102	106	107	106	104	100	096	089	083	076	067	060	050
	077	090	093	094	093	091	088	084	080	074	068	061	054	045
	067	076	079	079	078	076	074	070	067	062	057	051	p44	037
240 (120, 120)	079	087	089	088	084	085	082	076	074	068	063	056	048	039
	070	076	077	077	076	075	071	068	064	059	053	047	041	034
	062	067	067	067	066	064	061	059	056	051	047	041	035	030
	053	056	056	056	059	058	055	053	051	047	042	036	031	027
∞	0	0	0	0	0	0	0	0	0	0	0	0	0	0

Source: Stephens (1969a, Table 5); reproduced with kind permission of the author and the Biometrika Trust.

Appendix A

Let $N = n_1 + n_2 = 2n$, $\bar{R} = R/N$ and $Z = |R_1 - R_2|/N$. The table entries, preceded by a decimal point, give values, z_0 say, such that Prob $(Z > z_0) = \alpha$, given the values of $\bar{R}$ and N, for $\alpha = 0.01$, 0.025, 0.05 and 0.10 in descending order.

A23 Percentage points of Z for testing equality of the concentration parameters of r Fisher distributions: equal sample sizes $n, \ldots, n$.

Let $\nu = 2n$ when the mean directions are known, $\nu = 2n - 2$ when the mean directions are unknown. The table entries give values, z_0 say, such that Prob $(Z > z_0) = \alpha$, given the values of r and ν, for $\alpha = 0.05$ and 0.01.

Table A23

Upper 5% points

ν \ r	2	3	4	5	6	7	8	9	10	11	12
2	39·0	87·5	142	202	266	333	403	475	550	626	704
4	9·60	15·5	20·6	25·2	29·5	33·6	37·5	41·1	44·6	48·0	51·4
6	5·82	8·38	10·4	12·1	13·7	15·0	16·3	17·5	18·6	19·7	20·7
8	4·43	6·00	7·18	8·12	9·03	9·78	10·5	11·1	11·7	12·2	12·7
10	3·72	4·85	5·67	6·34	6·92	7·42	7·87	8·28	8·66	9·01	9·34
12	3·28	4·16	4·79	5·30	5·72	6·09	6·42	6·72	7·00	7·25	7·48
20	2·46	2·95	3·29	3·54	3·76	3·94	4·10	4·24	4·37	4·49	4·59
30	2·07	2·40	2·61	2·78	2·91	3·02	3·12	3·21	3·29	3·36	3·39
60	1·67	1·85	1·96	2·04	2·11	2·17	2·22	2·26	2·30	2·33	2·36
∞	1·00	1·00	1·00	1·00	1·00	1·00	1·00	1·00	1·00	1·00	1·00

Upper 1% points

ν \ r	2	3	4	5	6	7	8	9	10	11	12
2	199	448	729	1036	1362	1705	2063	2432	2813	3204	3605
4	23·2	37	49	59	69	79	89	97	106	113	120
6	11·1	15·5	19·1	22	25	27	30	32	34	36	37
8	7·50	9·9	11·7	13·2	14·5	15·8	16·9	17·9	18·9	19·8	21
10	5·85	7·4	8·6	9·6	10·4	11·1	11·8	12·4	12·9	13·4	13·9
12	4·91	6·1	6·9	7·6	8·2	8·7	9·1	9·5	9·9	10·2	10·6
20	3·32	3·8	4·3	4·6	4·9	5·1	5·3	5·5	5·6	5·8	5·9
30	2·63	3·0	3·3	3·4	3·6	3·7	3·8	3·9	4·0	4·1	4·2
60	1·96	2·2	2·3	2·4	2·4	2·5	2·5	2·6	2·6	2·7	2·7
∞	1·00	1·0	1·0	1·0	1·0	1·0	1·0	1·0	1·0	1·0	1·0

Source: Pearson & Hartley (1970, Table 31).

A24 Percentage points of $\hat{\rho}^*$ for uniform marginal distributions.

The tabulated values are percentage points ($100\alpha\%$) of the null distribution of the modified spherical correlation coefficient $\hat{\rho}^*$ when the marginal distributions are uniform. $100(1-\alpha)\%$ point $= -(100\alpha\%$ point).

Table A24

	Percentage point					
Statistic	20%	10%	5%	2.5%	1%	0.5%
$\hat{\rho}^*$	1.380	2.592	3.930	5.389	7.220	8.676

Source: Fisher & Lee (1986, Table 1).

Appendix B

Data sets

B1 **Data** Pole positions determined from the palaeomagnetic study of New Caledonian laterites.
    Coordinates Latitude, Longitude.
    Source We are most grateful to D.A. Falvey and R. Musgrave for permission to use these data.
    Analysis Examples 5.1, 5.7, 5.8, 5.10, 5.11.

	Lat. (°)	Long. (°)		Lat. (°)	Long. (°)		Lat. (°)	Long. (°)
1	−26.4	324.0	18	−49.0	65.6	35	−85.9	63.7
2	−32.2	163.7	19	−67.0	282.6	36	−84.8	74.9
3	−73.1	51.9	20	−56.7	56.2	37	−7.4	93.8
4	−80.2	140.5	21	−80.5	108.4	38	−29.8	72.8
5	−71.1	267.2	22	−77.7	266.0	39	−85.2	113.2
6	−58.7	32.0	23	−6.9	19.1	40	−53.1	51.5
7	−40.8	28.1	24	−59.4	281.7	41	−38.3	146.8
8	−14.9	266.3	25	−5.6	107.4	42	−72.7	103.1
9	−66.1	144.3	26	−62.6	105.3	43	−60.2	33.2
10	−1.8	256.2	27	−74.7	120.2	44	−63.4	154.8
11	−52.1	83.2	28	−65.3	286.6	45	−17.2	89.9
12	−77.3	182.1	29	−71.6	106.4	46	−81.6	295.6
13	−68.8	110.4	30	−23.3	96.5	47	−40.4	41.0
14	−68.4	142.2	31	−74.3	90.2	48	−53.6	59.1
15	−29.2	246.3	32	−81.0	170.9	49	−56.2	35.6
16	−78.5	222.6	33	−12.7	199.4	50	−75.1	70.7
17	−65.4	247.7	34	−75.4	118.6			

B2	Data	Measurements of magnetic remanence in specimens of Palaeozoic red-beds from Argentina.
	Coordinates	Declination, Inclination.
	Source	Embleton (1970).
	Analysis	Examples 5.2, 5.7, 5.8, 5.12, 5.13, 5.14, 5.15, 5.17, 5.21, 5.23, 5.24, 5.25, 5.26.

	Dec. (°)	Inc. (°)		Dec. (°)	Inc. (°)		Dec. (°)	Inc. (°)
1	122.5	55.5	10	147.5	54.5	19	141.0	57.0
2	130.5	58.0	11	142.0	51.0	20	143.5	67.5
3	132.5	44.0	12	163.5	56.0	21	131.5	62.5
4	148.5	56.0	13	141.0	59.5	22	147.5	63.5
5	140.0	63.0	14	156.0	56.5	23	147.0	55.5
6	133.0	64.5	15	139.5	54.0	24	149.0	62.0
7	157.5	53.0	16	153.5	47.5	25	144.0	53.5
8	153.0	44.5	17	151.5	61.0	26	139.5	58.0
9	140.0	61.5	18	147.5	58.5			

B3   *Data*   Arrival directions of low mu showers of cosmic rays.
     Coordinates Declination, Right Ascension.
     Source   Digitised from Figure 3 in Toyoda *et al.* (1965).
     Analysis   Examples 5.3, 5.7, 5.8, 5.35.

	Dec. (°)	RA (°)		Dec. (°)	RA (°)		Dec. (°)	RA (°)
1	−66	315	45	−23	342	89	−3	234
2	−66	198	46	−17	356	90	−3	185
3	−63	99	47	−19	347	91	−8	162
4	−63	50	48	−12	347	92	−15	144
5	−61	86	49	−15	342	93	−19	144
6	−61	221	50	−12	338	94	−19	108
7	−52	0	51	−17	329	95	−19	104
8	−50	14	52	−26	293	96	−19	86
9	−50	63	53	−32	284	97	−17	77
10	−52	176	54	−35	261	98	−12	68
11	−52	185	55	−21	252	99	−6	63
12	−52	207	56	−19	257	100	4	27
13	−52	221	57	−23	261	101	0	95
14	−50	230	58	−28	261	102	−3	99
15	−52	243	59	−39	252	103	−8	122
16	−48	347	60	−35	230	104	−6	140
17	−43	342	61	−32	230	105	−8	144
18	−48	311	62	−28	216	106	4	171
19	−48	293	63	−30	212	107	2	176
20	−48	284	64	−28	207	108	7	185
21	−48	216	65	−30	194	109	9	198
22	−46	207	66	−35	158	110	2	216
23	−48	144	67	−30	117	111	4	216
24	−46	149	68	−32	99	112	7	234
25	−46	77	69	−37	95	113	16	324
26	−46	41	70	−21	14	114	18	311
27	−46	27	71	−19	18	115	24	293
28	−46	14	72	−15	27	116	20	275
29	−43	5	73	−12	23	117	22	266
30	−32	41	74	−15	50	118	13	207
31	−41	36	75	−26	63	119	16	203
32	−39	50	76	−28	77	120	20	212
33	−35	63	77	−21	167	121	22	198
34	−41	77	78	−23	176	122	13	140
35	−41	113	79	−26	185	123	7	117
36	−41	153	80	−23	194	124	11	117
37	−43	176	81	−17	221	125	11	86
38	−37	203	82	−15	216	126	11	68
39	−41	216	83	−15	230	127	13	32
40	−41	252	84	−17	234	128	16	14
41	−41	279	85	−10	248	129	16	59
42	−41	288	86	−17	279	130	16	68
43	−37	311	87	−12	297	131	22	68
44	−30	320	88	24	324	132	24	72

B3 (*cont.*)

	Dec. (°)	RA (°)		Dec. (°)	RA (°)		Dec. (°)	RA (°)
133	18	86	139	40	338	144	42	198
134	20	104	140	−55	9	145	31	99
135	18	212	141	38	230	146	38	86
136	27	153	142	36	221	147	36	68
137	27	216	143	38	203	148	36	45
138	36	342						

B4 **Data** Facing directions of conically folded planes.
Coordinates Plunge, Plunge azimuth.
Source Cohen (1983); data kindly supplied to us on tape by P. H. Cohen.
Analysis Examples 5.4, 5.7, 5.8, 5.30.

	Pl. (°)	Pl. az. (°)		Pl. (°)	Pl. az. (°)		Pl. (°)	Pl. az. (°)
1	48	269	45	60	270	89	42	277
2	57	265	46	68	268	90	44	278
3	61	271	47	75	97	91	60	280
4	59	272	48	47	95	92	65	275
5	58	268	49	48	90	93	76	270
6	60	267	50	50	95	94	63	275
7	59	265	51	48	94	95	71	276
8	58	265	52	49	93	96	80	255
9	60	265	53	45	93	97	77	105
10	59	263	54	41	93	98	72	99
11	59	267	55	42	95	99	80	253
12	53	267	56	40	96	100	60	96
13	50	270	57	51	100	101	48	93
14	48	270	58	70	104	102	49	92
15	61	265	59	74	102	103	48	91
16	40	95	60	71	108	104	46	91
17	56	100	61	51	99	105	44	90
18	67	95	62	75	112	106	43	89
19	52	90	63	73	110	107	40	89
20	49	271	64	60	100	108	58	96
21	60	267	65	49	95	109	65	105
22	47	272	66	44	93	110	39	90
23	50	270	67	41	91	111	48	76
24	48	273	68	51	92	112	38	91
25	50	271	69	45	92	113	43	91
26	53	269	70	50	95	114	42	91
27	52	270	71	41	89	115	49	90
28	58	267	72	44	93	116	39	95
29	60	266	73	68	100	117	43	90
30	60	268	74	67	270	118	44	92
31	62	269	75	73	261	119	48	92
32	61	270	76	50	275	120	61	95
33	62	269	77	40	276	121	68	100
34	60	270	78	60	275	122	80	135
35	58	272	79	47	277	123	60	98
36	59	271	80	47	276	124	49	92
37	56	271	81	54	273	125	45	90
38	53	270	82	60	273	126	62	99
39	49	273	83	62	271	127	79	175
40	49	271	84	57	275	128	79	220
41	53	270	85	43	277	129	72	266
42	50	274	86	53	275	130	76	235
43	46	275	87	40	276	131	77	231
44	45	274	88	40	279	132	71	256

B4 (*cont.*)

	Pl. (°)	Pl. az. (°)		Pl. (°)	Pl. az. (°)		Pl. (°)	Pl. az. (°)
133	60	272	141	46	275	149	54	96
134	42	276	142	51	274	150	71	111
135	50	276	143	56	272	151	54	96
136	41	275	144	50	273	152	50	92
137	60	273	145	40	270	153	48	91
138	73	266	146	73	103	154	48	90
139	43	276	147	57	95	155	51	90
140	50	274	148	60	98			

Appendix B

B5 *Data* Measurements of magnetic remanence from specimens of red beds from the Bowen Basin, Queensland, after thermal demagnetisation to 670°C.
Coordinates Declination, Inclination.
Source We are most grateful to P. W. Schmidt for permission to use these data.
Analysis Examples 5.5, 5.7, 5.8, 5.9, 5.34.

	Dec. (°)	Inc. (°)		Dec. (°)	Inc. (°)		Dec. (°)	Inc. (°)
1	36.5	−70.5	19	338.2	−73.0	37	343.1	37.3
2	44.8	65.6	20	115.3	25.3	38	141.7	61.9
3	349.9	−17.2	21	293.9	−26.7	39	1.3	57.6
4	179.7	−3.3	22	21.9	−17.4	40	34.6	54.0
5	148.8	−37.3	23	115.7	14.2	41	271.9	−13.9
6	160.5	−62.0	24	342.3	24.4	42	338.4	14.9
7	6.7	−73.7	25	315.1	68.6	43	2.4	−37.3
8	178.2	−74.5	26	137.6	19.6	44	144.2	44.2
9	48.4	40.1	27	30.2	18.6	45	254.5	10.6
10	275.1	−3.7	28	43.8	−7.4	46	333.3	70.6
11	7.3	−40.2	29	274.1	−51.4	47	240.9	59.0
12	162.0	41.2	30	349.6	6.6	48	8.1	−4.9
13	188.7	−16.5	31	287.5	0.1	49	240.2	−13.5
14	159.0	−30.7	32	92.9	−32.6	50	339.2	−22.6
15	100.9	0.9	33	63.7	47.4	51	42.4	−8.8
16	150.0	38.3	34	341.1	48.5	52	260.8	29.9
17	168.8	46.8	35	16.9	82.6			
18	274.6	−39.6	36	270.5	67.0			

Data sets

B6 *Data* Measurement of magnetic remanence from specimens of Precambrian volcanics.
Coordinates Declination, Inclination.
Source Schmidt & Embleton (1985).
Analysis **§3.3.2, §3.3.3**, Examples 5.6, 5.7, 5.8, 5.31.

	Dec. (°)	Inc. (°)		Dec. (°)	Inc. (°)		Dec. (°)	Inc. (°)
1	319.1	30.7	37	323.2	62.6	73	89.4	22.5
2	347.4	−49.2	38	210.5	62.3	74	339.7	70.3
3	344.5	−37.3	39	316.0	−16.3	75	359.9	71.9
4	313.6	−11.4	40	306.9	−19.2	76	314.1	−45.1
5	177.0	31.3	41	331.0	−31.0	77	325.4	−24.4
6	3.0	−31.3	42	298.1	−12.3	78	1.9	−19.6
7	308.0	4.6	43	296.2	−21.7	79	177.3	−7.1
8	165.0	42.8	44	10.9	−14.5	80	324.5	−68.3
9	352.6	−22.2	45	249.0	28.8	81	346.9	−58.1
10	307.0	8.8	46	262.3	38.2	82	350.0	−57.5
11	190.3	19.3	47	270.0	31.8	83	321.4	−1.6
12	3.6	−28.2	48	338.5	−6.0	84	314.6	−5.7
13	306.1	−2.3	49	332.2	−2.8	85	320.6	−22.3
14	180.2	32.4	50	222.3	3.4	86	304.7	−17.3
15	9.1	−29.7	51	3.3	−20.6	87	314.2	−22.6
16	355.7	−39.3	52	325.9	−5.7	88	306.3	−16.5
17	286.8	21.2	53	345.6	4.6	89	271.6	−18.4
18	207.7	19.8	54	294.4	−34.4	90	316.0	−36.7
19	192.1	18.9	55	358.7	−8.2	91	309.8	−20.3
20	202.3	35.8	56	23.8	−44.3	92	313.0	−20.5
21	44.9	41.3	57	20.8	−43.5	93	329.4	−32.3
22	23.8	−41.4	58	280.5	8.7	94	334.9	−29.3
23	9.3	−25.9	59	347.4	−47.2	95	312.3	−15.7
24	18.3	−15.8	60	341.9	−48.3	96	269.8	5.9
25	186.2	27.3	61	10.2	−17.4	97	318.2	−24.9
26	31.1	−0.8	62	12.4	−2.7	98	0.4	−40.0
27	165.5	12.7	63	321.0	−17.6	99	309.0	−15.2
28	202.3	37.2	64	16.7	−45.5	100	310.4	−21.6
29	191.5	59.7	65	353.5	−22.7	101	348.0	−43.3
30	297.2	−16.4	66	264.8	26.8	102	291.9	41.0
31	248.7	6.3	67	258.5	39.1	103	272.2	51.4
32	155.8	−57.5	68	310.6	15.3	104	273.0	43.9
33	190.1	11.8	69	281.2	−12.0	105	338.2	−25.1
34	341.3	−11.4	70	7.0	−15.4	106	318.5	−4.7
35	346.3	−3.4	71	342.8	23.2	107	298.2	19.2
36	184.6	3.0	72	243.6	47.2			

B7   *Data*   Measurements of magnetic remanence from specimens collected from the Tumblagooda Sandstone, Western Australia.
Coordinates   Declination, Inclination.
Source   Embleton & Giddings (1974).
Analysis   Examples 5.16, 5.18, 5.20, 5.22.

	Dec. (°)	Inc. (°)		Dec. (°)	Inc. (°)		Dec. (°)	Inc. (°)
1	66	−11	7	25	−14	13	63	−21
2	73	−31	8	42	−36	14	63	−24
3	76	−38	9	51	−36	15	64	−33
4	78	−26	10	52	−24	16	81	−30
5	83	−58	11	56	−34	17	81	−43
6	122	−50	12	61	−35			

B8 *Data* Measurements of magnetic remanence in specimens of Mesozoic Dolerite from Prospect, New South Wales, after successive partial demagnetisation stages (200° and 350°C) for each of 62 specimens as part of an experiment to determine the blocking temperature spectrum of components of magnetisation. In Set A, observation No. 16 was originally entered incorrectly as (Dec. 194°, Inc. −69°) – see Example 5.19.

Coordinates Declination, Inclination.
Source We are most grateful to P.W. Schmidt for permission to use these data.
Analysis Examples 5.19, 5.22, 8.1, 8.5, 8.6, 8.7.

	A (200°C)		B (350°C)	
	Dec. (°)	Inc. (°)	Dec. (°)	Inc. (°)
1	1	−78	354	−75
2	11	−77	357	−75
3	334	−82	338	−79
4	300	−79	307	−79
5	319	−78	322	−78
6	347	−72	344	−70
7	347	−67	335	−67
8	2	−68	342	−69
9	355	−65	349	−66
10	31	−72	26	−72
11	27	−62	25	−63
12	356	−67	347	−70
13	350	−74	353	−75
14	303	−62	311	−63
15	338	−84	329	−75
16	294	−69	311	−69
	(194	−69	311	−69)
17	293	−67	301	−65
18	274	−87	289	−79
19	26	−75	11	−77
20	297	−76	300	−74
21	297	−76	300	−74
22	17	−77	359	−78
23	285	−76	290	−75
24	331	−80	326	−77
25	349	−81	341	−80
26	291	−72	288	−74
27	269	−89	303	−89
28	49	−85	39	−84

B8 (*cont.*)

	A (200°C)		B (350°C)	
	Dec. (°)	Inc. (°)	Dec. (°)	Inc. (°)
29	59	−83	56	−85
30	338	−67	332	−66
31	316	−76	310	−74
32	341	−79	338	−78
33	321	−70	316	−70
34	329	−63	324	−62
35	332	−68	329	−65
36	348	−79	349	−78
37	334	−81	346	−80
38	334	−81	346	−80
39	326	−70	315	−69
40	329	−68	307	−69
41	308	−70	294	−70
42	333	−73	319	−71
43	359	−76	358	−74
44	335	−77	322	−73
45	329	−74	320	−74
46	332	−72	312	−74
47	336	−73	318	−73
48	326	−72	310	−76
49	336	−73	320	−73
50	343	−78	329	−75
51	319	−70	318	−71
52	323	−72	318	−73
53	293	−73	289	−73
54	286	−81	289	−77
55	230	−85	331	−84
56	340	−70	335	−69
57	315	−79	334	−77
58	351	−71	320	−72
59	336	−81	343	−80
60	336	−82	316	−82
61	355	−79	346	−79
62	345	−74	344	−71

Data sets

B9 *Data* Measurements of the direction of magnetisation in specimens from the Great Whin Sill.
Coordinates Declination, Inclination.
Source Creer, Irving & Nairn (1959).
Analysis Examples 5.27, 5.28, 5.29, 7.12, 7.13.

	Dec. (°)	Inc. (°)		Dec. (°)	Inc. (°)		Dec. (°)	Inc. (°)
1	176	−10	13	187	−15	25	195	11
2	185	−6	14	203	−26	26	182	11
3	196	−5	15	196	−19	27	182	7
4	179	−4	16	205	6	28	190	−15
5	186	−5	17	182	9	29	183	−20
6	192	3	18	190	−13	30	194	−7
7	183	−1	19	183	1	31	186	−6
8	203	−11	20	190	15	32	192	−9
9	193	−28	21	183	7	33	192	0
10	171	−14	22	179	11	34	180	0
11	191	−14	23	190	13			
12	190	−20	24	177	−10			

B10 *Data* Measurements of the palaeomagnetic pole position from Tasmanian Dolerites.
Coordinates Latitude, Longitude.
Source Schmidt (1976).
Analysis Examples 5.32, 5.33.

	Lat. (°)	Long. (°)		Lat. (°)	Long. (°)		Lat. (°)	Long. (°)
1	−55.7	148.3	12	−55.6	124.4	23	−39.3	138.5
2	−47.2	176.5	13	−52.5	148.5	24	−48.9	107.8
3	−63.8	162.2	14	−45.7	75.2	25	−49.3	173.3
4	−49.1	196.5	15	−49.5	142.0	26	−44.1	170.8
5	−52.0	147.3	16	−59.5	94.9	27	−53.3	174.4
6	−56.7	178.5	17	−68.7	209.2	28	−51.1	174.6
7	−30.2	183.9	18	−47.8	159.6	29	−45.4	144.7
8	−33.5	184.1	19	−52.6	164.8	30	−40.5	125.6
9	−22.3	109.4	20	−63.0	168.6	31	−40.8	173.5
10	−42.9	133.4	21	−41.5	193.6	32	−53.3	165.7
11	−47.3	125.9	22	−51.9	164.4	33	−41.9	187.8

Appendix B

B11 *Data* Orientations of axial-plane cleavage surfaces of F_1 folds in Ordovician turbidites.
Coordinates Dip, Dip direction.
Source Powell, Cole & Cudahy (1985).
Analysis Examples 6.1, 6.6, 6.7, 6.11, 6.12, 6.13.

	Dip (°)	Dip dir. (°)		Dip (°)	Dip dir. (°)		Dip (°)	Dip dir. (°)
1	50	65	26	86	84	51	63	216
2	53	75	27	81	230	52	86	194
3	85	233	28	85	228	53	81	228
4	82	39	29	79	230	54	81	27
5	82	53	30	86	231	55	89	226
6	66	58	31	88	40	56	62	58
7	75	50	32	84	233	57	81	35
8	85	231	33	87	234	58	88	37
9	87	220	34	88	225	59	70	235
10	85	30	35	83	234	60	80	38
11	82	59	36	82	222	61	77	227
12	88	44	37	89	230	62	85	34
13	86	54	38	82	51	63	74	225
14	82	251	39	82	46	64	90	53
15	83	233	40	67	207	65	90	57
16	86	52	41	85	221	66	90	66
17	80	26	42	87	58	67	90	45
18	78	40	43	82	48	68	90	47
19	85	266	44	82	222	69	90	54
20	89	67	45	82	10	70	90	45
21	85	61	46	75	52	71	90	60
22	85	72	47	68	49	72	90	51
23	86	54	48	89	36	73	90	42
24	67	32	49	81	225	74	90	52
25	87	238	50	87	221	75	90	63

B12 *Data* Orientations of axial-plane cleavage surfaces of F_1 folds in Ordovician turbidites.
Coordinates Dip, Dip direction.
Source Powell, Cole & Cudahy (1985).
Analysis Examples 6.2, 6.6, 6.7, 6.24, 6.25, 6.26, 6.27, 6.28, 6.29, 6.30, 6.31, 6.32, 6.33, 7.19.

	Dip (°)	Dip dir. (°)		Dip (°)	Dip dir. (°)
1	80	122	*37	8	28
2	72	132	*38	20	310
3	63	141	39	21	310
4	51	145	40	18	331
*5	62	128	*41	25	326
*6	53	133	*42	28	332
7	53	130	**43	32	3
*8	52	129	44	32	324
*9	48	124	*45	32	308
10	45	120	*46	34	304
11	44	137	*47	38	304
*12	44	141	48	37	299
13	34	151	49	44	293
*14	37	138	*50	45	293
*15	38	135	51	48	306
16	40	135	52	42	310
*17	25	156	*53	47	313
*18	15	156	*54	45	319
*19	22	130	*55	43	320
20	63	112	56	45	320
21	35	116	57	50	330
*22	28	113	58	70	327
23	28	117	*59	59	312
*24	22	110	*60	66	317
*25	33	106	*61	65	314
26	37	106	*62	70	312
27	32	98	*63	66	311
28	27	84	64	67	307
*29	24	77	65	83	311
*30	8	111	*66	66	310
*31	6	122	67	69	310
*32	8	140	68	69	305
*33	11	48	*69	72	305
34	8	279	*70	67	301
35	6	19	*71	69	301
36	6	28	*72	82	300

B13 *Data* Measurements of the maximum susceptibility axis of magnetic fabric in specimens from ash strata.
Coordinates Plunge, Plunge azimuth.
Source Addison (1982).
Analysis Examples 6.3, 6.6, 6.7.

	Pl. (°)	Pl. az. (°)		Pl. (°)	Pl. az. (°)		Pl. (°)	Pl. az. (°)
1	11.0	114.9	7	−1.9	87.7	13	7.7	14.6
2	10.0	142.8	8	7.9	320.9	14	8.6	126.2
3	4.5	144.3	9	23.1	91.0	15	9.7	72.1
4	−3.5	154.7	10	5.5	33.9	16	1.5	354.1
5	−12.6	49.4	11	4.4	340.7	17	1.1	28.0
6	16.3	350.4	12	7.2	341.9	18	7.5	335.0

B14 *Data* Orientations of joint planes in Triassic sandstone at Wanganderry Lookout, New South Wales, collected as part of the Sydney Basin Fracture Analysis Project.
Coordinates Dip, Dip direction.
Source We are most grateful to J. F. Huntington, J. W. Creasey and A. Mauger for permission to use these data.
Analysis Examples 6.4, 6.6, 6.7.

	Dip (°)	Dip Dir. (°)		Dip (°)	Dip dir. (°)		Dip (°)	Dip dir. (°)
1	86	35	40	69	105	79	81	281
2	81	311	41	25	24	80	70	12
3	89	276	42	60	118	81	74	316
4	80	26	43	55	60	82	65	279
5	84	307	44	70	40	83	84	22
6	69	253	45	66	97	84	55	320
7	86	191	46	70	59	85	80	274
8	80	132	47	60	51	86	90	16
9	90	282	48	55	105	87	70	300
10	85	34	49	70	62	88	88	264
11	84	321	50	70	49	89	80	266
12	89	291	51	74	85	90	89	259
13	85	6	52	76	62	91	81	23
14	89	314	53	74	50	92	88	314
15	85	72	54	69	95	93	88	266
16	90	195	55	80	65	94	84	21
17	78	138	56	65	53	95	80	127
18	80	50	57	56	105	96	90	267
19	72	254	58	64	177	97	90	26
20	85	287	59	72	146	98	80	145
21	85	44	60	60	124	99	81	247
22	86	82	61	66	173	100	85	13
23	79	285	62	86	158	101	81	130
24	86	36	63	59	126	102	86	260
25	76	257	64	64	170	103	79	14
26	82	287	65	79	147	104	86	126
27	84	221	66	65	132	105	89	79
28	74	255	67	76	177	106	81	67
29	81	292	68	74	150	107	79	12
30	80	220	69	57	117	108	84	71
31	89	264	70	75	167	109	76	64
32	84	285	71	66	138	110	84	69
33	78	222	72	65	133	111	84	66
34	83	120	73	77	274	112	80	299
35	29	15	74	76	18	113	88	206
36	50	119	75	72	301	114	70	229
37	82	102	76	85	271	115	85	204
38	26	13	77	72	16	116	82	61
39	20	14	78	55	312	117	85	221

B14 (cont.)

	Dip (°)	Dip dir. (°)		Dip (°)	Dip dir. (°)		Dip (°)	Dip dir. (°)
118	75	64	153	84	2	188	88	281
119	69	59	154	86	150	189	87	249
120	80	66	155	83	246	190	90	125
121	74	356	156	84	183	191	84	22
122	81	191	157	80	219	192	86	326
123	74	66	158	75	134	193	86	210
124	80	29	159	80	210	194	80	160
125	90	92	160	80	138	195	76	66
126	80	174	161	84	218	196	90	111
127	89	84	162	77	146	197	84	310
128	83	67	163	79	208	198	81	35
129	86	65	164	70	149	199	85	234
130	88	168	165	86	52	200	89	335
131	81	65	166	88	210	201	88	147
132	86	20	167	86	172	202	87	127
133	86	8	168	86	125	203	90	211
134	88	359	169	84	119	204	89	104
135	75	137	170	86	260	205	86	206
136	80	198	171	81	94	206	77	112
137	80	116	172	78	153	207	84	106
138	86	76	173	82	70	208	81	108
139	84	288	174	82	137	209	74	114
140	85	128	175	85	291	210	86	232
141	86	70	176	82	335	211	86	34
142	86	66	177	88	156	212	85	34
143	86	300	178	88	127	213	74	347
144	74	11	179	79	216	214	84	34
145	89	304	180	84	154	215	86	220
146	90	306	181	81	212	216	83	247
147	90	304	182	89	25	217	77	273
148	88	18	183	90	132	218	83	66
149	89	129	184	85	105	219	20	11
150	89	310	185	90	212	220	80	162
151	90	330	186	60	62	221	80	50
152	77	310	187	85	17			

B15 **Data** Measurements of the orientation of the dendritic field at various sites in the retinas of 6 cats, in response to different visual stimuli.
Coordinates Latitude, Longitude.
Source Keilson, Petrondas, Sumita & Wellner (1983); the set constitutes the pooled data from Appendix C of this paper.
Analysis Examples 6.5, 6.6, 6.7, 6.34.

	Lat. (°)	Long. (°)		Lat. (°)	Long. (°)		Lat. (°)	Long. (°)
1	90	221	33	120	274	65	94	33
2	115	347	34	109	72	66	102	179
3	144	176	35	104	152	67	121	99
4	148	47	36	131	223	68	126	284
5	108	295	37	142	257	69	120	132
6	117	234	38	123	2	70	95	211
7	100	12	39	110	355	71	114	280
8	110	15	40	95	16	72	134	264
9	98	26	41	117	3	73	103	350
10	150	84	42	92	251	74	92	55
11	99	231	43	144	105	75	139	218
12	111	154	44	123	65	76	112	244
13	130	337	45	116	70	77	129	99
14	95	290	46	118	156	78	106	136
15	95	306	47	136	233	79	141	13
16	107	221	48	162	186	80	113	287
17	91	4	49	120	259	81	108	340
18	109	344	50	98	186	82	128	239
19	143	303	51	95	153	83	158	287
20	118	350	52	101	24	84	158	210
21	137	358	53	113	114	85	117	121
22	108	306	54	131	28	86	156	323
23	106	156	55	125	62	87	155	323
24	107	23	56	92	69	88	147	359
25	163	14	57	97	20	89	114	237
26	107	229	58	120	224	90	125	320
27	100	247	59	98	37	91	114	348
28	112	248	60	141	14	92	118	346
29	126	304	61	152	342	93	99	190
30	93	137	62	102	229	94	106	225
31	94	292	63	97	268			
32	99	183	64	96	265			

Appendix B

B16 *Data* Normals to the orbital planes of long-period comets with parabolic or nearly parabolic orbits.

Coordinates Colatitude, Longitude.

Source Marsden (1982, pp 55-56); data kindly supplied by P. E. Jupp. Marsden gives the orbital planes in terms of celestial longitude and latitude (L, B) of perihelion, and longitude (N say) of the ascending node. The direction cosines (x, y, z) of the colatitudes and longitudes of the *normals* to these planes (that is, of the data listed above), are given by

$$x = (\sin B \sin N)/r$$
$$y = -(\sin B \cos N)/r$$
$$z = -(\cos B \sin (N - L))/r$$

where

$$r^2 = \sin^2 B + \cos^2 B \sin^2(N - L).$$

Analysis Example 6.8.

	Colat. (°)	Long. (°)		Colat. (°)	Long. (°)		Colat. (°)	Long. (°)
1	96	123	28	119	105	55	95	307
2	104	191	29	117	77	56	132	100
3	110	173	30	107	309	57	122	341
4	129	12	31	166	268	58	138	162
5	106	119	32	143	16	59	142	51
6	148	136	33	143	296	60	148	68
7	125	11	34	93	106	61	107	317
8	159	71	35	146	311	62	129	241
9	159	63	36	153	350	63	121	311
10	147	5	37	95	269	64	171	202
11	132	19	38	148	292	65	156	168
12	127	33	39	147	348	66	161	353
13	118	17	40	97	163	67	142	231
14	176	359	41	108	204	68	96	160
15	145	51	42	120	350	69	155	31
16	173	277	43	107	110	70	96	142
17	139	12	44	136	328	71	94	138
18	136	338	45	140	103	72	114	138
19	146	307	46	123	226	73	149	185
20	143	170	47	106	296	74	91	146
21	124	89	48	143	68	75	104	142
22	102	96	49	114	304	76	154	136
23	174	133	50	151	87	77	148	88
24	92	284	51	130	287	78	145	92
25	106	136	52	92	157	79	104	174
26	90	105	53	125	54	80	125	107
27	99	284	54	108	305	81		95

B16 (cont.)

	Colat. (°)	Long. (°)		Colat. (°)	Long. (°)		Colat. (°)	Long. (°)
82	94	338	135	123	168	188	92	44
83	106	123	136	114	182	189	159	355
84	102	48	137	157	187	190	139	212
85	116	95	138	133	68	191	168	312
86	96	359	139	123	122	192	142	8
87	102	9	140	168	69	193	95	208
88	110	123	141	135	119	194	138	267
89	121	148	142	130	96	195	97	196
90	159	153	143	144	96	196	100	300
91	101	232	144	92	10	197	160	190
92	90	344	145	120	325	198	97	253
93	119	53	146	97	32	199	129	232
94	129	329	147	139	56	200	175	353
95	101	182	148	116	19	201	149	93
96	161	323	149	128	186	202	176	73
97	164	221	150	145	234	203	176	283
98	142	316	151	154	59	204	139	157
99	134	235	152	97	87	205	122	108
100	145	268	153	115	175	206	103	77
101	171	234	154	104	191	207	138	247
102	128	25	155	169	120	208	174	113
103	93	164	156	140	46	209	132	197
104	102	162	157	95	24	210	147	136
105	175	190	158	107	52	211	133	37
106	95	168	159	101	151	212	100	55
107	124	228	160	136	214	213	107	219
108	141	134	161	154	356	214	92	242
109	140	297	162	115	114	215	139	268
110	116	36	163	113	181	216	171	165
111	92	169	164	147	183	217	100	58
112	148	117	165	147	176	218	160	341
113	121	121	166	94	31	219	147	268
114	154	147	167	173	190	220	155	355
115	170	323	168	156	157	221	139	248
116	100	162	169	106	222	222	128	46
117	110	350	170	139	359	223	134	237
118	131	65	171	140	289	224	138	282
119	124	175	172	107	169	225	103	239
120	158	198	173	136	208	226	94	81
121	146	24	174	116	32	227	151	173
122	100	175	175	101	176	228	124	269
123	122	73	176	108	207	229	157	293
124	135	105	177	167	307	230	164	252
125	102	123	178	149	41	231	151	228
126	124	120	179	105	224	232	105	301
127	117	105	180	102	309	233	157	352
128	110	357	181	106	192	234	164	241
129	178	160	182	99	216	235	152	346
130	119	214	183	133	194	236	166	330
131	133	203	184	93	42	237	138	56
132	117	2	185	130	26	238	100	85
133	147	5	186	111	223	239	112	150
134	103	189	187	171	282	240	97	227

B17   *Data*   Orientations of axial-plane cleavage surfaces of F_1 folds in Ordovician turbidites.
       Coordinates Dip, Dip direction.
       Source   Powell, Cole & Cudahy (1985).
       Analysis   Examples 6.9, 6.10.

	Dip(°)	Dip dir.(°)		Dip(°)	Dip dir.(°)		Dip(°)	Dip dir.(°)
1	80	108	23	83	110	45	86	118
2	37	113	24	67	293	46	85	78
3	32	111	25	85	295	47	85	130
4	58	119	26	78	302	48	45	138
5	58	119	27	88	302	49	79	294
6	52	101	28	85	124	50	85	123
7	52	107	29	78	124	51	54	123
8	31	65	30	88	113	52	80	108
9	45	115	31	85	110	53	69	104
10	83	117	32	62	121	54	57	120
11	84	300	33	77	116	55	85	89
12	44	115	34	75	120	56	65	111
13	63	297	35	66	133	57	55	112
14	72	120	36	84	127	58	86	87
15	41	132	37	70	110	59	58	117
16	88	300	38	75	107	60	55	252
17	68	120	39	79	112	61	80	265
18	83	287	40	62	109	62	70	120
19	87	297	41	80	120	63	60	133
20	14	112	42	48	133	64	85	109
21	70	292	43	60	180	65	63	118
22	68	95	44	83	330			

Data sets

B18   *Data*   Measurements of L_0^1 axes (intersections between cleavage and bedding planes of F_1 folds) in Ordovician turbidites.
      Coordinates Plunge, Plunge azimuth.
      Source Powell, Cole & Cudahy (1985).
      Analysis Examples 6.14, 6.15, 6.16, 6.17, 6.18, 6.19, 6.20, 6.21, 7.3, 7.17.

	Pl. (°)	Pl. az. (°)		Pl. (°)	Pl. az. (°)		Pl. (°)	Pl. az. (°)
1	−5	12	13	−1	2	25	0	12
2	−1	17	14	−12	353	26	1	13
3	−10	9	15	11	350	27	3	353
4	8	342	16	3	9	28	10	3
*5	31	348	17	15	355	29	12	4
6	0	12	18	3	344	30	15	347
7	8	0	**19	−13	329	31	−6	2
8	−7	4	20	0	2	32	−13	7
9	−13	2	21	−9	359	33	−17	354
10	5	4	22	0	12	34	25	4
11	−7	4	23	−11	10	35	0	4
12	5	15	24	−13	14			

B19 *Data* Orientations of bedding planes of F_1 folds in Ordovician turbidites.
Coordinates Dip, Dip direction.
Source Powell, Cole & Cudahy (1985).
Analysis Examples 6.22, 6.23, 7.19.

	Dip (°)	Dip dir. (°)		Dip (°)	Dip dir. (°)		Dip (°)	Dip dir. (°)
1	16	185	23	39	116	45	7	320
2	12	154	24	70	116	46	66	317
3	50	145	25	87	118	47	85	314
4	62	141	26	14	115	48	29	314
5	43	141	27	21	115	49	59	311
6	83	140	28	24	113	50	62	311
7	28	138	29	64	112	51	83	310
8	22	136	30	22	111	52	79	309
9	40	136	31	30	102	53	75	309
10	52	133	32	16	80	54	20	309
11	2	156	33	13	71	55	81	306
12	52	130	34	3	32	56	24	306
13	32	129	35	6	20	57	69	304
14	52	129	36	9	14	58	87	303
15	34	127	37	29	6	59	81	301
16	72	127	38	14	351	60	90	297
17	28	125	39	35	343	61	90	296
18	32	125	40	15	338	62	44	293
19	89	125	41	28	333	63	12	239
20	47	124	42	30	323	64	9	1
21	18	118	43	87	323			
22	27	116	44	70	320			

Data sets   301

B20   *Data*   Measurements of occupational judgments (each
                     response being transformed to a unit vector)
                     according to 4 different criteria (Earnings, Social
                     Status, Reward, Social Usefulness); see Example
                     7.1 for further details.
      Coordinates Colatitude, Longitude.
      Source   Holguin (1980, Table A.1.4); we are most
                     grateful to C. L. Jones for supplying us with
                     these data.
      Analysis   Examples 7.1, 7.4, 7.5.

	A (Earnings)		B (Social Status)	
	Colat. (°)	Long. (°)	Colat. (°)	Long. (°)
1	97.4	344.9	107.4	45.5
2	124.4	322.6	84.7	4.2
3	92.2	323.2	114.9	0.0
4	112.7	339.6	75.1	83.8
5	91.0	326.4	89.1	34.3
6	87.9	327.6	92.8	340.0
7	173.2	152.1	80.4	89.4
8	98.0	334.8	87.5	339.2
9	89.3	333.5	79.6	343.8
10	93.1	334.7	96.1	10.0
11	98.5	328.4	97.6	10.5
12	100.8	340.4	91.5	28.6
13	107.5	323.5	87.7	346.6
14	97.1	329.2	81.8	6.0
15	86.7	318.6	91.1	24.9
16	84.3	338.2	83.1	12.9
17	88.1	333.1	90.2	1.9
18	91.4	336.0	108.7	353.5
19	80.4	337.8	96.2	357.2
20	83.8	327.9	98.6	21.2
21	83.5	331.5	88.7	48.4
22	88.6	331.1	92.8	26.8
23	84.6	324.9	95.0	19.5
24	114.3	335.8	83.9	353.3
25	87.7	320.3	90.4	8.2
26	83.7	297.5	82.1	29.1
27	78.6	309.3	79.5	8.4
28	127.7	25.9	98.2	37.0
29	84.0	339.0	81.7	24.1
30	108.5	9.3	95.6	38.3
31	85.2	2.6	83.7	14.6
32	82.1	341.0	94.4	6.3
33	83.3	334.3	93.5	356.8
34	90.7	332.7	90.5	26.7
35	109.8	329.3	87.8	31.0

B20 (*cont.*)

	A (Earnings)		B (Social Status)	
	Colat. (°)	Long. (°)	Colat. (°)	Long. (°)
36	96.5	347.3	79.6	15.5
37	77.1	322.5	86.3	24.8
38	86.2	331.3	86.0	333.9
39	86.2	332.8	86.3	334.0
40	85.1	331.3	69.3	97.1
41	81.2	326.0	114.0	352.3
42	86.5	350.0	78.5	24.7
43	94.3	345.2	104.3	51.7
44	87.4	10.1	100.2	53.6
45	93.7	355.3	83.9	2.0
46	84.2	327.1	104.5	33.8
47	82.7	326.3	51.8	124.8
48	114.4	36.0	72.0	27.1

Data sets

B20 (*cont.*)

	C (Reward)		D (Social Usefulness)	
	Colat. (°)	Long. (°)	Colat. (°)	Long. (°)
1	99.4	37.1	84.6	92.2
2	104.5	49.3	136.5	54.8
3	73.7	82.0	135.5	27.3
4	77.8	63.2	85.4	38.5
5	115.0	40.4	103.5	87.2
6	85.0	2.9	104.6	62.0
7	85.0	74.6	70.6	103.9
8	116.2	32.9	122.1	93.5
9	98.0	29.5	86.2	30.4
10	108.1	75.3	121.3	113.1
11	102.8	46.7	88.1	94.2
12	87.2	343.6	82.2	35.8
13	89.4	3.2	105.6	52.8
14	71.5	23.8	89.0	70.7
15	84.3	22.2	89.8	87.1
16	92.5	19.3	128.2	59.6
17	92.6	11.5	92.2	24.0
18	65.2	24.3	119.8	15.2
19	90.7	342.4	125.0	94.8
20	110.7	81.3	114.5	75.9
21	76.6	31.9	85.0	73.4
22	105.4	27.2	102.0	41.2
23	86.5	5.9	88.4	17.7
24	89.4	6.5	136.9	57.2
25	76.3	19.6	103.6	43.7
26	101.5	31.4	82.9	42.5
27	71.9	86.9	72.2	68.0
28	99.6	38.5	86.3	76.8
29	76.2	8.5	83.5	40.4
30	72.4	232.7	74.0	98.7
31	94.2	7.0	91.5	15.7
32	93.0	5.9	91.2	66.0
33	73.0	91.6	92.1	102.1
34	85.9	30.3	82.2	66.4
35	100.1	29.8	78.2	88.1
36	95.4	14.7	103.5	49.9
37	86.8	23.0	96.3	51.7
38	86.0	333.9	111.9	53.0
39	86.3	334.0	69.2	105.4
40	68.8	100.6	69.8	107.3
41	64.3	90.4	109.2	41.2
42	79.0	13.5	71.9	35.1
43	107.8	45.0	95.9	43.8
44	108.5	64.5	119.4	32.3
45	89.2	348.5	72.6	341.7
46	81.3	34.3	91.2	27.8
47	76.1	98.3	70.2	107.6
48	87.4	41.7	89.4	62.0

Appendix B

B21 *Data* Measurements of remanent magnetisation in red silts and claystones, made at 4 locations in Eastern New South Wales.
Coordinates Declination, Inclination.
Source The data were reported in summary form only, in Embleton & McDonnell (1980).
Analysis Examples 7.2, 7.6, 7.7, 7.8, 7.9, 7.10, 7.11, 7.14, 7.15, 7.16.

	A		B		C		D	
	Dec. (°)	Inc. (°)	Dec. (°)	Inc. (°)	Dec. (°)	Inc. (°)	Dec. (°)	Inc. (°)
1	32.7	−79.3	13	−75	1	−74	313.3	−85.4
2	18.9	−77.3	27	−72	23	−73	7.7	−68.2
3	25.3	−72.6	20	−74	356	−62	2.1	−71.0
4	359.3	−73.6	0	−73	44	−81	337.5	−81.1
5	12.5	−72.7	23	−70	70	−75	53.1	−78.9
6	14.7	−76.0	44	−72	315	−71	49.8	−71.9
7	1.3	−73.8	334	−78	58	−78	25.0	−80.8
8	6.4	−74.2	51	−75	30	−50	59.3	−80.9
9	17.3	−71.0	31	−73	350	−72	13.6	−77.5
10	27.3	−82.2	10	−75	356	−80	353.2	−78.8
11	42.1	−72.6	16	−73	31	−73	338.7	−81.2
12	339.4	−70.8	41	−81	344	−73	25.7	−81.8
13	347.9	−75.8	42	−75	41	−71	319.7	−76.9
14	27.9	−73.8	29	−75	15	−82	126.0	−81.4
15	24.7	−74.6	52	−80	342	−77	331.9	−83.0
16	11.6	−73.6	356	−78	334	−74	316.7	−76.6
17	37.2	−74.4	21	−84				
18	7.9	−79.7	330	−80				
19	6.5	−69.1	18	−77				
20	355.3	−74.3	344	−80				
21	359.2	−80.1	340	−78				
22	23.5	−84.2	350	−74				
23	60.4	−73.4	12	−70				
24	351.0	−78.2	316	−76				
25	297.9	−73.8	334	−76				
26	23.8	−65.3	11	−79				
27	34.2	−61.0	354	−82				
28	22.4	−79.7	29	−73				
29	21.2	−73.3	32	−74				
30	49.7	−79.1	359	−75				
31	24.0	−69.4	339	−79				
32	3.5	−82.8	357	−75				
33	38.2	−82.8	2	−77				
34	0.4	−82.5	14	−76				
35	43.9	71.7	357	−80				
36	49.4	−78.1	17	−70				
37			355	−77				
38			24	−83				
39			342	−78				

Data sets

B22 *Data* Measurements of L_0^1 axes (intersections between cleavage and bedding planes of F_1 folds) in Ordovician turbidites.
  Coordinates Plunge, Plunge azimuth.
  Source Powell, Cole & Cudahy (1985).
  Analysis Examples 7.3, 7.17.

	Pl. (°)	Pl. az. (°)		Pl. (°)	Pl. az. (°)		Pl. (°)	Pl. az. (°)
1	−9	26	15	−20	20	29	−4	30
2	−8	16	16	−1	35	30	25	16
3	0	24	17	−7	12	31	15	18
4	−26	12	18	−12	22	32	0	13
5	−12	11	19	−4	21	33	0	9
6	−4	33	20	−7	24	34	−18	11
7	1	340	21	−14	26	35	−1	30
8	−4	336	22	−14	29	36	0	23
9	5	28	23	0	16	37	−17	26
10	6	18	24	−4	31	38	0	98
11	0	27	25	−5	44	39	0	17
12	5	19	26	7	35	40	−35	12
13	10	16	27	0	33	41	12	12
14	0	15	28	0	22	42	19	24

B23 *Data* Measurements of natural remanent magnetisation in Old Red Sandstone rocks from two sites in Pembrokeshire, Wales.
Coordinates Declination, Inclination.
Source Set A – Creer (1957; Table 2A);
Set B – Creer (1957; Table 2C, Manorbier Bay site only).
Analysis Examples 7.12, 7.13.

	A		B	
	Dec. (°)	Inc. (°)	Dec. (°)	Inc. (°)
1	202	−25	194	17
2	197	6	197	28
3	190	−13	190	22
4	194	−8	197	14
5	192	0	202	17
6	200	7	202	17
7	201	4	200	40
8	202	9	200	11
9	197	0	197	12
10	197	−9	215	43
11	200	0	209	17
12	187	−1	197	26
13	192	3	209	37
14	191	−16		
15	184	−15		
16	193	−16		
17	197	−11		
18	193	16		
19	198	−21		
20	190	−8		
21	197	−13		
22	199	−9		
23	197	−7		
24	193	−3		
25	210	−6		
26	189	−16		
27	195	−18		
28	194	−6		
29	198	−6		
30	217	24		
31	177	−20		
32	214	−16		
33	167	52		
34	204	4		
35	235	15		

B24

Data Two samples of measurements of L_0^1 axes (intersections between cleavage and bedding planes of F_1 folds) in Ordovician turbidites, collected in the same sub-domain.
Coordinates Plunge, Plunge azimuth.
Source Powell, Cole & Cudahy (1985).
Analysis Example 7.18.

	A		B	
	Pl. (°)	Pl. az. (°)	Pl. (°)	Pl. az. (°)
1	−12	44	2	39
2	−6	48	10	43
3	−25	18	1	54
4	−21	41	−5	33
5	−3	36	73	106
6	10	40	−7	44
7	−6	26	0	57
8	0	35	29	30
9	−6	57	−10	42
10	0	35	−10	92
11	−5	54	4	36
12	−9	0	−2	49
13	−5	34	−2	19
14	−1	67	−5	38
15	2	41	−7	36
16	2	35	−9	40
17	−1	39	−10	44
18	2	39	1	36
19	−7	45	14	49
20	−22	39	−4	47
21	1	50	31	36
22	−14	42	21	40
23	−5	41	3	34
24	−2	38	−19	32
25	−12	34	−22	31

B25 *Data* Measurements of wind direction on consecutive days near Singleton, New South Wales, during March 1981. The measurements were made between 0600 h and 0630 h at a height of 300 m, using Doppler shift acoustic radar.
Coordinates Colatitude, Longitude.
Source We are most grateful to R. Hyde for permission to use these data.
Analysis Examples 8.2, 8.9.

	Colat. (°)	Long. (°)		Colat. (°)	Long. (°)		Colat. (°)	Long. (°)
1	90.0	132.2	11	88.8	131.5	21	89.3	95.8
2	90.0	107.2	12	88.7	108.6	22	88.3	112.1
3	90.0	282.5	13	91.8	282.7	23	90.0	225.0
4	87.0	213.3	14	90.0	296.6	24	89.4	120.2
5	87.6	167.7	15	92.7	243.4	25	88.6	124.9
6	90.0	158.3	16	89.1	115.3	26	89.0	144.1
7	88.6	182.8	17	92.2	283.5	27	88.2	95.4
8	90.0	135.0	18	91.3	336.3	28	90.0	67.8
9	92.7	300.3	19	92.1	155.8	29	88.8	83.6
10	91.3	240.3	20	89.4	98.3			

Data sets 309

B26 *Data* Measurements of the longest axis and shortest axis orientations of tabular stones on a slope at Windy Hills, Scotland.
Coordinates Plunge, Plunge azimuth.
Source We are most grateful to E. A. Fitzpatrick for permission to use these data, and to R. M. Cormack for assistance in obtaining them.
Analysis Example 8.8.

	A (Longest Axis)		B (Shortest Axis)	
	Pl. (°)	Pl. az. (°)	Pl. (°)	Pl. az. (°)
1	10	150	0	60
2	40	160	40	70
3	30	140	60	50
4	50	160	40	70
5	60	170	0	80
6	20	140	0	50
7	30	150	50	60
8	20	150	0	60
9	10	160	20	250
10	50	170	20	80
11	30	350	70	80
12	0	110	50	200
13	40	120	0	30
14	0	180	40	90
15	60	340	30	250
16	30	170	70	260
17	0	160	10	70
18	50	140	20	50
19	60	50	20	140
20	0	140	90	50
21	30	150	60	60
22	10	160	40	250
23	20	160	30	250
24	10	170	70	260
25	20	150	0	60
26	20	150	0	60
27	0	160	10	70
28	40	170	0	80
29	10	150	20	240
30	20	340	0	70
31	50	340	60	70
32	0	140	80	230
33	20	180	10	90
34	50	170	10	260
35	30	160	20	250
36	30	150	70	240
37	30	160	10	70

Appendix B

B26 (*cont.*)

	A (Longest Axis)		B (Shortest Axis)	
	Pl. (°)	Pl. az. (°)	Pl. (°)	Pl. az. (°)
38	10	170	0	80
39	10	150	10	240
40	30	170	90	260
41	30	130	20	220
42	10	160	10	250
43	30	180	0	90
44	50	130	50	40
45	40	150	20	240
46	30	130	0	40
47	10	180	20	90
48	10	200	10	290
49	10	300	20	210
50	10	120	50	210
51	0	150	40	240
52	30	150	0	240
53	30	140	0	230
54	10	140	10	230
55	30	150	30	60
56	30	150	10	60
57	0	140	0	50
58	40	170	0	80
59	40	160	10	70
60	20	150	10	240
61	20	130	10	220
62	20	180	10	90
63	40	180	20	270
64	10	310	30	220
65	60	90	20	180
66	20	140	30	230
67	20	180	10	270
68	40	280	20	10
69	0	180	20	270
70	20	150	60	60
71	40	130	10	220
72	0	130	60	220
73	20	180	10	90
74	40	150	0	60
75	30	180	20	270
76	50	150	0	60
77	0	170	0	80
78	0	130	20	40
79	30	310	50	220
80	40	170	0	80
81	10	140	70	230
82	20	0	50	90
83	50	110	0	20

B26 (*cont.*)

	A (Longest Axis)		B (Shortest Axis)	
	Pl. (°)	Pl. az. (°)	Pl. (°)	Pl. az. (°)
84	0	170	10	80
85	40	140	70	50
86	0	150	10	60
87	20	220	30	130
88	30	170	20	260
89	50	170	10	260
90	40	150	10	60
91	0	170	10	260
92	30	210	10	120
93	40	160	20	230
94	20	30	80	120
95	70	120	20	30
96	10	160	50	70
97	40	130	0	40
98	30	80	10	170
99	30	180	30	90
100	40	200	20	290
101	50	180	40	270

References

Abramowitz, M. & Stegun, I. A. (1970). *Handbook of Mathematical Functions.* New York: National Bureau of Standards. (51)

Addison, F. T. (1982). *A Magnetic Study of Diagenesis in Carbonate Sediments.* Ph.D. Thesis, University of Newcastle upon Tyne, U.K. (292)

Adler, R. K. *See* Richardus *et al.* (1972).

Anderson, T. W. & Stephens, M. A. (1972). Tests for randomness against equatorial and bimodal alternatives. *Biometrika* **59**, 613-621. (162, 179)

Arnold, K. J. (1941). *On spherical probability distributions.* Ph.D. Thesis, Massachusetts Institute of Technology. (12, 14, 97)

Barnett, V. & Lewis, T. (1984). *Outliers in Statistical Data.* 2nd Edition. New York: John Wiley. (128)

Batschelet, E. (1981). *Circular Statistics in Biology.* London: Academic Press. (12, 82, 83, 86, 88, 92)

Beasley, J. D. & Springer, S. G. (1977). The percentage points of the normal distribution. *Appl. Statist.* **26**, 118-121. (51)

Beran, R. (1968). Testing for uniformity on a compact homogeneous space. *J. Appl. Prob.* **5**, 177-195. (149, 150)

Beran, R. (1979). Exponential models for directional data. *Ann. Statist.* **7**, 1162-1178. (98)

Beran, R. J. *See* Watson *et al.* (1967).

Berry, L. G. & Mason, B. (1959). *Mineralogy: Concepts, Descriptions, Determinations.* San Francisco: W. H. Freeman & Co. (36)

Best, D. J. & Fisher, N. I. (1981). The bias of the maximum likelihood estimators of the von Mises-Fisher concentration parameters. *Commun. Statist. - Simula. Computa.* **B10(5)**, 493-502. (133)

Best, D. J. & Fisher, N. I. (1986). Goodness-of-fit and discordancy tests for samples from the Watson distribution on the sphere. *Austral. J. Statist.* **28**, 13-31. (60, 128, 170, 175, 177, 186, 189, 190, 263, 272, 273)

Best, D. J. *See* Fisher *et al.* (1984).

Bingham, C. (1964). *Distributions on the sphere and on the projective plane.* Ph.D. Thesis, Yale University. (14, 92, 97, 150, 177)

Bingham, C. (1974). An antipodally symmetric distribution on the sphere. *Ann. Statist.* **2**, 1201-1225. (92, 97, 150, 177)

Bingham, C. & Mardia, K. V. (1978). A small circle distribution on the sphere. *Biometrika* **65**, 379-389. (99, 143)

References

Bomford, G. (1980). *Geodesy*. 4th Edition. Oxford: Clarendon Press. (37)

Bradt, H. *See* Toyoda *et al.* (1965).

Breitenberger, E. (1963). Analogues of the normal distribution on the circle and the sphere. *Biometrika* **50**, 81-88. (14)

Briden, J. C. & Ward, M. A. (1966). Analysis of magnetic inclination in bore-cores. *Pure & Appl. Geophys.* **63**, 133-145. (135)

Brown, E. T. *See* Hoek *et al.* (1980).

Chambers, J. M., Cleveland, W. S., Kleiner, B., & Tukey, P. A. (1983). *Graphical Methods for Data Analysis*. Boston: Duxbury Press. (40)

Chayes, F. (1949). Pp 297-326 in Fairbairn, H. W., Ed., *Structural Geology of Deformed Rocks*. 2nd Edition. Reading, Massachusetts: Addison-Wesley. (10)

Clark, G. *See* Toyoda *et al.* (1965).

Clark, R. M. (1983). Estimation of parameters in the marginal Fisher distribution. *Austral. J. Statist.* **25**, 227-237. (135)

Clark, R. M. (1985). A FORTRAN program for constrained sequence-slotting based on minimum combined path length. *Computers and Geosciences* **11**, 605-617. (244)

Clark, R. M. & Thompson, R. (1978). An objective method for smoothing palaeomagnetic data. *Geophys. J. R. astr. Soc.* **52**, 205-213. (245, 247)

Clark, R. M. & Thompson, R. (1979). A new statistical approach to the alignment of time series. *Geophys. J.R. astr. Soc.* **58**, 593-607. (245, 247)

Clark, R. M. & Thompson, R. (1984). Statistical comparison of palaeomagnetic records from lake sediments. *Geophys. J.R. astr. Soc.* **76**, 337-368. (245, 247)

Clark, R. M. *See* Thompson *et al.* (1981)

Clark, R. M. *See* Thompson *et al.* (1982).

Cleveland, W. S. *See* Chambers *et al.* (1983).

Cochran, W. G. (1977). *Sampling Techniques*. 3rd Edition. New York: John Wiley. (64)

Cohen, P. H. (1983). *A Computer Based Trend Surface Study of Structural Analysis: Test Case from Lipson Cove, South Australia*. Ph.D. Thesis, University of Adelaide, South Australia. (282)

Cole, J. P. *See* Powell *et al.* (1985).

Collett, D. & Lewis, T. (1981). Discriminating between the von Mises and Wrapped Normal distributions. *Austral. J. Statist.* **23**, 73-79. (84)

Creasey, J. W. (1985). *See* Fisher, Huntington, Jackett, Willcox & Creasey (1985).

Creer, K. M. (1957). Palaeomagnetic investigations in Great Britain IV. The natural remanent magnetization of certain stable rocks from Great Britain. *Phil. Trans. R. Soc., London*, **A, 974**, 111-129. (306)

Creer, K. M., Irving, E. & Nairn, A. E. M. (1959). Palaeomagnetism of the Great Whin Sill. *Geophys. J.R. astr. Soc.* **2**, 306-323. (289)

Cudahy, T. J. *See* Powell *et al.* (1985).

Daniels, H. E. (1975). Contribution to Discussion of Mardia (1975a). (14, 15)

Denham, C. R. *See* Parker *et al.* (1979).

Diggle, P. J. & Fisher, M. I. (1985). SPHERE: A contouring program for spherical data. *Computers and Geosciences* **11**, 725-766. (34, 41, 42, 43, 46)

Diggle, P. J., Fisher, N. I. & Lee, A. J. (1985). A comparison of tests of uniformity for spherical data. *Austral. J. Statist.* **27**, 53-59. (57, 111, 150, 270)

Dimroth, E. (1962). Untersuchungen zum Mechanismus von Blastesis und Syntexis in Phylliten und Hornfelsen des südwestlichen Fichtelgebirges. I. Die statistische Auswertung einfacher Gürteldiagramme. *Tscherm. Min. Petr. Mitt.* **8**, 248-274. (14, 89)

Domingo, V. *See* Toyoda *et al.* (1965).

Dudley, R. M., Perkins, P. C. & Evarist Giné, M. (1975). Statistical tests for preferred orientation. *J. Geol.* **83**, 685-705. (10)

Efron, B. (1982). *The Jackknife, the Bootstrap and Other Resampling Plans.* SIAM (Regional Conference Series in Applied Mathematics, 38). Philadelphia. (57)

Efron, B. & Gong, G. (1983). A leisurely look at the bootstrap, the jackknife, and cross-validation. *Am. Statist.* **37**, 36-48. (57)

Embleton, B. J. J. (1970). Palaeomagnetic results for the Permian of South America and a comparison with the African and Australian data. *Geophys. J.R. astr. Soc.* **21**, 105-118. (279)

Embleton, B. J. J., Fisher, N. I. & Schmidt, P. W. (1983). Analytic comparison of apparent polar wander paths. *Earth Planet. Sci. Lett.* **64**, 276-282. (244)

Embleton, B. J. J. & Giddings, J. W. (1974). Late Precambrian and Lower Palaeozoic palaeomagnetic results from South Australia and Western Australia. *Earth Planet. Sci. Lett.* **22**, 355-365. (286)

Embleton, B. J. J. & McDonnell K. L. (1980). Magnetostratigraphy in the Sydney Basin, Southeastern Australia. *J. Geomag. Geoelectr.* **32**, Suppl. III, SIII1-SIII10. (304)

Embleton, B. J. J. *See* Schmidt *et al.* (1985).

Epp, R. J., Tukey, J. W. & Watson G. S. (1971). Testing unit vectors for correlation. *J. geophys. Res.* **76**, 8480-8483. (243)

Escobar, I. *See* Toyoda *et al.* (1965).

Fara, H. D. & Scheidegger, A. E. (1963). An eigenvalue method for the statistical evaluation of fault plane solutions of earthquakes. *Bull. seism. Soc. Am.* **53** (4), 811-816. (181)

Fisher, N. I. (1982a). Goodness-of-fit and outlier detection procedures for samples from Fisher's distribution on the sphere. *Contemporary Mathematics* **9**, 377-380. (125)

Fisher, N. I. (1982b). Robust estimation of the concentration parameter of Fisher's distribution on the sphere. *Appl. Statist.* **31**, 152-154. (133)

Fisher, N. I. (1985). Spherical medians. *J.R. Statist. Soc.* **B47**, 342-348. (113, 114, 201)

Fisher, N. I. & Best, D. J. (1984). Goodness-of-fit tests for Fisher's distribution on the sphere. *Austral. J. Statist.* **26**, 142-150. (125, 263)

Fisher, N. I., Huntington, J. F., Jackett, D. R., Willcox, M. E. & Creasey, J. W. (1985). Spatial analysis of two-dimensional orientation data. *J. Math. Geol.* **17**, 177-194. (248)

Fisher, N. I. & Lee, A. J. (1986). Correlation coefficients for random variables on a unit sphere or hypersphere. *Biometrika* **73**, 159-164. (232, 234, 277)

Fisher, N. I. & Lewis, T. (1983). Estimating the common mean direction of several circular or spherical distributions with differing dispersions. *Biometrika* **70**, 333-341. (116, 208, 213)

Fisher, N. I., Lewis, T. & Willcox, M. E. (1981). Tests of discordancy for samples from Fisher's distribution on the sphere. *Appl. Statist.* **30**, 230-237. (59, 128)

Fisher, N. I. *See* Best *et al.* (1981).

Fisher, N. I. *See* Best *et al.* (1985).

Fisher, N. I. *See* Diggle & Fisher (1985).

Fisher, N. I. *See* Diggle, Fisher & Lee (1985).

Fisher, N. I. *See* Embleton *et al.* (1983).

Fisher, N. I. *See* Lewis *et al.* (1982).

Fisher, R. A. (1953). Dispersion on a sphere. *Proc. R. Soc.* **A 217**, 295-305 (9, 11, 12, 13, 88, 132, 133)

Fisher, R. A. & Yates, F. (1974). *Statistical Tables for Biological, Agricultural and Medical Research.* 6th Edition. Edinburgh: Oliver & Boyd. (75)

Flinn, D. (1958). On tests of significance of preferred orientation in three-dimensional fabric diagrams. *J. Geol.* **66**, 526-539. (10)
Gadsden, R. J. *See* Mardia & Gadsden (1977).
Geiser, J. R. *See* Gray et al. (1980).
Geiser, P. A. *See* Gray et al. (1980).
Giddings, J. W. *See* Embleton et al. (1974).
Giné, Evarist M (1975). Invariant tests for uniformity on compact Riemannian manifolds based on Sobolev norms. *Ann. Statist.* **3**, 1243-1266. (149, 150)
Giné, Evarist M. *See* Dudley et al. (1975).
Gong, G. *See* Efron et al. (1983).
Gould, A. L. (1969). A regression technique for angular variates. *Biometrics* **25**, 683-700. (241)
Gray, N. H., Geiser, P. A. & Geiser, J. R. (1980). On the least-squares fit of small and great circles to spherically projected orientation data. *J. Math. Geol.* **12**, 173-184. (143)
Hartley, H. O. *See* Pearson et al. (1970).
Hartley, H. O. *See* Pearson et al. (1972).
Hasegawa, H. *See* Toyoda et al. (1965).
Hawkins, D. M. (1980). *Identification of Outliers.* London: Chapman & Hall. (128)
Hill, I. D. *See* Wichmann et al. (1982).
Hoek, E., & Brown, E. T. (1980). *Underground Excavations in Rock.* London: The Institute of Mining & Metallurgy. (36)
Holguin, J. (1980). *The application of directional methods in p dimensions.* M.Sc. Thesis. Simon Fraser University, Burnaby, Canada. (301)
Holmes, D. *See* Mardia et al. (1984).
Hospers, J. (1951). Remanent magnetism of rocks and the history of the geomagnetic field. *Nature* **168**, 1111-1112. (11)
Huntington, J. F. *See* Fisher, Huntington, Jackett, Willcox & Creasey (1985).
IMSL (1982). *The IMSL Library.* Edition 9. International Mathematical and Statistical Libraries, Inc., Houston, Texas. (111, 146)
Irving, E. (1964). *Paleomagnetism and Its Application to Geological and Geophysical Problems.* New York: John Wiley. (14, 66)
Irving, E. (1977). Drift of the major continental blocks since the Devonian. *Nature* **270**, 304-309. (245)
Irving, E. *See* Creer et al. (1959).
Irving, E. *See* Watson et al. (1957).
Jackett, D. R. *See* Fisher, Huntington, Jackett, Willcox & Creasey (1985).
Jones, D. L. *See* McFadden et al. (1981).
Jupp, P. E. & Mardia, K. V. (1980). A general correlation coefficient for directional data and related regression problems. *Biometrika* **67**, 163-173. (235, 237, 238, 240)
Jupp, P. E. & Spurr, B. D. (1983). Sobolev tests for symmetry of directional data. *Ann. Statist.* **11**, 1225-1231. (151)
Kamata, K. *See* Toyoda et al. (1965).
Keilson, J., Petrondas, D., Sumita, U. & Wellner, J. (1983). Significance points for tests of uniformity on the sphere. *J. Statist. Comput. Simul.* **17**, 195-218. (149, 295)
Kelker, D. & Langenberg, C. W. (1982). A mathematical model for orientation data from macroscopic conical folds. *J. Math. Geol.* **14**, 289-307. (98)
Kendall, D. G. (1974). Pole-seeking Brownian motion and bird navigation (with Discussion). *J.R. Statist. Soc.* **B 36**, 365-417. (84)

Kent, J. T. (1982). The Fisher-Bingham distribution on the sphere. *J.R. Statist. Soc.* **B 44**, 71-80. (15, 92, 93, 94, 98, 136, 137, 140)

Kent, J. T. *See* Mardia *et al.* (1984).

Kinderman, A. J. & Monahan, J. F. (1977). Computer generation of random variables using the ratio of uniform deviates. *ACM Trans. Math. Software* **3**, 257-260. (58)

Kleiner, B. *See* Chambers *et al.* (1983).

Kraus, J. D. (1966). *Radio Astronomy.* New York: McGraw-Hill Book Company. (21, 23)

Lafeber, D. (1965). The graphical representation of planar pore patterns in soils. *Austral. J. Soil Res.* **3**, 143-164. (37)

Langenberg, C. W. *See* Kelker *et al.* (1982).

Langevin, P. (1905). Magnétisme et théorie des électrons. *Ann. de Chim. et de Phys.* **5**, 70-127. (12)

La Pointe, M. *see* Toyoda *et al.* (1965).

Laycock, P. J. *See* Morris *et al.* (1974).

Lee, A. J. *See* Fisher & Lee (1985).

Lewis, T. (1974). Contribution to Discussion of Kendall (1974). (84)

Lewis, T. (1986). A simple improved-accuracy normal approximation for χ^2. *Biometrika.* To appear.

Lewis, T. & Fisher, N. I. (1982). Graphical methods for investigating the fit of a Fisher distribution to spherical data. *Geophys. J.R. astr. Soc.* **69**, 1-13. (51, 125)

Lewis, T. *See* Barnett *et al.* (1984).

Lewis, T. *See* Collett *et al.* (1981).

Lewis, T. *See* Fisher *et al.* (1981).

Lewis, T. *See* Fisher *et al.* (1983).

Lindley, D. V. & Scott, W. F. (1984). *New Cambridge Elementary Statistical Tables.* Cambridge: Cambridge University Press. (249, 251, 253, 254, 256)

Mackenzie, J. K. (1957). The estimation of an orientation relationship. *Acta. Cryst.* **10**, 61-62. (235)

Macleod, I. D. G. (1970). Pictorial output on a line printer. *IEEE Transactions on Computers* **19**, 160-162. (46)

McDonnell, K. L. *See* Embleton *et al.* (1980).

McDougall, I. *See* Schmidt *et al.* (1977).

McElhinny, M. W. (1964). Statistical significance of the fold test in palaeomagnetism. *Geophys. J.R. astr. Soc.* **8**, 338-340. (211)

McElhinny, M. W. (1973). *Palaeomagnetism and Plate Tectonics.* Cambridge: Cambridge University Press. (66)

McFadden, P. L. & Jones, D. L. (1981). The fold test in palaeomagnetism. *Geophys. J.R. astr. Soc.* **67**, 53-58. (211)

Mardia, K. V. (1972). *Statistics of Directional Data.* London: Academic Press. (14, 39, 47, 69, 83, 88, 91, 92, 97, 99, 133, 150, 177, 190, 191, 210, 211, 216, 218, 220, 223, 224)

Mardia, K. V. (1975a). Statistics of directional data (with Discussion). *J.R. Statist. Soc.* **B 37**, 349-393. (97, 133)

Mardia, K. V. (1975b). Distribution theory for the von Mises-Fisher distribution and its application. In *Statistical Distributions in Scientific Work*, Vol. 1 edited by G. P. Patil, S. Kotz & J. K. Ord, *pp* 113-130. Dordrecht, Holland: D. Reidel Publishing Company. (133)

Mardia, K. V. & Gadsden, R. J. (1977). A small circle of best fit for spherical data and areas of vulcanism. *Appl. Statist.* **26**, 238-245. (100, 143)

Mardia, K. V. & Zemroch, P. J. (1977). Table of maximum likelihood estimates for the Bingham distribution. *J. Statist. Comput. Simul.* **6**, 29-34. (97)

Mardia, K. V. & Zemroch, P. J. (1978). *Tables of the F- and Related Distributions with Algorithms.* New York; Academic Press. (81)

Mardia, K. V., Holmes, D. & Kent, J. T. (1984). A goodness-of-fit test for the von Mises-Fisher distribution. *J.R. Statist. Soc.* **B 46**, 72-78. (125)

Mardia, K. V. *See* Bingham *et al.* (1978).

Mardia, K. V. *See* Jupp *et al.* (1980).

Maritz, J. S. (1981). *Distribution-free Statistical Methods.* London: Chapman & Hall. (63)

Marsden, B. G. (1982). *Catalogue of Cometary Orbits.* 4th Edition. International Astronomical Union. (296)

Mason, B. *See* Berry *et al.* (1959).

Mead, R. *See* Nelder *et al.* (1965).

Monahan, J. F. *See* Kinderman *et al.* (1977).

Moran, P. A. P. (1976). Quaternions, Haar measure, and the estimation of a palaeomagnetic rotation. In *Perspectives in Probability and Statistics,* pp 295-301, J. Gani, Editor. Sheffield: Applied Probability Trust. (235)

Morris, J. E. & Laycock, P. J. (1974). Discriminant analysis of directional data. *Biometrika* **61**, 335-341. (224)

Murakami, K. *See* Toyoda *et al.* (1965).

NAG Fortran Library Manual (1977). Mark 6. Numerical Algorithms Group, Oxford, England. (111, 146)

Nairn, A. E. M. *See* Creer *et al.* (1959).

Naylor, M. A. *See* Woodcock *et al.* (1983).

Nelder, J. A. & Mead, R. (1965). A simplex method for function minimisation. *Comput. J.* **7**, 308-313. (111, 146)

Nijenhuis, A. & Wilf, H. S. (1978). *Combinatorial Algorithms for Computers and Calculators.* 2nd Edition. New York: Academic Press. (61)

Owen, D. B. (1962). *Handbook of Statistical Tables.* Reading, Massachusetts: Addison-Wesley. (75, 81)

Parker, R. L. & Denham, C. R. (1979). Interpolation of unit vectors. *Geophys. J.R. astr. Soc.* **58**, 685-687. (245)

Pearson, E. S. & Hartley, H. O. (1970). *Biometrika Tables for Statisticians,* Vol. I. 3rd Edition. Cambridge: Cambridge University Press. (75, 79, 81, 276)

Pearson, E. S. & Hartley, H. O. (1972). *Biometrika Tables for Statisticians,* Vol. II. Cambridge: Cambridge University Press. (269, 271, 272, 274)

Perkins, P. C. *See* Dudley *et al.* (1975).

Petrondas, D. *See* Keilson *et al.* (1983).

Pettijohn, F. J. *See* Potter *et al.* (1977).

Phillips, F. C. (1971). *Stereographic Projections in Structural Geology.* 3rd Edition. London: Edward Arnold (Publishers) Ltd. (36)

Potter, P. E. & Pettijohn, F. J. (1977). *Paleocurrents and Basin Analysis.* 2nd Edition. Berlin: Springer-Verlag. (66)

Powell, C. McA., Cole, J. P. & Cudahy, T. J. (1985). Megakinking in the Lachlan Fold Belt. *J. Struct. Geol.* **7**, 281-300. (290, 291, 298, 299, 300, 305, 307)

Prentice, M. J. (1978). On invariant tests of uniformity for directions and orientations. *Ann. Statist.* **6**, 169-176. (150)

Prentice, M. J. (1984). A distribution-free method for unsigned directional data. *Biometrika* **71**, 147-154. (163, 165, 166, 181, 182, 192)

Rayleigh, *Lord* (1919). On a problem of vibrations, and of random flights in one, two and three dimensions. *Phil. Mag.* (6) **37**, 321-347. (111)

Richardus, P. & Adler, R. K. (1972). *Map Projections for Geodesists, Cartographers & Geographers*. London: North-Holland Publishing Company. (36)

Saw, J. G. (1978). A family of distributions on the m-sphere and some hypothesis tests. *Biometrika* 65, 69–73. (99)

Saw, J. G. (1983). Dependent unit vectors. *Biometrika* 70, 665–671. (235)

Schaeben, H. (1984). A new cluster algorithm for orientation data. *J. Math. Geol.* 16, 139–154. (144)

Scheidegger, A. E. (1964). The tectonic stress and tectonic motion direction in Europe and Western Asia as calculated from earthquake fault plane solutions. *Bull seism. Soc. Am.* 54 (5A), 1519–1528. (14)

Scheidegger, A. E. (1965). On the statics of the orientation of bedding planes, grain axes and similar sedimentological data. *U.S. Geol. Survey Prof. Paper.* 525c, 164–167. (14, 89)

Scheidegger, A. E. *See* Fara *et al.* (1963).

Schmidt, P. W. (1976). The non-uniqueness of the Australian Mesozoic palaeomagnetic pole position. *Geophys. J.R. astr. Soc.* 47, 285–300. (147, 289)

Schmidt, P. W. & Embleton, B. J. J. (1985). Pre-folding and overprint magnetic signatures in Precambrian (~2.9–2.7ga) igneous rocks from the Pilbara Craton and Hamersley Basin, N. W. Australia. *J. geophys. Res.* 90 (B4), 2967–2984. (285)

Schmidt, P. W. & McDougall, I. (1977). Palaeomagnetic and Potassium–argon dating studies of the Tasmanian Dolerites. *J. geol. Soc. Aust.*, 25 (6), 321–328. (149)

Schmidt, P. W. *See* Embleton *et al.* (1983).

Scott, W. F. *See* Lindley *et al.* (1984).

Schou, G. (1978). Estimation of the concentration parameter in von Mises–Fisher distributions. *Biometrika* 65, 369–377. (133)

Selby, B. (1964). Girdle distributions on the sphere. *Biometrika* 51, 381–392. (97)

Shibata, S. *See* Toyoda *et al.* (1965).

Smart, W. M. (1977). *Textbook on Spherical Astronomy*. 6th Edition, revised by R. M. Green. London: Cambridge University Press. (21)

Springer, S. G. *See* Beasley *et al.* (1977).

Spurr, B. D. *See* Jupp *et al.* (1983).

Stegun, I. A. *See* Abramowitz *et al.* (1970).

Stephens, M. A. (1962). Exact and approximate tests for directions. II. *Biometrika* 49, 547–552. (133, 136, 268)

Stephens, M. A. (1964). The testing of unit vectors for randomness. *J. Am. Statist. Assoc.* 59, 160–167. (111)

Stephens, M. A. (1965). The distributions of $S = \sum_{i=1}^{N} \cos^2 \theta_i$: 3 dimensions. Appendix to Watson (1965). (162)

Stephens, M. A. (1966). Statistics connected with the uniform distribution: percentage points and application to testing for randomness of directions. *Biometrika* 53, 235–240. (162, 180)

Stephens, M. A. (1967). Tests for the dispersion and for the modal vector of a distribution on the sphere. *Biometrika* 54, 211-223. (133, 266)

Stephens, M. A. (1969a). Multi-sample tests for the Fisher distribution for directions. *Biometrika* 56, 169–181. (209, 218, 224, 275)

Stephens, M. A. (1969b). Techniques for directional data, Technical report #150, Dept. of Statistics, Stanford University, 29 pp. (144)

Stephens, M. A. (1972). Analysis of directions on a circle and sphere. Section IX in E. S. Pearson & H. O. Hartley (1972), *Biometrika Tables for Statisticians.* (162, 180)

Stephens, M. A. (1974). EDF statistics for goodness of fit and some comparisons. *J. Amer. Statist. Assoc.* **69**, 730-737. (262, 263)

Stephens, M. A. (1975). A new test for the modal vector of the Fisher distribution. *Biometrika* **62**, 171-174. (131)

Stephens, M. A. (1979). Vector correlation. *Biometrika* **66**, 41-48. (234, 235)

Stephens, M. A. *See* Anderson *et al.* (1972).

Suga, K. *See* Toyoda *et al.* (1965).

Sumita, U. *See* Keilson *et al.* (1983).

Thompson, R. & Clark, R. M. (1981). Fitting polar wander paths. *Physics Earth Planet. Int.* **27**, 1-7. (245, 247)

Thompson, R. & Clark, R. M. (1982). A robust least-squares Gondwanan apparent polar wander path and the question of palaeomagnetic assessment of Gondwanan reconstructions. *Earth Planet. Sci. Lett.* **57**, 152-158. (245, 247)

Thompson, R. *See* Clark *et al.* (1978).

Thompson, R. *See* Clark *et al.* (1979).

Thompson, R. *See* Clark *et al.* (1984).

Toyoda, Y., Suga, K., Murakami, K., Hasegawa, H., Shibata, S., Domingo, V., Escobar, I., Kamata, K., Bradt, H., Clark, G. & La Pointe, M. (1965). Studies of primary cosmic rays in the energy region 10^{14} eV to 10^{17} eV (Bolivian Air Shower Joint Experiment). *Proc. Int. Conf. Cosmic Rays*, Vol. 2, London, September 1965, pp 708-711. London: The Institute of Physics and The Physical Society. (280)

Tukey, J. W. *See* Epp *et al.* (1971).

Tukey, P. A. *See* Chambers *et al.* (1983).

Turner, F. J. & Weiss, L. E. (1963). *Structural Analysis of Metamorphic Tectonites*. New York: McGraw-Hill Book Company. (36, 66)

Ward, M. A. *See* Briden *et al.* (1966).

Watson, G. S. (1956a). Analysis of dispersion on a sphere. *Mon. Not. R. astr. Soc. geophys. Suppl.* **7**, 153-159. (13, 133, 224)

Watson, G. S. (1956b). A test for randomness of directions. *Mon. Not. R. astr. Soc. geophys. Suppl.* **7**, 160-161. (111, 211)

Watson, G. S. (1960). More significance tests on the sphere. *Biometrika* **47**, 87-91. (216, 218)

Watson, G. S. (1965). Equatorial distributions on a sphere. *Biometrika* **52**, 193-201. (14, 92, 179, 180, 190, 191, 228)

Watson, G. S. (1966). The statistics of orientation data. *J. Geol.* **74**, 786-797. (13, 14, 177)

Watson, G. S. (1967). Some problems in the statistics of directions. *Bull. Int. Statist. Inst.* **42**, 374-385. (133)

Watson, G. S. (1970). The statistical treatment of orientation data. In *Geostastics. A Colloquium*, D. F. Merriam, Editor. New York: Plenum Press. (2, 64, 99, 133)

Watson, G. S. (1983a). *Statistics on Spheres*. University of Arkansas Lecture Notes in the Mathematical Sciences, Volume 6. New York: John Wiley. (14, 16, 21, 33, 42, 64, 84, 86, 116, 125, 133, 150, 162, 163, 165, 166, 167, 177, 180, 181, 182, 183, 192, 206, 208, 213)

Watson, G. S. (1983b). Large sample theory of the Langevin distributions. *Journal of Statistical Planning and Inference* **8**, 245-256. (125, 133, 211)

Watson, G. S. (1983c). Large sample theory for distributions on the hypersphere with rotational symmetrics. *Ann. Inst. Statist. Math.* **35**, Part A, 303-319. (167, 182, 183, 226, 227, 228, 229)

Watson, G. S. (1983d). Interpolation and smoothing of directed and undirected line data. In *Multivariate Analysis* VI, P. R. Krishnaiah, Editor. New York; Academic Press. (245, 247, 248)

Watson, G. S. (1984a). The calculation of confidence regions for eigenvectors. *Austral. J. Statist.* **26**, 272-276. (163, 181)

Watson, G. S. (1984b). The theory of concentrated Langevin distributions. *J. Multivar. Anal.* **14**, 74-82. (211)

Watson, G. S. & Beran, R. J. (1967). Testing a sequence of unit vectors for randomness. *J. geophys. Res.* **72**, 5655-5659. (243)

Watson, G. S. & Irving, E. (1957). Statistical methods in rock magnetism. *Mon. Not. R. astr. Soc. geophys. Suppl.* **7**, 289-300. (66, 136, 224)

Watson, G. S. & Williams, E. J. (1956). On the construction of significance tests on the circle and the sphere. *Biometrika* **43**, 344-352. (14, 133, 211, 224)

Watson, G. S. *See* Epp *et al.* (1971).

Weiss, L. E. *See* Turner *et al.* (1963).

Wellner, J. A. (1979). Permutation tests for directional data. *Ann. Statist.* **7**, 929-943. (201)

Wellner, J. *See* Keilson *et al.* (1983).

Wichmann, B. A. & Hill, I. D. (1982). Algorithm AS183: An efficient and portable pseudo-random number generator. *Appl. Statist.* **31**, 188-190. (58)

Wilf, H. S. *See* Nijenhuis *et al.* (1978).

Willcox, M. E. *See* Fisher *et al.* (1981).

Willcox, M. E. *See* Fisher, Huntington, Jackett, Willcox & Creasey (1985).

Williams, E. J. *See* Watson *et al.* (1956).

Winchell, H. (1937). A new method of interpretation of petrofabric diagrams. *Am. Mineralogist* **22**, 15-36. (10)

Wood, A. (1982). A bimodal distribution on the sphere. *Appl. Statist.* **31**, 52-58. (94, 96, 147, 148, 149)

Wood, A. (1986). Some notes on the Fisher-Bingham family. *Commun. Statist. - Theor. Meth.* To appear. (97, 98)

Woodcock, N. H. (1977). Specification of fabric shapes using an eigenvalue method. *Bull. geol. Soc. Am.* **88**, 1231-1236. (47, 48, 49)

Woodcock, N. H. & Naylor, M. A. (1983). Randomness testing in three-dimensional orientation data. *J. Struct. Geol.* **5**, 539-548. (48, 49)

Yates, F. *See* Fisher *et al.* (1957).

Zemroch, P. J. *See* Mardia & Zemroch (1977).

Zemroch, P. J. *See* Mardia & Zemroch (1978).

Figure A1
Figure A2
Figure A3(a) (The diagram is for illustrative purposes. In fact the above shape applies for $\nu \geq 3$ only; when $\nu < 3$ the mode is at the origin as may be seen in Figure 4.5, Chapter 4).

Index

alternative hypothesis *see* null hypothesis
altitude 21-2
animal physiology 1
antimode 94, 96
apparent polar wander 3, 15, 244-7
Arnold distribution 97
association *see* correlation, serial association
astronomical coordinates 21-2
   ecliptic system 22
   equatorial system 21
   galactic system 22
   horizon system 21
   super-galactic system 22
astronomy 1, 37
astrophysics 1
axes of tabular stones (B26) 309-11
axial data 1
   analysis of a single sample 152-93
   analysis of two or more samples 197-8, 225-9
   see also bipolar data, girdle data, small-circle
axial distribution 22
   see also bipolar distribution, girdle distribution
axis 1, 18, 22
   see also axial data, axial distribution, random axis
azimuth
   of altitude 21
   of dip *see* dip direction
   of plunge *see* plunge azimuth

bedding planes (B4), (B19) 282-3, 300
Bessel function 82, 83
bimodal data *see* Wood distribution
bimodal distribution *see* Fisher-Bingham distribution, Fisher-Watson distribution, general Fisher-Bingham distribution, Kent distribution, Wood distribution

Bingham distribution 97, 155-6
Bingham-Mardia distribution 99
bipolar data
   analysis of a single sample 152-78
   analysis of two or more samples 225-7
   estimation of common principal axis of two or more samples 226-7
   estimation of principal axis of sample 162-3, 165-6
   test for common principal axis of two or more samples:
      formal 225-6
      graphical 225-6
   test for rotational symmetry 164-5
   test for serial association 243
   test for specified principal axis 163-4, 167
   see also principal axis, Watson distribution
bipolar distribution *see* Arnold distribution, Bingham distribution, Selby distribution, Watson distribution
block test for outliers *see* Fisher distribution, Watson distribution
bootstrap method 23, 57-8, 111, 115, 142-3, 147, 162, 166, 192, 201, 206, 234-5, 241, 245-7

cats 1, 153, 193, 295
celestial latitude 22
celestial longitude 22
central limit theorem 74
central projection 38-9
   see also gnomonic projection
centre of mass 29, 31
characteristic vector 34
   see also eigenvector
chi-squared (χ^2) distribution 76-9
   tables 78-9, 251-5
   chi-squared test 79

321

322 *Index*

circular confidence cone
  for common mean direction of two or more unimodal distributions 115–16
  for common polar axis of two or more girdle distributions 182–3
  for common principal axis of two or more bipolar distributions 165–6
  for mean direction 115–16
  for polar axis 182–3
  for principal axis 16, 165–6
  see also Fisher distribution, Watson distribution
circular data 81, 83
circular uniform distribution *see* uniform distribution on the circle
cleavage surfaces (B11), (B12), (B17) 290, 291, 298
colatitude 18
  analysis of 135
colatitude plot *see* Fisher distribution, Watson distribution
combined path length 244
cometary orbital planes (B16) 296–7
colatitude test *see* Fisher distribution, Watson distribution
concentration parameter
  for Arnold distribution 97
  for Fisher distribution 86
  for Kent distribution 93
  for Saw distribution 99
  for Selby distribution 97
  for von Mises distribution 82
  for Watson distribution 89
  for Wood distribution 94–5
  inference for *see under* heading for individual distribution
confidence cone
  definition 23
  see also circular confidence cone, elliptical confidence cone
confidence interval 23
  see also correlation, involving spherical variables
confidence level 23
contour
  choice of level 43–6
  definition 24
  method of display for spherical data 37
  of spherical density 43–6
conventional display of spherical data 37
coordinate system 17–22
  astronomical 21–2
  geographical 18–19
  geological 19–21
  polar 18
coplanar mean directions 213–8
  estimation of common plane 216–18
  test for coplanarity 213–14
  test for specified normal to common plane 214–16

correlation, involving linear variables
  permutation test for 60–1
correlation, involving spherical variables 230–8
  between a unit vector or axis and another quantity 237–8
  estimate of correlation between two unit random axes 235–7
  estimate of correlation between two unit random vectors 232–5
  test for correlation between two unit random axes 235–6
  test for correlation between two unit random vectors 232–5
  see also serial association
cosmic rays (B3) 280–1
cross-validation *see* nonparametric density estimation
critical region 24
  for permutation test for correlation 61–2
crystallography 1, 37–8
cumulative distribution function 24

data collection 63–6
data display 35–46
  see also projections
data rotation 32
  see also rotation matrix
data sets (B1–B26) 278–311
  bedding planes (B4), (B19) 282–3, 300
  cleavage surfaces (B11), (B12), (B17) 290, 291, 298
  cometary orbital planes (B16) 296–7
  cosmic rays (B3) 280–1
  dendritic field orientations (B15) 295
  joint planes (B14) 293–4
  L_0^1 axes (B18), (B22), (B24) 299, 305, 307
  magnetic remanence (B2), (B5), (B6), (B7), (B8), (B9), (B21) 279, 284, 285, 286, 287–8, 289, 304
  natural remanent magnetisation (B23) 306
  occupational judgments (B20) 301–3
  palaeomagnetic pole positions (B1), (B10) 278, 289
  susceptibility axes (B13) 292
  tabular stones (B26) 309–11
  wind directions (B25) 308
declination
  definition for equatorial coordinates 21–2
  definition for geological coordinates 20–1
degenerate distribution *see* "point" distribution
degrees of freedom 76, 80
dendritic field orientations (B15) 295

Index

Dimroth-Watson distribution *see* Watson distribution
dip 19-21
dip direction 19-21
direction cosines 18, 29
direction of dip 19-21
directional-spatial plots 46
discordancy 128-9
  test for *see* Fisher distribution, Watson distribution
discordant observation 24, 128-9
  see also Fisher distribution, Watson distribution
discriminant analysis 224
display convention for spherical data 37
distribution 67-100
  on the circle 81-4
  on the line 72-81
  on the sphere 84-100
  type of *see* bimodal distribution, girdle distribution, great circle distribution of unit vectors, multimodal data, small circle distribution, unimodal distribution
  see also under specific name of distribution
distribution function 24
double-exponential distribution 233

Earth Sciences 2
ecliptic coordinates 22-3
eigenvalue
  of orientation matrix 33-4
  of symmetric 2×2 matrix 34-5
  see also exploratory analysis
eigenvector
  of orientation matrix 33-4
  of symmetric 2×2 matrix 34-5
elemental area 67-8
elemental range 67
elemental surface region 68
elliptical confidence cone
  calculation of 34-5
  for common median direction of two or more unimodal distributions 202-3
  for median direction 111-12
  for polar axis 180-1
  for principal axis 162-3
  for second axis 192
  see also Kent distribution
empty space test 10
equal-angle projection 36
equal-area projection 36-8
equatorial coordinates 21-3
equatorial distribution 24
  see also Girdle distribution
equatorial projection 39
estimate 4
exploratory analysis
  for samples of unit axes 155-9
  for samples of unit vectors 103-4, 107-9
  techniques for 46-9
exponential distribution 75-6
  goodness of fit of exponential model 53
  simulation of 58

F-distribution 79-81
  tables 80-1, 255-61
Fisher distribution 4, 6-9, 117-36, 208-24
  analysis of a single sample from 117-36
  analysis of colatitudes 135
  analysis of two or more samples from Fisher distributions 208-24
  bias in estimation of concentration parameter 130, 133
  choice of sample size for estimating mean direction 136
  circular confidence cone for mean direction 131-2, 133
  confidence interval for concentration parameter 132-3
  definition 86-8
  discordancy tests 125-9
  discriminant analysis 224
  estimation of common concentration parameter of several distributions 224
  estimation of common mean direction of two or more distributions 211-13
  estimation of common plane for mean directions of several distributions 216-18
  estimation of concentration parameter 129-30, 132-3: graphical estimate 118-19, robust estimate 133
  estimation of mean direction 129, 132-3
  for errors in regression model 241
  goodness-of-fit tests: formal 122-5, graphical 117, 122, omnibus test 125, versus Kent alternative 139-40, versus Wood alternative 147
  graphical 117-22, 208, 218, 220-1
  historical 11-13
  mixture of Fisher distributions 144
  notation for 86
  outliers: accommodation 129, 130, block tests for discordancy of several outliers 127-8, detection 117-22, discussion of how to handle 128-9, 130, test for discordancy of single outlier 125-9
  parameters of 86
  pdf of colatitude distribution 68-9
  profile graph 69, 87

Fisher distribution (*cont.*)
  properties of 86–8
  pseudo-random values from 59
  robust estimation 129, 130, 133
  simulation of 59
  tables 88
  test against alternative with elliptical contours 139–40
  test for common concentration parameter: several samples – formal 221–4, graphical 220–1; two samples – formal 218–20, graphical 218, 220
  test for common mean direction: two samples – graphical 208, formal 208–211, several samples, graphical 208, formal 208–11
  test for coplanarity of several mean directions 213–14
  test for specified concentration parameter 134–5
  test for specified mean direction 133–4
  test for specified normal to common plane of several mean directions 214–16
  test for uniformity against Fisher alternative 117–22
  test that mean direction lies in a specified plane 135–6
Fisher–Bingham distribution 15, 98
Fisher–Dimroth–Watson distribution *see* Fisher–Watson distribution
Fisher–Watson distribution 98
fold test 211

galactic coordinates 22
gamma function 77
geographical coordinates 18–19
geological coordinates 19–21
Gaussian distribution *see* normal distribution
general Fisher–Bingham distribution 98
girdle data 152–6, 178–93
  analysis of a single sample 178–93
  analysis of two or more samples 225–9
  estimation of common polar axis of two or more distributions 229
  estimation of polar axis of distribution 180–1, 182
  test for common polar axis of two or more distributions: formal 228, graphical 228
  test for rotational symmetry 182
  test for serial association 235
  test for specified polar axis 181–2, 183
  see also polar axis, Watson distribution
girdle distribution 24
  see also Girdle data, Watson distribution
gnomonic projection 38–39

goodness-of-fit 55–6
  chi-squared test 79
  definition 24
  graphical method for assessing 50–5
  Kolmogorov–Smirnov test for 55–6
  Kuiper test for 55–6
  see also Fisher distribution, Kent distribution, probability plot, Watson distribution, Wood distribution
graphical method *see* contour, directional-spatial plots, goodness-of-fit, Fisher distribution, mean direction, nonparametric density estimation, polar axis, principal axis, probability plots, projection, quantile–quantile plots, shade plots, sunflower plots, Watson distribution
great circle 24
great circle distribution
  see girdle distribution, great circle distribution of unit vectors
great circle distribution of unit vectors
  estimation of pole to plane 141
  general 140–1
  parametric models for 143
Greengate xiv

hierarchical error structure 64–5
horizon coordinates 21–3

inclination 19–20
  analysis of 135
independent 24, 26
inverse
  of 2×2 symmetric matrix 33
  of 3×3 symmetric matrix 33
isotropic 25
  see also uniform, uniform distribution on the circle, uniform distribution on the sphere

Jackknife method
  definition 25
  general 60
  used in estimating correlation between two random unit axes 236–7
  used in estimating correlation between two random unit vectors 233
joint planes (B14) 293–4

Kelker–Langenberg distribution 97, 98, 191
Kent distribution 6
  bimodal form 93–4
  comparison with Wood bimodal distribution 95–6
  definition and properties 92–4
  elliptical confidence cone for mean direction 138–9
  estimation of parameters 136–8

Index

notation 92
parameters of 92-3
test as alternative to Fisher
   model 139-40
see also Fisher distribution, Watson
   distribution
Kolmogorov-Smirnov test 56, 122-3,
   125, 169, 185-86
Kuiper test 56, 114, 123
see also Fisher distribution

L_0^1 axes (B18), (B22), (B24) 299, 305,
   307
Lambert projection 10, 36, 38-9
Langevin distribution 86
latent vector 34
see also eigenvector
latitude 18
see also celestial latitude, galactic,
   super-galactic
level of significance 25
linear (or real) random variable 25
longitude 18
see also celestial longitude, galactic,
   super-galactic
longitude plot
   to assess rotational symmetry *see*
      quantile-quantile plots
see also Fisher distribution, Watson
   distribution

magnetic remanence (B2), (B5), (B6),
   (B7), (B8), (B9), (B21) 279, 284,
   285, 286, 287-8, 289, 304
Mardia-Gadsden distribution 99-100
matrix
   eigenvalues and eigenvectors of: 2×2
      symmetric matrix 34-5, 3×3
      symmetric matrix 33-4
   inverse of 2×2 symmetric matrix 33
   orientation matrix 33-4
   orthogonal matrix 33
   reflection matrix 27
   rotation matrix 32
   square root of symmetric 2×2
      matrix 33
   see also rotation
mean 25
see also central limit theorem
mean direction 29-32
   calculation of 30
   circular confidence cone for 115-16
   comparison with median direction 111
   estimation of 115-16
   estimation of common mean direction
      of two or more distributions 206-8
   test for common mean direction of two
      or more distributions: formal 204-6,
      graphical 204
   test for rotational symmetry about mean

direction; formal 114,
   graphical 114
test for specified mean direction 116
see also Fisher distribution, Kent
   distribution, Wood distribution
mean resultant length 31-2
   definition 31
   use in exploratory analysis 47-9
median direction 111-14
   comparison with mean direction 111
   definition 111
   elliptical confidence cone for 111-12
   estimation of 111-13
   estimation of common median direction
      of two or more distributions 201-3
   test for specified value 113-14
   test of common median direction of two
      or more distributions 200-1
meteorology 1, 231, 243
minimum combined path length 244
minor axis 34
see also polar axis
mode 144-9
model 67-100
   choice of 2, 4, 10-11
   see also distribution
modelling 6-7
moment 25
   of inertia 33
multimodal data
   analysis for vector data 144-9
   analysis for axial data 192
   display of 40-1
   exploratory methods 47-9
   see also Wood distribution

natural remanent magnetisation
   (B23) 306
nets 36
nonparametric density estimate
   for spherical data 41-6
   use of cross-validation 42-3
normal distribution 73-5
   goodness of fit of model 51-2
   simulation of 58
   tables 75, 249-51
normalized eigenvalue 34
notation xiii
   for probability distributions 70
   for summary statistics in several-sample
      problems 199
   see also under specific name of
      distribution
null model 81, 84-5
null hypothesis 25

observation 25
occupational judgments (B20) 1, 301-3
order statistic 25, 50
optimisation routines 111, 146

326 Index

orientation matrix
  definition 33
  use in exploratory analysis 47-9
  see also Watson distribution
orthogonal matrix 25
orthographic projection 38-9
outlier, outlying observation, outlying point 3, 26
  see also discordancy, Fisher distribution, Watson distribution
ovalness parameter 93

palaeocurrents 64-5
palaeomagnetic pole positions (B1), (B10) 245-7, 278, 289
palaeomagnetism 1, 11, 13, 38, 211
parameter 26
  notation for 70
percentile 26
permutations, random 61
permutation (or randomisation) distribution 26
  see also randomisation (or permutation) test
permutation (or randomisation) test see randomisation (or permutation) test
Physical Oceanography 1
pitch 20-1
plots see contour, directional-spatial plots, goodness-of-fit, Fisher distribution, nonparametric density estimation, probability plots, quantile-quantile plots, shade plots, sunflower plots, Watson distribution
plunge 20-1
point "distribution" 84
polar axis (of a girdle or equatorial distribution) 153, 154
  circular confidence cone for 182-3
  elliptical confidence cone for 180-1
  estimation of 180-1, 182
  estimation of common polar axis of two or more girdle distributions 229
  test for common polar axis for two or more girdle distributions:
    formal 228, graphical 228
  test for rotational symmetry:
    formal 182, graphical 182
  test for specified polar axis 181-2, 183
  see also Watson distribution
polar coordinates 18
  in the plane 35-6
polar projection 39
polar wander see apparent polar wander pole 19
population 26
principal axis (of a bipolar distribution) 34
  circular confidence cone for 16, 165-6
  elliptical confidence cone for 162-3
  estimation of 162-3, 165-6
  estimation of common principal axis of two or more bipolar distributions 226-7
  test for common principal axis for two or more bipolar distributions:
    formal 225-6, graphical 225-6
  test for rotational symmetry:
    formal 164-5, graphical 164-5
  test for specified principal axis 163-4, 167
  see also Watson distribution
probability density element (pde) 67-9
  basic form for spherical distributions 70
  see also under specific name of distribution
probability density function (pdf) 67-9
  definition 26
probability plot
  definition 26
  description 50-5
  for exponential model 53
  for normal model 51-2
  for uniform model 53
  two-sample 53-5
  see also Fisher distribution, quantile-quantile plots, Watson distribution
profile graph 69
projection 35-9
  equal-angle (Wulff/stereographic) 36
  equal-area (Schmidt) 36, 38-9
  equatorial 39
  gnomonic or central 38-9
  orthographic 38-9
pseudo-random number 26
  see also simulation
pseudo-values 60

quantile 26
quantile-quantile (Q-Q) plot
  description 50-5
  for assessing rotational symmetry 150-151; about a mean direction 114, polar axis 182, principal axis 164-5
  for comparison of mean directions of two or more distributions 204
  for comparison of polar axes of two or more girdle distributions 228
  for comparison of principal axes of two or more bipolar distributions 225
  two-sample 53-5
  see also Fisher distribution, probability plots, Watson distribution

random (unit) axis 22
random sample 26
  see also sample
random variable 27

Index

random (unit) vector 28
randomisation (or permutation)
   distribution 27
   see also randomisation test
randomisation (or permutation) test
   definition 27
   description of use 60-3
   examples of 210, 219, 221, 233, 236, 238, 243
randomness, tests for
   see uniform distribution on the sphere
Rayleigh tests 110-11
reflection matrix
   definition 27
   in model for negative correlation of random unit vectors 232-5
regression
   general 238
   of a unit vector on a circular variable 239-40
   of a unit vector on a set of linear variables 240-2
resultant length 29-31
resultant vector 30-1
Right Ascension 21-3
robust (statistical) procedure 11, 27
   see also Fisher distribution, median direction, Watson distribution
robustness (of median direction) 113
rotation
   as model for spherical correlation between two random unit axes 235
   as model for spherical correlation between two random unit vectors 232-5
   estimate of rotation between two random unit vectors 234-5
   of unit vector or axis 25
rotation matrix
   calculation of 32
   definition 27
   see also rotation
rotational symmetry
   test of rotational symmetry for axial data 164-5, 182
   test of rotational symmetry for vector data 114, 150-1
   see also Fisher distribution, Kent distribution, polar axis, principal axis, Watson distribution

sample 2
   definition 27
   discussion of statistical sample vis-à-vis scientific sample (specimen) 17
sampling problems 63-6
saw distribution 99
Scheidegger-Watson distribution see Watson distribution
Schmidt projection 36, 38-9
   see also projection
sedimentology
Selby distribution 97
serial association
   test for unit axes 243
   test for unit vectors 242
serial correlation see serial association
several-sample procedures 194-229
   estimation of common mean direction of two or more distributions 206-8
   estimation of common median direction of two or more distributions 201-3
   estimation of common polar axis of two or more distributions 229
   estimation of common principal axis of two or more distributions 226-7
   test for common mean direction of two or more distributions: formal 204-6, graphical 204
   test for common median direction of two or more distributions 200-1
   test for common polar axis of two or more distributions; formal 228, graphical 228
   test for common principal axis of two or more distributions; formal 225-6, graphical 225-6
   see also Fisher distribution, probability plot, two-sample procedure, Watson distribution
shade plot 43-6
shape
   empirical shape criterion 48; examples of use 108, 159
   parameters of Bingham distribution 97
   parameters of Bingham-Mardia distribution 99
   parameters of Fisher-Bingham distribution 98
   parameters of Fisher-Watson distribution 98
   parameters of Kent distribution 92
   parameter of wrapped normal distribution 83
significance level 27
"simple" methods 5
   for samples of axial bipolar data 161-7, 225-7
   for samples of axial girdle data 179-183, 228-9
   for samples of vectorial data 110-17, 199-208
simulation
   examples of use 56-8
   general 56-60
   of particular distributions:
      exponential 58, Fisher 59, normal 58, uniform on the circle or line 58, uniform on the sphere 59, Watson 59-60

Index

simulation (*contd.*)
  see also bootstrap
small circle
  definition 27
  distribution of unit axes: estimation of parameters 192, general 191-2
  distribution of unit vectors: estimation of parameters 141-3, general 140-1
  see also Bingham-Mardia distribution, Kelker-Langenberg distribution, Mardia-Gadsden distribution
Social Science 1, 194-6, 203-4
Soil Science 37
spatial analysis 231, 242, 248
  graphical display 46
  sampling problems 63-6
spherical data
  definition 1
spherical distribution see distribution
spherical median see median direction
spherical standard error
  of pooled estimate of mean direction 207-8
  of pooled estimate of polar axis 229
  of pooled estimate of principal axis 227
  of sample mean direction 115-16
  of sample polar axis 182
  of sample principal axis 165-6
  use in sampling 65
spherical uniform distribution see uniform distribution on the sphere
spherical variance 206
spline
  definition 27
  in temporal analysis 244-5
standard deviation 27
standard error see jackknife method, spherical standard error
statistic 27
stereographic projection 36-8
  see also projection
strength
  empirical strength criterion 48; examples of use 108, 159
strike 19-21
Structural Geology 1, 18, 36, 38, 140
Structural Petrology 10
sunflower plot 40-1
super-galactic coordinates 22
susceptibility axes (B13) 292
symmetry
  test for rotational symmetry: about mean direction 114, about polar axis 182, about principal axis 164-5
  see also Fisher distribution, rotational symmetry, Watson distribution

tabular stones (B26) 309-11

temporal analysis
  analysis of time-ordered sequences of unit vectors 244-7
  general 231
  test for serial association: for sequence of unit axes 243, for sequence of unit vectors 242
  see also apparent polar wander
terminology 17-28
time series
  see temporal analysis
trace 27
trend 20-1
two-sample procedure
  graphical method for linear data 53-5
  see also Fisher distribution, probability plot, several-sample procedure, Watson distribution
two-sample test see two-sample procedure

uniform distribution see uniform distribution on the circle, uniform distribution on the line, uniform distribution on the sphere
uniform distribution on the circle 81
  simulation of 58
uniform distribution on the line
  definition and properties 72
  goodness of fit of uniform model 53
  simulation of 58
uniform distribution on the sphere 10
  definition and properties 84-6
  simulation of 59
  test against general alternative: for a sample of unit axes 192-3, for a sample of unit vectors 149-50
  test against bipolar alternative 160-2
  test against girdle alternative 179-80
  test against unimodal alternative 110-11
  see also empty space test, Fisher distribution, Watson distribution
unimodal distribution
  analysis of a sample of data from a unimodal distribution of unit vectors 101-40
  analysis of two or more samples of data from unimodal distributions 194-224
  see also Fisher distribution, general Fisher-Bingham distribution, Kent distribution, mean direction, median direction, Wood distribution
unit axes
  mathematical methods for 29-35
unit vectors
  mathematical methods for 29-35

variance
  of a linear random variable 27

Index

spherical 206
variance ratio distribution *see*
  F-distribution
variate 27
vector 1
  see also random unit vector, vectorial data
vectorial data 1
  analysis of a single sample of 101–51
  analysis of two or more samples of 194–224
von Mises distribution 81–3
  tables 82–3
von Mises-Arnold-Fisher distribution *see* Fisher distribution
von Mises-Fisher distribution *see* Fisher distribution
vulcanology 100

Watson distribution 85, 89–92, 167–78
  analysis of a single sample from: bipolar distribution 167–78, girdle distribution 178–91
  circular confidence cone for: polar axis of girdle distribution 190 principal axis of bipolar distribution 176–7
  confidence interval for concentration parameter: of bipolar distribution 177, of girdle distribution 190
  definition 89
  discordancy tests: bipolar distribution 170–5, girdle distribution 184–9
  estimation of concentration parameter of bipolar distribution 175–7; graphical estimate 168–9, robust estimate 175
  estimation of concentration parameter of girdle distribution 189–90; graphical estimate 185, robust estimate 190
  estimation of polar axis of girdle distribution 189
  estimation of principal axis of bipolar distribution 175
  goodness-of-fit tests for bipolar distributions: graphical 168–9, formal 169–70
  goodness-of-fit tests for girdle distribution: formal 185–6, graphical 184–5

  graphical methods: for bipolar distribution 168–9, girdle distribution 184–5
  historical 14
  notation for 89
  outliers in samples from bipolar distributions: accommodation 176, block test for discordancy of several outliers 173–5, detection 168–9, discussion 170, test for discordancy of single outlier 170–3
  outliers in samples from girdle distributions: accommodation 187, block test for discordancy of several outliers 187–9, detection 184–5, discussion 185, 187, test for discordancy of a single outlier 186–9
  parameters of 89
  polar graph of 89
  properties of 89–92
  simulation of 59–60
  test of uniformity against Watson bipolar alternative 168
  test of uniformity against Watson girdle alternative 184
  test for specified concentration parameter: of bipolar distribution 178, of girdle distribution 191
  test for specified polar axis of girdle distribution 190–1
  test for specified principal axis of bipolar distribution 177
  two-sample test for girdle data 228
Watson bipolar distribution *see* Watson distribution
Watson girdle distribution *see* Watson distribution
wind directions (B25) 308
Wood distribution
  comparison with Kent distribution 95–6
  definition and properties 94–6
  estimation of parameters 145–8
  notation 94–5
  parameters of 94–5
  test as alternative to Fisher model 147
  test for specified modal directions 148–9
wrapped normal distribution 83–4
Wulff projection 36–8
  see also projection